Bonenberger

The First Snap-Fit Handbook

Paul R. Bonenberger

The First Snap-Fit Handbook

Creating and Managing Attachments for Plastics Parts

3rd Edition

HANSER

Hanser Publishers, Munich

Hanser Publications, Cincinnati

The Author:

Paul R. Bonenberger, 1572 Pebble Creek Dr., Rochester MI 48307-1765

Distributed by:
Carl Hanser Verlag
Postfach 86 04 20, 81631 München, Germany
eMail: info@hanser.de
www.hanser-fachbuch.de

The use of general descriptive names, trademarks, etc., in this publication, even if the former are not especially identified, is not to be taken as a sign that such names, as understood by the Trade Marks and Merchandise Marks Act, may accordingly be used freely by anyone. While the advice and information in this book are believed to be true and accurate at the date of going to press, neither the authors nor the editors nor the publisher can accept any legal responsibility for any errors or omissions that may be made. The publisher makes no warranty, express or implied, with respect to the material contained herein.

The final determination of the suitability of any information for the use contemplated for a given application remains the sole responsibility of the user.

Cataloging-in-Publication Data is on file with the Library of Congress

Editor: Cheryl Hamilton
Production Management: Jörg Strohbach
Coverconcept: Marc Müller-Bremer, www.rebranding.de, München
Coverdesign: Stephan Rönigk
Typesetting: Kösel Media GmbH, Krugzell

ISBN: 978-1-56990-595-1
E-Book ISBN: 978-1-56990-596-8

Foreword to Third Edition

Globalization resulted in the off-shoring of American manufacturing to low labor rate countries. In order to compete or just survive, the manufacturers of plastic products were forced to improve quality and reduce cost.

All aspects of the manufacturing process were scrutinized. Most of the plastics molding processes were or could be automated. The only manufacturing operations that were still labor intense were tool making and assembly.

This realization resulted in a new technology that came to be called *Design for Manufacturability* or DFM. This technology encompassed all aspects of the manufacturing process. However, the easiest and quickest savings were realized by improving assembly. Almost overnight the trade magazines were full of case studies and articles extolling the savings to be had by designing for manufacturability (or assembly). Conference speakers and seminar teachers begin explaining the advantages of replacing fasteners with molded-in attachment features.

In the midst of that frenzy, the University of Wisconsin recruited me to join Paul Bonenberger's multispeaker *Snap-Fits and Product Design* seminar. I remember telling the recruiter that I was not an expert on snap-fits. He replied they wanted me to talk about how to improve the design of the two plastic parts required for a snap-fit. Paul and the other speakers would cover the details of designing the actual snap-fit structures.

The first seminar was held in 1998. By that time I had been designing plastic parts for over forty years, and I had designed my share of snap-fits. I thought I already knew what I needed to know. In spite of that, I sat in on Paul's lecture. It quickly become evident that Paul and the other snap-fit speaker knew far more than I did about the design and development of snap-fits. They explained concepts and details that had never occurred to me. How could this be? I had far more experience than either of them in designing and developing plastic products. The answer to that question was that snap-fits are just one of the hundreds and hundreds of details that I and other designers have to take into account in the design and development of a new plastic product. Most designers will have only an occasional need to design a snap-fit and cannot devote a lot of time to that one detail.

Paul Bonenberger, on the other hand, worked at General Motors. That giant company generates an endless stream of potential snap-fit applications. Paul was not only there but was commissioned to do something about the *too many loose fasteners* in GM products. That was the beginning of his analysis of various types of snap-fits. GM provided the opportunity to try different methodologies and learn how they performed over time. Being in the right place at the right time allowed Paul to perfect the engineering that determines how snap-fits function.

This work also led to Paul's development of the Attachment Level Construct (ALC) concept that provides a proven method of managing the design and development of a successful snap-fit application. The ALC concept is the basis for this book.

Most designers have only an occasional need for a snap-fit. If the resulting structure does not function as required, the part design and the mold are modified to overcome the assembly's failure. They do not design enough snap-fits to develop a true understanding of how they function and how they fail. Fortunately Paul has done that work for us. His *The First Snap-Fit Handbook* contains what he has learned by concentrating on snap-fit design and development work. All of which has been fine-tuned by his teaching programs and tempered by his many years of hands-on experience.

If you already own an assembly book or two, you will be surprised by the *The First Snap-Fit Handbook*. It does not make the usual attempt to include all of the many assembly techniques and all of the different metal fasteners. Like the name says, this book concentrates on only snap-fits. If you have this book, you possess the best of what Paul Bonenberger has learned about snap-fits. I have no hesitation in recommending the third edition of this book to anyone interested in optimizing the design of snap-fit plastic assemblies.

Libertyville, Illinois *Glenn L. Beall*
2016

Preface to Third Edition

This third edition of *The First Snap-Fit Handbook* contains some content additions and clarifications and major organizational changes. These are the result of 12 years of teaching a class based on earlier editions of the book. Participant comments and questions influenced many changes to the class itself and many of those changes have found their way into this edition. Thanks to all who asked the tough questions and pointed out areas for improvement. I owe a special thanks to David Schattner who once arranged for me to teach a class at Lexmark, Inc. Since then, he has gone far beyond the call of duty in continuing to offer valuable suggestions for improvements to the book.

Content changes include:

- More emphasis on why the cantilever hook style lock should *not* be used in some applications. This topic was addressed indirectly in previous editions, but deserves more direct and emphatic attention.
- More discussion of the feature analysis procedures for other beam-based lock features, not just cantilever hooks.
- Elimination of unnecessary content.

Format changes include:

- Some graphic content has been made more understandable with supporting text.
- Side notes have been added for special or incidental information.
- Chapters have been divided and some subchapters have been rewritten and rearranged to improve content flow, reduce redundancy, and to better accomodate the use of the book as a design reference.

I believe these changes are improvements to the book's usability and I think the reader will agree. Because my goal is to continue to improve this book as a practical reference tool for plastic part and snap-fit developers; I am dedicating this effort to them.

I also hope to use the *fasteningsmart.net* website for occasional updates of snap-fit information and topics in this book. The reader might want to visit it occasionally.

Suggestions and comments are welcome and can be sent directly to the publisher and/or to the author at paulrb@fasteningsmart.net.

Many thanks to Cheryl Hamilton, my editor, and the rest of the Hanser team for their patience and help throughout this project.

Rochester, Michigan *Paul Bonenberger*
2016

Foreword to Previous Editions

Over the past decade we have seen a complete redefinition of the expected outcome of design for manufacture in the product development process. The term, design for manufacture (DFM), was often applied to a process of using rules or guidelines to assist in the design of individual parts for efficient processing. For this purpose the rule sets, or lists of guidelines, were often made available to designers through company specific design guides. Such information is clearly valuable to design teams who can make very costly decisions about the design of individual parts if these are made without regard to the capabilities and limitations of the required manufacturing processes. However, if DFM rules are used as the main principles to guide a new design in the direction of manufacturing efficiency, then the result will usually be very unsatisfactory. The end result of this guidance towards individual part simplicity will often be a product with an unnecessarily large number of individual functional parts, with a corresponding large number of interfaces between parts, and with a large number of associated items for connecting and securing. At the assembly level, as opposed to the manufactured part level, the resulting product will often be very far from optimal with respect to total cost or reliability.

Dr. Dewhurst of Boothroyd-Dewhurst, Inc. and the University of Rhode Island is a pioneer in Design for Assembly (DFA) practices.

The alternative approach to part-focused DFM, is to concentrate initially on the structure of the product and try to reach team consensus on the design structure which is likely to minimize cost when assembly as well as part manufacturing costs are considered. With this goal in mind, Design for Assembly (DFA) is now most often the first stage in the design for manufacture evaluation of a new product concept. The activity of DFA naturally guides the design team in the direction of part count reduction. DFA challenges the product development team to reduce the time and cost required for assembly of the product. Clearly, a powerful way to achieve this result is to reduce the number of parts which must be put together in the assembly process. DFA is a vehicle for questioning the relationship between the parts in a design and for attempting to simplify the structure through combinations of parts or features, through alternative choices of securing methods, or through spatial relationship changes.

An important role of DFA is to assist in the determination of the most efficient fastening methods for the necessary interfaces between separate items in a design. This is an important consideration since separate fasteners are often the most labor-intensive group of items when considering mechanical assembly work. To reduce the assembly cost of dealing with separate fasteners, fastening methods, which are an integral part of functional items, should always be considered. For plastic molded parts, well-designed snap fits of various types can provide reliable high-quality fastening arrangements, which are extremely efficient for product assembly. It is not an overstatement to claim that snap-fitted assembly structures have revolutionized the manufacturing efficiency of almost all categories of consumer products.

In this context, *The First Snap-Fit Handbook* by Paul Bonenberger provides an extremely valuable resource for product development teams. The concept of complete snap-fit attachment systems, rather than isolated analyses of the mechanics of the snap-fit elements, represents a major advance in the design of integral plastic attachment methods. This concentration on "attachment level" rather than snap-fit "feature level" design has been developed and tested by Paul Bonenberger through years of solving attachment problems with product development teams at General Motors Corporation. This handbook contains the best blend of analysis and real-world design experience.

Wakefield, Rhode Island *Peter Dewhurst*
1999

Prefaces to Previous Editions

Preface to First Edition

This book is a reference and design handbook for the attachment technology called snap-fits or sometimes, integral attachments. Its purpose is to help the reader apply snap-fit technology effectively to plastic applications. To do this, it arranges and explains snap-fit technology according to an Attachment Level™ knowledge construct. The book is intended to be a user-friendly guide and practical reference for anyone involved with plastic part development and snap-fits. It is called The First Snap-Fit Handbook for two reasons: I believe it is the first book written that is devoted exclusively to snap-fits. I also hope it leads to increased interest and more books on the subject.

The reader should consider this book to be a "good start" in the ongoing process of understanding and organizing snap-fit technology. There is much more to be done, but one must begin somewhere. Although the original "attachment level" construct (created in 1990 and 1991) has proven to be fairly robust and complete, many details have evolved over the years as I learned more about the topic. The construct will continue to evolve and I encourage and welcome reader's comments on the subject; they will certainly help in the process.

My interest in the subject of snap-fits grew out of a very real need at General Motors. As a long-time fastening expert, I had typically been involved with threaded fasteners and traditional mechanical attachments. In the late 1980s and early 90s, as GM embraced design for manufacturing and assembly, the philosophies of Dr. Geoffrey Boothroyd and Dr. Peter Dewhurst [*Product Design for Manufacture and Assembly*, 1988, G. Boothroyd and P. Dewhurst, Department of Industrial and Manufacturing Engineering, University of Rhode Island, Kingston, RI] were formally adopted as the corporate direction, and were rolled out in a series of intensive training/workshop sessions. As a result, product designers and engineers began looking for alternatives to traditional loose fasteners, including threaded fasteners. Snap-fit attachments immediately became popular but we soon discovered that there was little design information available in the subject. Calculations for cantilever hook performance could be found in many supplier design guides or as software, but beyond that, no general snap-fit attachment expertise was captured in design or reference books. GM needed to bootstrap itself to a level of snap-fit expertise that was not written down anywhere. An intensive study of snap-fit applications resulted and eventually patterns of good design practices began to emerge. A "systems level" understanding of snap-fit attachments began to grow.

I called this systems level organization of snap-fits *attachment level* to emphasize its focus on the interface as a whole and to distinguish it from the traditional *feature level* approach. I have been teaching about snap-fits according to this attachment level model since 1991. The reaction after each class has been that attendees had indeed reached a new or better understanding of snap-fits. I trust and hope this book will have the same results for the reader.

The Attachment Level Construct (ALC) was only a personal vision in 1990. I believed it had potential and that it represented a unique approach to understanding snap-fit applications but I needed much more than that to make it reality. I needed verification that I was not just reinventing or paraphrasing some existing but obscure snap-fit design practices; an extensive literature search verified that systems-level snap-fit practices were not documented anywhere. I also needed impartial validation that the model was indeed useful and worth pursuing. A colleague, Mr. Dennis Wiese, who was Manager of the Advanced Product Engineering Body Components Group at that time, provided that initial validation. He also gave moral support and generously provided resources including his own engineers and significant amounts of his own time for debate and discussion of the fledgling snap-fit design methodology. Those discussions, sometimes lively and always useful, drove the insights that helped shape the original attachment level model. Dennis was certainly the midwife of the attachment level approach and I cannot thank him enough for his help. Other GM people involved with the infant methodology included Florian Dutke, Tom Froling, Daphne Joachim, Colette Kuhl, Chris Nelander, Tom Nistor, Tim Rossiter, and Teresa Shirley.

Finally, Mike Carter, of GM University, deserves special thanks because in the early 1990s he asked me, *what are you fastening guys going to do about too many loose fasteners in our products?* That question was the beginning of my involvement with design for assembly. Mike, this book is your answer.

As pressure of other work grew, the development team dwindled back to one (me). In 1992, Tony Luscher, the project manager of a planned snap-fit program at Rensselaer Polytechnic Institute (RPI), and I learned of each other's work and made contact (once again, thanks to Mike Carter). The RPI program was originally designed around feature level research but Tony enthusiastically embraced the concept of attachment level thinking. Tony, with the concurrence of Dr. Gary Gabrielle, the project leader, modified the RPI program to include some aspects of the attachment level method. Tony's technical insights, contributed during many hours of personal discussion and through exchange of correspondence, helped drive more refinements to the method. Under his guidance, some work to apply and extend the methodology occurred under the RPI program. Tony is now a professor at the Ohio State University and he has carried his interest and enthusiasm for the subject to his new position. Tony and I shared a long-term vision for snap-fit technology: that attachment level thinking will lead to evolution of the snap-fit design and development process from an art to an engineering science.

The original motivation for the attachment level work was to provide support for Design for Manufacturing and Design for Assembly initiatives at General Motors. Joe Joseph, then the Director of the GM DFM Knowledge Center, supported my early efforts by providing a site for snap-fit training classes. This also gave the kind of validation needed to justify continued efforts to develop the methodology. Joe is now Dean of the Engineering College of the GM University and he continues to provide valued moral support. The patience and support of Jim Rutledge, Dave Bubolz, and Roger Heimbuch is also greatly appreciated. They provided an environment in which ongoing development work could flourish and gave me much encouragement. Tony Wojcik also deserves thanks because he first sent a publisher my way. That marked the beginning of the snap-fit book project.

I must also acknowledge the creative people who designed and developed the numerous snap-fit applications that I have studied. In products from around the world, the level of cleverness and creativity evident in many snap-fits is truly impressive. My admiration for and fascination with these designs helped to drive the original ideas behind the Attachment Level Construct in the following manner:

- *Observation*: There are many clever, well-designed, and complex snap-fit applications in existence; there are also many poor snap-fits.
- *Hypothesis*: Many snap-fit designers must possess tacit knowledge that allows them to develop good snap-fits; others do not.
- *Problem*: Snap-fit application design information could not be found as documented knowledge. Principles of good snap-fit application design were not written down anywhere.
- *Solution*: Discover the information and define it. Study successful snap-fit applications and look for patterns of good design practices. Capture and organize the concepts behind good snap-fit design.
- *Result*: A deep understanding of snap-fit concepts and principles organized in a knowledge construct.

I cannot claim credit for the clever snap-fit applications I describe here. Most were found on existing products or inspired by products. I simply interpreted them, inferred a logical process by which they could have been developed, and organized everything I found into a knowledge structure. The only new "invention" here is the construct itself. Hopefully, it will inspire readers to create their own product inventions.

My wife and son have provided endless encouragement and understanding through the long process of writing this book, putting up with my long hours at the computer, and tolerating (barely) my monopolization of same.

With thanks and appreciation to all.

Rochester, Michigan *Paul Bonenberger*
1999

Preface to Second Edition

The first edition of this book introduced a systematic way of thinking about snap-fit attachments. By intent, it did not spend a lot of time or space on calculations of feature behavior because this information was available elsewhere. That information is still available in various resources, including online sources; therefore, no new calculations have been added. However, equations for locking feature analysis are available online. The reader should check Appendix A for resources providing snap-fit feature calculations.

This second edition is an opportunity to add clarification and more detail in some areas. Most significantly, a new chapter, "Creating a Snap-Fit Capable Organization - Beyond Individual Expertise" has been added. This chapter is targeted primarily toward engineering executives and managers. It explains how engineering organizations can and should leverage their individual snap-fit expertise into organizational capability for competitive advantage.

After publication of the first edition of *The First Snap-Fit Handbook*, I was approached by the Automotive Learning Center of the American Chemistry Council and asked to create a class based on the book. That was the start of a very satisfying relationship, one which has given me the opportunity to teach the subject of snap-fits to many individuals from a variety of industries. The interaction with class attendees, answering their questions and being required to clarify my thinking in response to their challenges, has been extremely valuable to me. This second edition is dedicated to them.

Rochester, Michigan *Paul Bonenberger*
2005

Contents

1 Introduction

This book presents information about snap-fit technology in a logical format for learning and understanding. Once the reader understands snap-fit technology, this book will provide design guidance as a reference handbook.

The book has multiple purposes:

- Teach the reader a practical method of thinking about and using snap-fit technology.
- Be a comprehensive product development reference for snap-fit solutions.
- Provide a place for readers to record their own snap-fit lessons-learned.
- Provide guidance for managers wishing to develop a sustainable culture of snap-fit expertise in their product development organizations.

Any scientific discipline has a need for a specific language for describing and summarizing the observations in that area [1].

Experience without theory teaches ... nothing [2].

This book captures both the *language* and *theory* of snap-fits in a unique knowledge model that explains the snap-fit interface as a system. Readers with some snap-fit experience will find this model allows them to integrate their existing knowledge with new snap-fit information. Snap-fit novices will find the model makes understanding snap-fit technology easier. All readers will learn a practical way of thinking about and, most importantly, *using* snap-fits in product applications.

Snap-fit attachments are a *system*. It's time to start treating them that way.

The task of developing snap-fits generally falls on product engineers, designers, and developers (referred to collectively in this book as *developers*). A developer with little or no snap-fit experience can quickly find calculations in the literature for determining snap-fit lock behavior. However, next they will learn that while calculating lock feature behavior is important, it is not enough. Their learning will then go through a trial-and-error process during product testing and redesign. Sometimes design flaws are not discovered until a product is in the consumer's hands. In any case, product development through trial-and-error is time-consuming and potentially quite expensive. We want to avoid that.

Product developers may have access to someone with snap-fit experience, but their usefulness is generally limited to what they too have learned through trial and error.

A couple of bad experiences with snap-fits may cause a product developer or an entire organization to decide that snap-fits are not worth the trouble. This is unfortunate; to remain competitive, companies must utilize all possible design strategies. To ignore snap-fits as a legitimate attachment option is a mistake.

Reasons for using snap-fits include appearance, packaging, and tamper resistance. However, the most compelling reason is economic. When snap-fits replace loose fasteners and the associated assembly tools and tightening operations, significant cost savings are possible.

The increasing use of snap-fit technology parallels the growing use of plastics in products. Processing technologies have made production of complex shapes economically feasible. The advantages of ease of assembly and disassembly and the ever-increasing engineering capabilities of plastic materials now make snap-fit technology a serious candidate for applications once considered the domain of threaded or other mechanical fasteners.

The growth and advancement of rapid-prototyping technology has made the creation of accurate part models possible. These models provide early and meaningful evaluation of attachment concepts for more potential snap-fit applications.

While toys and small appliances have long made extensive use of snap-fits, the technology is now applied in virtually every product field including medical devices, automotive components, small and large appliances, electronics, and numerous consumer goods. Snap-fit technology is also being extended to structural applications [3–5].

Although commonly associated with plastic parts, snap-fits are also possible in metal-to-metal and plastic-to-metal applications. Keep this in mind as you read this book, and look for opportunities to use snap-fits in metal as well as plastic applications.

1.1 Reader Expectations

This book is not what a reader is likely to expect in a book about snap-fits.

Because snap-fit technology has traditionally been viewed as nothing more than lock feature calculations, readers may expect this book to be full of equations for calculating snap-fit lock behavior. *It is not.* This book includes those calculations but there is much more to snap-fit application development than just calculations.

Material property and part processing information is presented here only to the extent needed to support understanding of the snap-fit development process. Many excellent books and references are available on those topics and this book would serve no purpose repeating that material.

The reader must understand that experience with threaded fasteners, the most common method of mechanical attachment, is not transferable to understanding or developing snap-fit attachments. New ways of thinking about the attachment must be learned. There is more discussion of this subject in the next section.

The reader should expect to acquire a deep intuitive or gut-level understanding of snap-fits. You will learn how to *think* about snap-fits to solve routine as well as unique snap-fit design issues during product development.

After studying some sophisticated snap-fit applications, one cannot help being impressed and maybe intimidated. It's OK to be impressed, but do not be intimidated. With the knowledge in this book and through experience, every reader will gain the knowledge needed to create world-class snap-fits.

The reader will find that, occasionally, information may appear more than once in different chapters. This is intentional; information is repeated because of its importance or because it is being presented in a different context. Sometimes repetition is unavoidable because of the multiple interactions between elements and design concepts, and repetition is needed to ensure clarity and understanding of these interactions.

1.2 Harmful Beliefs

Seven common beliefs about snap-fit technology are described here. In this book, you will learn why these beliefs are wrong and how these beliefs interfere with developing cost-effective and reliable snap-fit attachments. You, the reader, may hold some of these beliefs. You will also find that your peers, management, and suppliers may likely hold some of these beliefs as well. Some of these beliefs will manifest themselves as a fear of using snap-fits. Other beliefs can have the opposite effect, leading to the misconception that snap-fits are so simple they require little or no thought at all. The harmful beliefs are:

- **The battery cover syndrome.**

  Most people are familiar with snap-fits thanks to their usage on common applications like remote control battery covers and toys. This can lead to two common and erroneous beliefs: (1) Snap-fits are only appropriate for simple or noncritical applications and (2) Snap-fits are trivial and easy to design.

- **Snap-fits are a materials technology.**

  Because snap-fits are generally found in products made from polymers, there is a belief that polymer experts (including resin suppliers) can be the design resource for snap-fit applications. Polymer experts should certainly be a primary resource for material properties, but they should not necessarily be expected to be the primary source for product design. Many polymer suppliers do have a wealth of experience in product design, and there is no reason not to use them as a secondary resource. Even when a supplier is, by contract, providing the primary design work, it is still up to you, the customer, to ensure the design, including the snap-fits, is done properly.

  This author would be very pleased to find the attachment level design principles appearing in plastic supplier design guides, but it hasn't happened yet.

- **Cantilever hooks represent snap-fit technology.**

  The cantilever hook style locking feature seems to be everywhere, but it is not representative of all snap-fit technology. When asked to create a snap-fit attachment, many developers will default to this style because of its familiarity. Many other lock feature styles exist as attachment options *and are often a better choice.*

- **All I need to do is design the locking feature.**

  A snap-fit attachment is an interface system and it must be developed as such. Many well-designed lock features fail to perform as expected because the systemic aspects of the part-to-part interface have been ignored.

- **Experience in other fastening methods transfers to snap-fits.**

  No, that experience does not transfer. Snap-fit attachments are *fundamentally different* from all other fastening methods. New and different knowledge is required to understand and apply snap-fit technology to product development.
- **Every snap-fit application is a new invention.**

  With snap-fits, the same fundamental rules of design are true for a finite number of common part-to-part combinations. Once those basic combinations are understood, a new application can be designed around existing and well-understood basic principles and rules.
- **I can do the attachment after I do everything else.**

  The attachment concept must be developed simultaneously with the parts that are being attached. Certain design details can wait until later, but getting the basic snap-fit concept right early in the development process is critical to the attachment's success.

These beliefs are discussed in more detail in Chapter 15.

1.3 Snap-Fit Technology

A snap-fit is the entire part-to-part interface.

The terms *snap-fit* and *integral attachment* are often used interchangeably because snap-fit lock features are molded or formed as integral features of parts. To avoid confusion, we will stick with the term *snap-fit.*

In the traditional meaning of the term, snap-fit referred to only the lock features.

In this book, the term snap-fit refers to the *entire attachment interface* (see Fig. 1.1), of which the lock feature(s) is only one element.

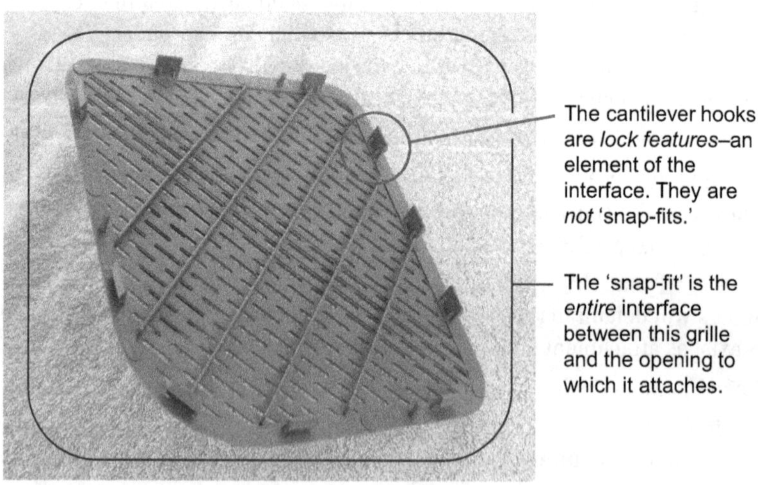

Figure 1.1 A snap-fit is the entire attachment interface, not just the locks

Snap-fit applications range from the very simple to the very complex. Some snap-fits hold one part to another and little or no force is transmitted across the interface. In other applications, snap-fit attachments must be strong and extremely reliable, see Fig.1.2.

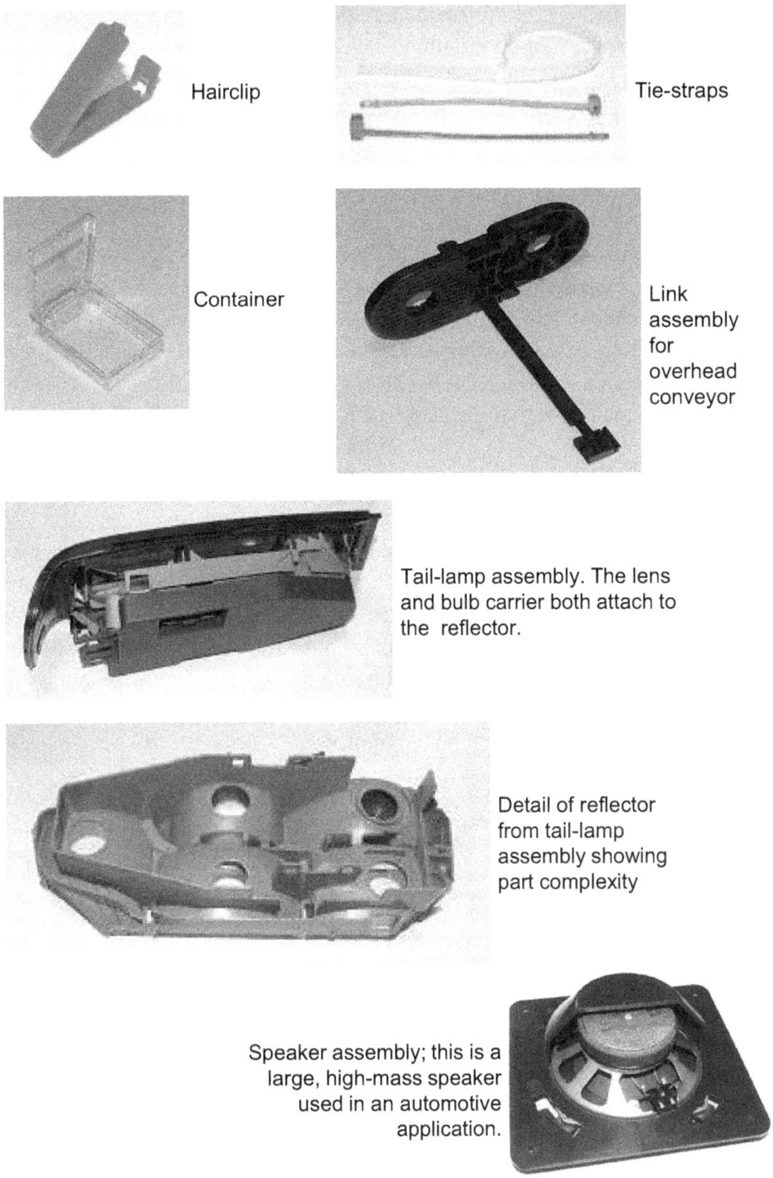

Figure 1.2 Snap-fit application examples

1.4 Snap-Fits and Loose Fasteners

A snap-fit is different from loose threaded fasteners and other mechanical or chemical attachment methods in that it requires no additional pieces, materials, tools, or operations to carry out the part joining function.

The choice between snap-fits or loose fasteners is a major decision point in product development. Chapter 3, Section 3.3, discusses this decision in depth. Neither snap-fit nor threaded fastener technology is inherently good or bad; both have their place in product design based on informed decisions about the best attachment for the application.

Without intending insult to threaded fastener technology (the author spent 30 years as a threaded fastener subject matter expert), we can think of a threaded attachment as a *brute force* approach to connecting parts. The fastener's strength makes it easy to ignore or forget some of the finer points of interface design and behavior. A retention problem can often be fixed by simply using a higher strength material for the fastener, tightening it to a higher clamp load, specifying a larger fastener, or adding more fasteners. Indeed, a major advantage of a loose fastener is that its strength is independent of the joined components. This is not the case with snap-fits.

Experience with threaded fasteners **does not** transfer to snap-fits.

With a snap-fit application, we do not have the luxury of selecting a fastener material and strength that is independent of the joined components. Most of the time, material selection is driven by other application considerations, not by attachment requirements. One must work with the material(s) selected for the parent components. Processing requirements can also restrict design options because the attachment features must be formed with the part. The subtleties of interface design and behavior must be well understood and reflected in the design. A snap-fit application, therefore, must be a more *elegant* method of attachment than a bolted joint.

1.5 Snap-Fits as Interface Systems

The key word here is *system*. In any assembly of individual components, part-to-part attachment occurs across an interface. A successful product development process must treat that interface as a system and it must be developed as the parts themselves are being developed. To start, we will define two major areas of snap-fit technology: *feature level* and *attachment level*.

1.5.1 Feature Level

In a snap-fit application, *locks* are flexible features that deflect for assembly and also latch parts together. The cantilever hook lock, for example, is a very common lock feature used in snap-fits. Calculations for assembly force, assembly strain, and retention strength are necessary to ensure the lock design will meet application requirements. This information is available in numerous publications and design guides. Feature level decisions and calculations will be discussed in Chapters 11, 12, and 13. The traditional lock analysis approach to snap-fit development is reflected in this feature level definition:

A snap-fit is an integral latching mechanism that deflects for assembly then returns to its original position to cause interference that will latch one part to another.

Feature level is the traditional approach to snap-fit development. Feature level lock calculations are necessary and important, but locks do not exist in isolation. They are part of an interface system and the entire system must work. This is the *attachment level* of snap-fit technology and it is the unique snap-fit concept described in this book.

1.5.2 Attachment Level

In contrast to feature level thinking, at the attachment (or systems) level we treat the interface as a system where we will fully comprehend the mechanical environment in which the *entire* interface, including the lock features, functions.

Using attachment level thinking, the developer establishes a fundamentally sound attachment concept. Once a good concept is established, feature level analysis is used to evaluate feature performance and provide feature dimensions. If a good attachment concept is not established first, even well-designed features may fail. Furthermore, if the causes of a problem at the attachment level are not understood, an attempt to fix that problem at the feature level can be more expensive than necessary and possibly doomed to failure.

Attachment level definitions of a snap-fit:

Short version: *A snap-fit is a system of compatible locators, locks, and enhancements forming a mechanical attachment between parts.*

Long version: *A snap-fit is a mechanical joining system where part-to-part attachment occurs using constraint features, that is, locators and locks, which are homogenous with one or the other of the components being joined.*

The attachment level definition includes the feature level but goes beyond it.

Locators are strong, inflexible features providing strength and stability in the attachment.

Locks are flexible features on one part that must deflect for engagement with locators on a second part, followed by return of the lock feature toward its original position to create interference and latch the parts together.

Enhancements complete the snap-fit system, adding robustness and user-friendliness to the attachment.

These definitions highlight significant differences between the traditional feature level approach described above and the more comprehensive attachment level approach to snap-fits. Some new terms introduced here are defined in Section 1.6.1.

Figure 1.3 shows how thinking of a snap-fit as a system rather than just as a lock feature moves the development process closer to the realities of a product application. Look again at the complexity of some of the interfaces shown in Fig. 1.2. It is clear that much more than feature level calculations are required to develop these kinds of applications.

We will discuss lock design shortly and the reader will learn that the locks in this application, while adequate, do not represent good cantilever hook feature design.

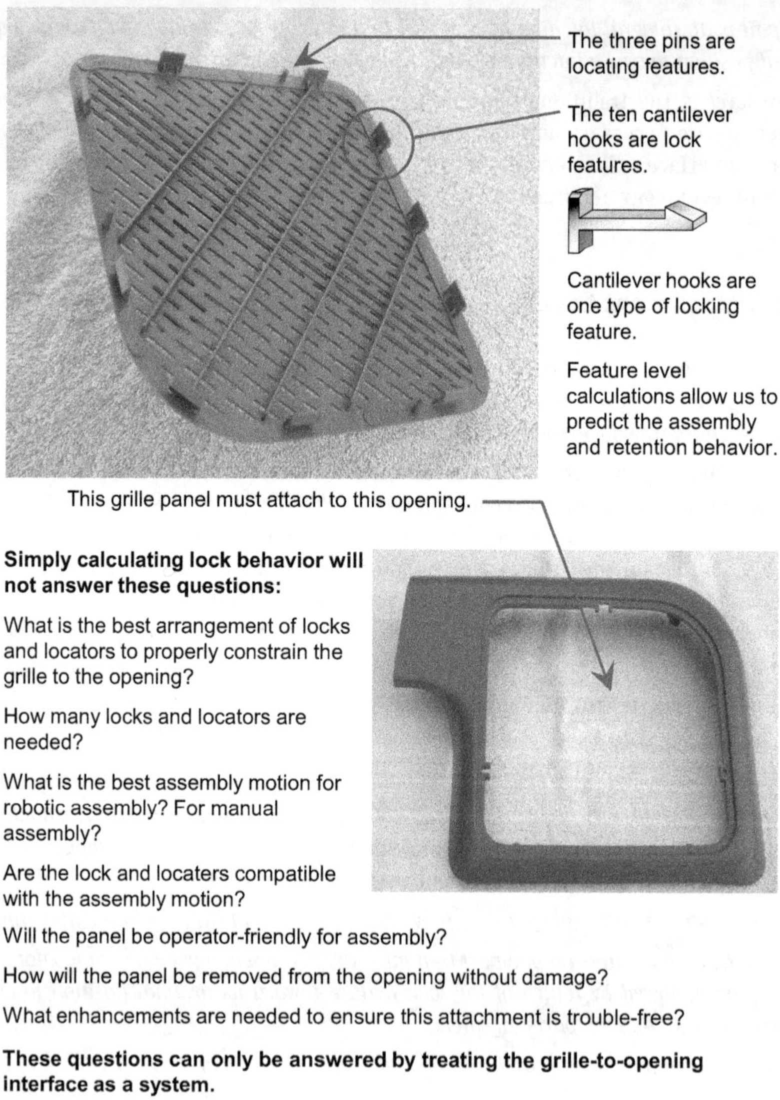

Figure 1.3 A systems or attachment level approach is closer to reality

1.6 The Attachment Level Construct© (ALC)

The snap-fit features, designs, and technology described in this book have evolved and accumulated over many years thanks to the efforts of countless product developers. The author's contribution to the field is in trying to capture, understand, and describe snap-fit technology and in creating the *Attachment Level Construct©* (ALC) to provide an organizing structure and language for snap-fit knowledge.

1.6.1 Attachment Level Terminology

Figure 1.4 shows the Attachment Level Construct©. Its primary sections are:

- **Key Requirements**: In the top row, common technical characteristics shared by all fundamentally sound snap-fits.
- **Elements**: In the second row, attributes used to describe or characterize the snap-fit application as well as physical features or design attributes that make up the snap-fit attachment. Elements are used at specific times during the development process to make decisions about and to build the snap-fit interface.
- **Development Process**: In the third row, steps for creating a fundamentally sound attachment concept and then moving that concept into a successful snap-fit attachment.

In the *Elements* row in Fig. 1.4, some common terms are:

- **Locators**: Constraint or interface features that position or locate the mating part to the base part.
- **Locks**: Constraint or interface features that perform the final latching of the mating part to the base part.
- **Constraint**: The concept of fixing one part to another using locators and locks in such a manner that relative motion between the parts is prevented or controlled.
- **Enhancements**: Additional interface features or attributes that contribute to the attachment's performance and/or robustness in important ways.

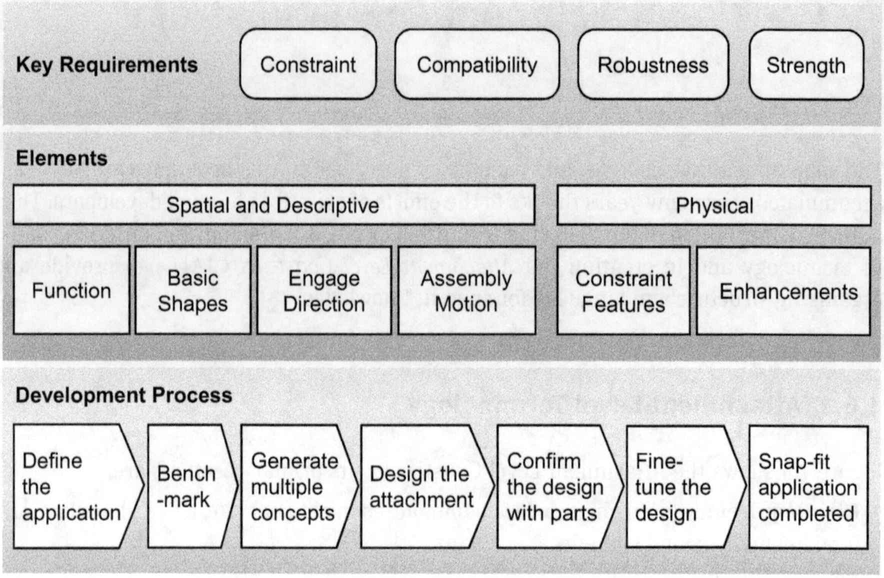

Figure 1.4 The Attachment Level Construct (ALC)

1.6.2 Applying the ALC to Other Attachment Methods

Snap-fit design principles also apply to other fastening methods.

Many of the design principles presented here can, and should, be applied to all mechanical attachments and interface designs, not just to snap-fits. For example, by applying attachment level principles to applications requiring threaded fasteners, it is sometimes possible to reduce the number of loose fasteners used (cost savings) and/or increase the strength and reliability of the attachment (improved quality). This is further discussed in Chapter 10, Section 10.4.

Interestingly, while the use of snap-fit principles can improve other attachments, design principles for threaded fastener joints *do not* apply to snap-fits. This is one reason for problems that occur when developers familiar with threaded fasteners make their first attempt at a snap-fit application.

The reader may want to revisit the brief discussion of snap-fits and threaded fasteners in Section 1.4.

1.6.3 Required Capabilities for Snap-Fit Development

The ALC supports five capabilities necessary for successful snap-fit application development. They are:

- **Communication**: The ALC provides a vocabulary for exchanging ideas and information about snap-fits. Every technical discipline requires a common language if it is to be understood and used effectively.

- **Technical Understanding**: The ALC organizes existing knowledge about snap-fits for easy understanding and use. It also supports capture and transfer of useful snap-fit knowledge and lessons-learned from one application to another. The organizing structure of the ALC also helps the user to grow in knowledge and add to their own technical understanding of snap-fits. Technical understanding includes analytical capability for evaluating feature performance, which is the traditional feature level of snap-fit technology.
- **Spatial Reasoning**: Snap-fit development is enhanced when the developer can visualize the interactions and behaviors of the parts to be joined as well as the features of the parts. The ALC provides a logical set of generic shapes and motions to enable this visualization and quantifies the spatial concept *constraint*.
- **Creativity**: The snap-fit development process, introduced in Chapter 3 and explained in detail in Chapter 10, encourages creativity by supporting rapid generation of multiple attachment concepts for consideration by the developer.
- **Attention to Detail**: It can be easy to forget some of the many design rules and options available when developing a snap-fit application. These details are captured in a logical structure for the developer's recollection and consideration.

1.6.4 Justifying the ALC

A systematic approach to the subject should appeal to anyone wanting to develop snap-fit attachments. But some people have pointed out that the attachment level approach is *too basic*. This author's response is,

Yes it is basic, just as arithmetic is basic to all the higher levels of mathematics. Because it is so fundamental to good snap-fits, it must be understood. Furthermore, it must be written down somewhere.

The author has seen far too many snap-fit applications in which very simple, even trivial errors of commission or omission caused product issues requiring costly redesign.

Some design principles included in this book will seem so obvious that the reader may think, *no one would ever make that mistake*. In fact, every one of those principles is the result of the author having seen that mistake multiple times in products where somebody failed to recognize its importance. That design information is captured here so readers can learn from these mistakes and not repeat them.

Dr. W. Edwards Deming said, "*Experience without theory teaches ... nothing*" [2]. The theory provided by the ALC can greatly improve learning and understanding of snap-fit technology.

The ALC is a tool for organizing and capturing information and concepts: ...*we create constructs by combining concepts and less complex constructs into purposeful patterns... Constructs are useful for interpreting empirical data and building theory. They are used to account for observed regularities and relationships. Constructs are created in order to summarize observations and to provide explanations* [1].

Other comments have mentioned the *jargon* in the ALC, with the term having a negative connotation. To go back to the statement at the beginning of this chapter: *Any scientific discipline has a need for a specific language for describing and summarizing the observations in that area* [1]. Before the ALC was created, there was no consistent and organized terminology and no structured design knowledge for snap-fit technology. Consistency and organization are necessary for accurate communication, understanding, and growth of a subject.

To draw a historical parallel: In the 1700s, Carl Linnaeus, a Swedish botanist and physician, developed his revolutionary taxonomy for classification of species. The organization it provided to the complex plant and animal kingdoms contributed to the proliferation of scientific discovery that followed [6]. Scientists finally had a language and a structure for organizing and understanding their subjects. Linnaeus' classification scheme remains in use today.

1.7 Using This Book

After reading this chapter, if you have not already done so, go back and read the preface to the first edition. This will help you understand the foundations and evolution of the attachment level technology and the how and why of this book.

Engineering managers should read Chapter 15.

Figure 1.5 shows the book's chapters. They are organized around the ALC shown above in Fig. 1.4. Most chapters conclude with a summary of important points introduced in that chapter. Refer to these end sections as quick reviews of the chapter content or use them as an overview before reading the chapter.

Blank space for recording notes is provided at the end of most chapters.

Chapter 1 – Introduction

You are here.

Chapter 2 – Key Requirements

Key Requirements are high-level technical requirements shared by all fundamentally sound snap-fits.

Chapter 3 – Introduction to the Snap-Fit Development Process

This introduction to the development process supports discussions in the chapters that follow. Chapter 10 describes the process in more detail.

Chapter 4 – Descriptive Elements

These are generic terms and concepts for describing a snap-fit application. They also support transfer of snap-fit knowledge between applications.

Chapter 5 – Physical Elements: Locators

Styles of locator features are described. Locators are the strong, inflexible constraint features in an interface.

Chapter 6 – Physical Elements: Locks

Styles of lock features are described in Chapter 6, and their strengths and weaknesses are discussed. Locks are the latching features in an interface.

Chapters 7 and 8 explain important concepts related to the physical elements, locators, and locks, which are introduced in Chapters 5 and 6.

Chapter 7– Lock Strength and Decoupling

Decoupling explains why some lock features are far superior to others for assembly and part retention.

Chapter 8 – Constraint

The most fundamental of the key requirements. Constraint describes and quantifies how the joined parts are properly positioned and latched together.

Chapter 9 – Physical Elements: Enhancements

Enhancements are physical features or attributes of other features in the interface. They are often the kind of design tricks or details an experienced developer may know to use but the novice will not.

Figure 1.5 Book contents

Chapter 10 – Applying the Snap-Fit Development Process
The snap-fit concepts, elements and design rules described in the previous chapters are applied to product development.

Chapters 11, 12, and 13 discuss feature analysis topics.

Chapter 11 – Feature Design: Material Properties
The material properties used in feature calculations are explained.

Chapter 12 – Feature Design: Rules of Thumb
Some general design rules are useful for preliminary lock feature development.

Chapter 13 – Feature Design: Calculations
Beam-based lock calculations are discussed in detail, and modifications to the classic beam calculations are introduced. Calculations for other lock styles are also provided without detailed discussion.

Chapter 14 – Diagnosing Snap-Fit Problems
Just as it guides development, the ALC provides the basis for diagnosing common snap-fit application issues.

Chapter 15 – Gaining a Competitive Advantage in Snap-Fit Technology
An organization can go beyond individual snap-fit expertise and create a sustainable culture of competence to gain a competitive business advantage.

Appendix: Resources – Sources of additional snap-fit information and data.

Figure 1.5 Book contents, continued

1.71 Sample Parts

Snap-fits are a highly spatial and visual topic. The best way, by far, to understand them is to hold parts in your hands. The reader should have snap-fit applications available to study for reinforcement of the design rules and concepts presented here. As you read, use these parts to identify and understand the principles and rules being discussed.

Snap-fit applications are everywhere: find them in toys, electronics, small appliances, vacuum cleaners, etc. They can be found in products as diverse as patio lamps, chemical sprayers, slot-car tracks, and toilet tank shut-off valves. An excellent product for studying a wide variety of snap-fit applications are the old Polaroid One-Step© cameras. They are no longer in production but may be found online and at garage sales. They are 100 % snap-fit and the variety and cleverness of the attachments is impressive.

A suggested exercise as you learn about snap-fits is to critique them on toys, cars, household products, appliances, small electronics, and so on, every chance you get. After a while, you will find yourself noticing how just about every application you study can be improved. Many of the improvements are no-cost; they are simply doing the right thing during concept development and final design.

Study snap-fits outside of your work environment.

The applications and examples in this book were collected over a period of many years from a wide variety of products. They are provided here as idea starters and to illustrate various principles. Use the information in this book to create your own unique products.

Do not simply copy designs out of this book or from parts you study. You must fully understand why and how every feature of a design works. A feature or arrangement of features may work in a particular application or design but not in another. Subtle differences in the product's working environment, materials, assembly methods, use of enhancements, and feature interactions may not be apparent and simply copying a design can result in problems in your application.

Product benchmarking is Step 2 in the snap-fit development process and is discussed in Chapter 10, Section 10.2.

1.7.2 Snap-Fit Novices

A team approach to learning about snap-fits is extremely effective; a group of people can study parts and, using attachment level terminology, discuss the design's good and bad characteristics and its assembly/separation behavior. This will encourage attachment level thinking and reinforce understanding.

A team-based and hands-on approach will enhance the learning process.

If you are a novice in snap-fit development, I suggest that you read the book in order through Chapter 9. You will learn the snap-fit development concepts without getting into any feature calculations. Then, while referring to some real products that use snap-fits, step through the development process described in Chapter 10. Critique the attachments and identify good design practices and discuss how the parts could be improved. Do this with a variety of products - see the above discussion of sample parts. By this time, you should have a good understanding of the concept of the snap-fit as a system. Jump ahead to Chapter 14 and read about diagnosing snap-fit problems. You will likely have already seen some problems if you have done the critiquing exercise suggested above and studied parts on your own.

Finally, read Chapters 11, 12, and 13. These chapters describe material considerations and feature level calculations for feature design.

While Chapter 15 is intended for engineering managers, it contains useful information for everyone and reading it will encourage deeper thinking about the subject. Read it after having some snap-fit experience so that you can better relate to the discussion and ideas presented.

Encourage your management to read Chapter 15 too.

1.7.3 Experienced Product Developers

Most experienced snap-fit developers have learned about snap-fits through intuition and trial-and-error. You will find that your existing knowledge fits well into the snap-fit knowledge model. The ALC will also help you to mentally organize both your existing as well as new snap-fit knowledge.

After finishing this chapter, consider jumping ahead and reading Chapters 7 and 8 where two very important snap-fit concepts are discussed. You have probably already encountered the effects of these concepts in your work. Then come back and continue with Chapter 2.

Chapter 15 discusses engineering business strategies. Experienced developers should find it interesting because of their own experiences. The help and support of the product development community will be critical to success of these strategies in the organization. Encourage your management to read Chapter 15 too.

1.7.4 Design for Assembly/Manufacturing Practitioners

Design for Assembly (DFA) and Design for Manufacturing (DFM) practitioners will be pleased to know the attachment level approach supports and is compatible with those philosophies. The original motivation for creating this material was to support DFA and DFM efforts.

After reading Chapter 1, DFA/DFM practitioners interested in encouraging wise use of snap-fits should read Chapter 3, *Introduction to the Snap-Fit Development Process,* to understand its compatibility with recommended DFA/DFM practices.

Chapter 10 will be of particular interest because it shows how snap-fit development principles can be applied to all fastened interfaces to optimize an attachment interface and reduce the number of loose parts and fastening operations.

Read Chapter 15, it discusses business strategies for embedding snap-fit expertise in the engineering culture. DFA and DFM practitioners can play an important role in causing this to happen. Section 15.3 describes the seven harmful beliefs (briefly introduced in Section 1.2) that can negatively impact implementation of snap-fit technology in an organization. DFA/DFM practitioners are in a unique position to watch for these harmful beliefs and work with management to address them.

Encourage product engineering management to read Chapter 15 too.

DFA and DFM trainers will find the entire book useful as they discuss fastening alternatives and lead product development exercises.

1.7.5 Executives and Engineering Managers

Leaders of companies engaged in designing, manufacturing, and assembling plastic parts, and managers of engineering/product development groups, should read Chapter 15. This chapter explains how to minimize snap-fit issues, reduce engineering structural costs, and gain competitive advantage by implementing snap-fit expertise at both the individual and departmental/corporate levels.

While this chapter is written from the perspective of a large engineering organization, any size product engineering organization can pick and choose the strategies and initiatives that will work for them.

In any case, all executives and managers must read Section 15.3 in Chapter 15. This section discusses beliefs that can slow or even stop implementation of snap-fit technology in an organization. Engineering executives and managers can watch for these harmful beliefs and address them so that snap-fit application decisions are made with a proper understanding of the technology.

Section 15.4 should also be read for an awareness of resource decisions related to snap-fit development.

1.8 Summary

Chapter 1 introduced the idea of a new *attachment level* approach to snap-fit design to supplement the (traditional) feature level snap-fit design practices. Benefits of this systems approach to snap-fit development and design were discussed.

By learning and applying the principles in this book, the reader will:

- Gain valuable insights into exactly how snap-fits work. An additional benefit is an increased and practical understanding of how all mechanical attachments work.
- Be able to design better, more effective snap-fit applications and do it in less time.
- Save product cost and support Design for Assembly/Manufacturing through proper use of snap-fits.
- Learn how to think about snap-fits.

Important points in Chapter 1 include:

- The Attachment Level Construct© (ALC) is a *knowledge model* that organizes snap-fit technology at both the attachment (or systems) level and feature level.
- The term *snap-fit* does not refer to a locking feature; it refers to the entire interface system.
- Experience with popular mechanical attachment methods (loose fasteners across an interface) is not suitable experience for developing snap-fit interfaces. New ways of thinking about function, component interfaces, and attachments must be learned.

Rivets, nuts, bolts, and screws are not snap-fits; therefore, the knowledge does NOT transfer.

- Snap-fit knowledge *does* transfer to other mechanical attachments. Applying attachment level principles can help improve development and design of all interfaces and support Design for Assembly and Design for Manufacturing.
- The root causes of many problems with snap-fit applications are at the attachment level, not at the feature level. Therefore, prevention, diagnosis, and the solution of application problems must start with an understanding of the attachment level.
- Snap-fits involve a level of detail and creativity that can require a certain amount of attachment evolution into its final form. That evolution should occur during *development*, not after part production begins. The snap-fit development process, introduced in Chapter 3 and described in detail in Chapter 10, will reduce the number of design iterations required to reach a superior snap-fit attachment.
- The ALC defines and organizes the design space for snap-fits, explaining it in terms of *key requirements, elements,* and a logical *development process.*

References

[1] Ary, D., Jacobs, L.C., Razavieh, A., *Introduction to Research in Education*, 5th ed., Harcourt Brace College Publishers, Orlando, FL (1996) pp. 27–28

[2] Deming, W.E., *Out of the Crisis*, Massachusetts Institute of Technology, Center for Advanced Engineering Study, Cambridge, MA (1982) p. 19

[3] Goldsworthy, W.B., Hiel, C., *Composite Structures are a Snap*, SAMPE J. (1998) v34 n1, pp. 24–30

[4] Lee, D.E., Hahn, H.T., *Composite Additive Locking Joint Elements (C-Locks) for Standard Structural Components*, Proc. ASC Twelfth Ann. Tech. Conf. (1997) pp. 351-360

[5] Lee, D.E., Hahn, H.T., *Assembly Modeling and Analysis of Integral Fit Joints for Composite Transportation Structures, 93-DETC/FAS-1362*, Proc. 1996 ASME Des. Eng. Tech. Conf., Irvine, CA (1996)

[6] Warne, K., *Organization Man*, Smithsonian, May (2007) pp. 105–111

2 Key Requirements

Chapter 2 introduces the *key requirements* for snap-fit applications. These are common technical characteristics shared by all fundamentally sound snap-fits and satisfying them is the goal of snap-fit application development. These key requirements are the top level of the Attachment Level Construct (ALC), see Fig 2.1.

Meeting specific application requirements like durability, reliability, quality, and ease of assembly will be difficult, costlier, or impossible unless the key requirements are satisfied.

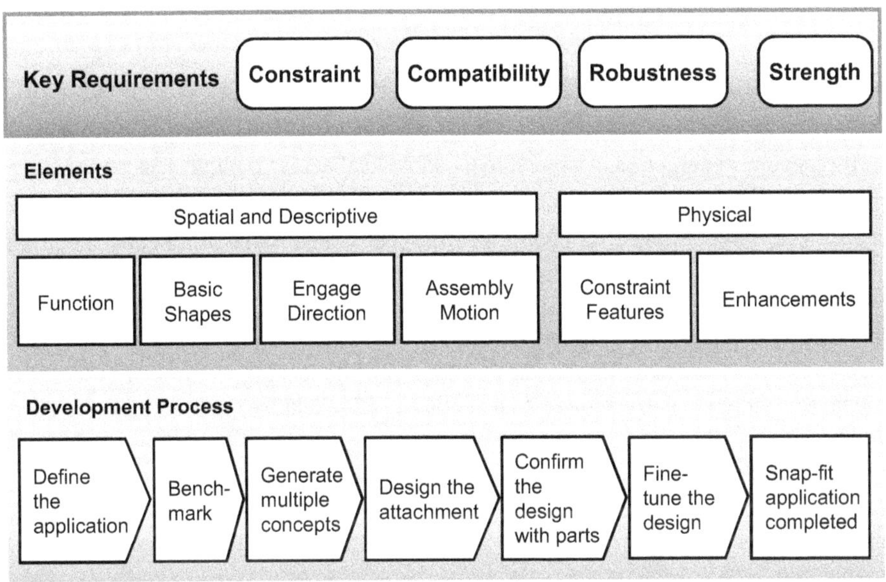

Figure 2.1 Key requirements in the Attachment Level Construct (ALC)

2.1 Constraint

Proper *constraint* is the foundation for a good snap-fit attachment. This is a brief introduction; Chapter 8 discusses the subject in detail.

In a Cartesian coordinate system, linear motion of a free object in space is described by ± translational movement along the three axes and ± rotational movement around the axes. To fix an object in a given location, each of those motions must be *constrained.*

In any mechanical attachment, one part is held in a specific location to another part across an interface. We'll refer to them as the *mating part* and the *base part*, respectively, see Fig. 2.2.

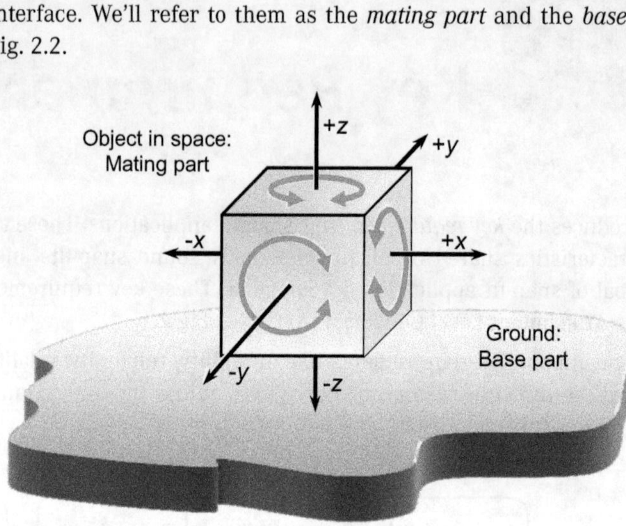

Figure 2.2 Mating and base parts and a Cartesian coordinate system

In threaded fastener joints, friction due to clamp-load across the interface and the fastener's tensile strength provide much, if not all of the constraint to hold the parts together. With threaded fasteners, we usually do not even need to think about constraint, it just happens.

In a snap-fit attachment, there is no real clamp-load. Relative movement of the mating and base parts is prevented by interacting features designed into the parts (Fig. 2.3).

Locating features or *locators* provide positioning while locking features or *locks* latch the mating and base parts in their located relationship. Relative movement is controlled and all forces on the parts are transmitted across the interface through the locator and lock constraint features.

Locks and locators are used in *constraint pairs.* In a locator pair, a locator engages another locator. In a lock pair, a lock engages a locator, although there can be exceptions to this rule.

Success in satisfying the other key requirements depends on a properly constrained snap-fit. Because it describes part-to-part and feature interactions, *constraint* is strongly tied to the concept of a snap-fit as a system.

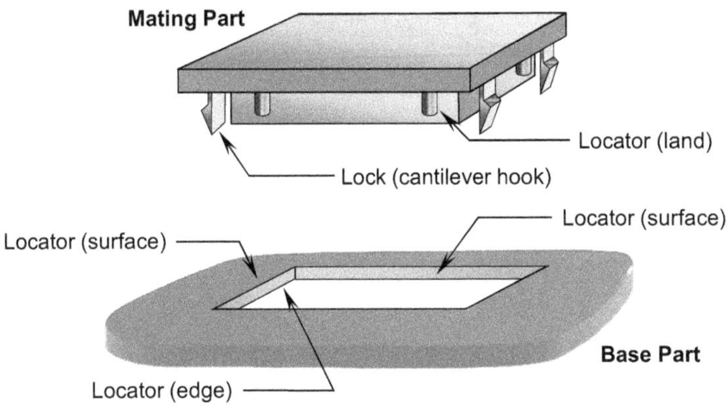

Figure 2.3 Constraint features

2.2 Compatibility

Compatibility is harmony between the elements of the snap-fit interface. Some combinations of part shapes, constraint features, assembly/disassembly motions, and directions can cause difficult assembly or feature damage and should be avoided.

Incompatibility can be a subtle mistake, not easily recognized until symptoms and problems occur. One reason for this may be that decisions affecting compatibility can be made at different times during the development process, sometimes by different individuals.

For example, the door handle application in Fig. 2.4 requires a tipping motion for assembly. But, with this motion, the rigid lugs cannot deflect for engagement with an edge on the mating surface. This causes assembly difficulties in the form of high assembly force, a high scrap rate due to broken lugs on the handles, and the possibility of handles with damaged, but not fully broken lugs, not being discovered until they literally end up in the customer's hands.

In this design, the lug style and locations are not compatible with the assembly motion. This can be fixed by redesigning and relocating the lugs.

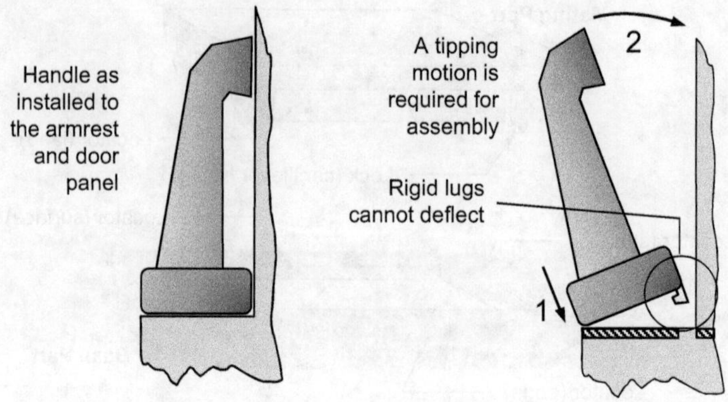

Figure 2.4 Incompatible locator features and assembly motion

Figure 2.5 shows an access panel covering an opening. The application is designed for a linear (push) assembly motion (although a tip assembly motion is also possible, which could result in damage for the same reason the tip disassembly motion can cause damage). This is a second shortcoming of the design.

The access panel can only be removed with a tipping motion by using the finger assist feature at one end of the panel.

The lock features (cantilever hooks) are designed for a push assembly motion where the deflection required for engagement is shared across the opposing hooks. A tipping motion during disassembly results in most of the hook deflection occurring at the finger assist end of the panel. Thus, those hooks can be weakened or completely broken because the deflection is not distributed across both sets of locks.

The locking features are not compatible with the tipping motion required for assembly or disassembly. This can be fixed by redesigning the interface so only a tipping motion is possible for both assembly and separation, and then ensuring the lock features are properly selected and designed for the required deflection.

Important compatibility rules are:

- All physical features in the interface must be compatible with the assembly motion.
- The selected assembly motion must be compatible with the basic shapes.
- The assembly and disassembly motions should be the same, but opposite in direction, unless special design provisions are made.
- Allow sufficient clearance for lock feature deflection during both assembly and disassembly.

These are simple and seemingly obvious rules, yet they are violated. Both examples of compatibility violations shown were found on product applications.

Compatibility violations may not prevent assembly or cause feature damage, but they may make assembly/disassembly more difficult than necessary or they may limit design options as part development proceeds.

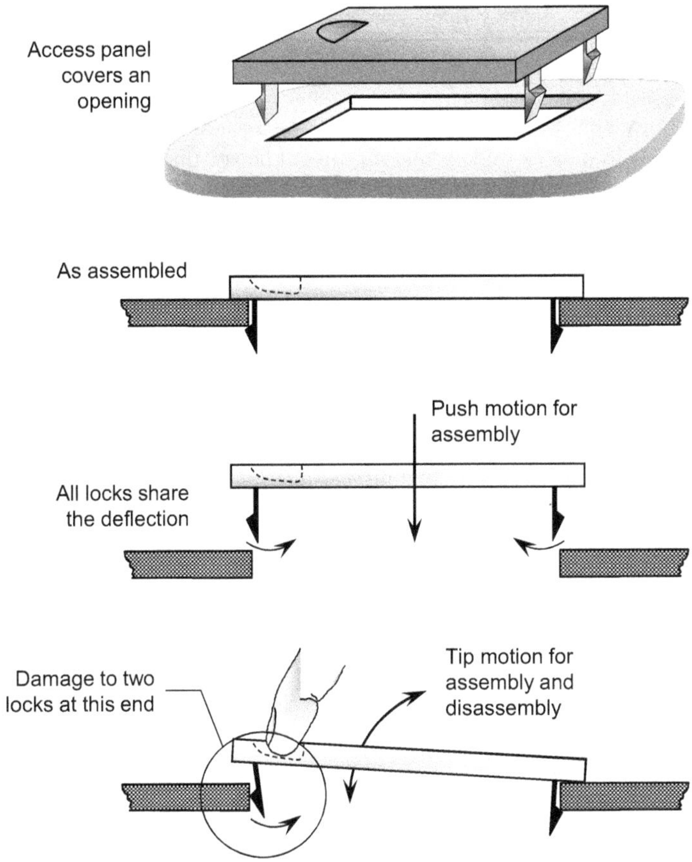

Figure 2.5 Incompatible lock features and assembly/disassembly motions

IMPORTANT:

Do not consider every figure in this book to represent a complete snap-fit attachment design. Many figures support a specific discussion and to keep the figure simple, details are omitted. For example, Fig. 2.5 supports a discussion about incompatibility so only the relevant features and motions are shown. Other important interface features like locators and enhancements are omitted. Likewise, design details like radii, draft angle, and others are not included in the figures. When a figure is intended to represent a complete and correct snap-fit attachment, it will be identified as such.

2.3 Robustness

Robustness in snap-fits is more than just tolerance to dimensional variation.

As a key requirement for snap-fit applications, the term *robustness* means tolerance to many kinds of variation as well as tolerance to unknowns that can affect the reliability, durability, and quality of an attachment in many ways. Variables and unknowns in the life of a snap-fit can include

- Design of the mold(s) used to make the parts. This includes dimensional variation but also cooling and flow capability.
- Variation in the manufacturing process, including cycle time, mold wear, and raw material variation.
- The conditions under which the parts are assembled or disassembled, including temperature extremes.
- The customer's ability to use or operate the snap-fit as intended.
- The possibility and extent of misuse.
- Applied forces that are unanticipated in the design.
- A service technician's or customer's ability to disassemble and reassemble the attachment for service without damage.

Many of these variables and unknowns require more than simply part strength to ensure robustness. *Enhancements* are very important to an application's robustness and are discussed in Chapter 9.

The concept of *robustness* will be illustrated with the discussion of strength in the next section.

2.4 Strength

Strength alone will not guarantee a robust or a reliable attachment.

We are familiar with strength because it is the basis for the feature level approach to snap-fit design. But strength is a potential. The attachment must be robust in many ways to ensure it can meet its strength requirements.

Assembly strength is the ability of the interface, particularly flexible lock features, to survive part assembly without damage.

Separation strength means maintaining part-to-part constraint without looseness, breakage, or squeaks and rattles throughout the product's *useful life.* Useful life includes part handling, operation by a user, and disassembly and reassembly for maintenance or repair.

Strength and robustness are sometimes confused, the thinking being that if an attachment is strong, it is also robust; this is not necessarily true. The following example illustrates the difference between strength and robustness.

Figure 2.6 shows an application in which a small reflector attaches to a recess in an automotive application: an inner door panel. The reflector must stay in place if bumped or kicked, it must resist the occasional jarring and vibration that can occur in a vehicle, and it must resist the deceleration forces of door closure. Since the reflector's mass is very low, these forces are also low and holding the reflector in place should be easy.

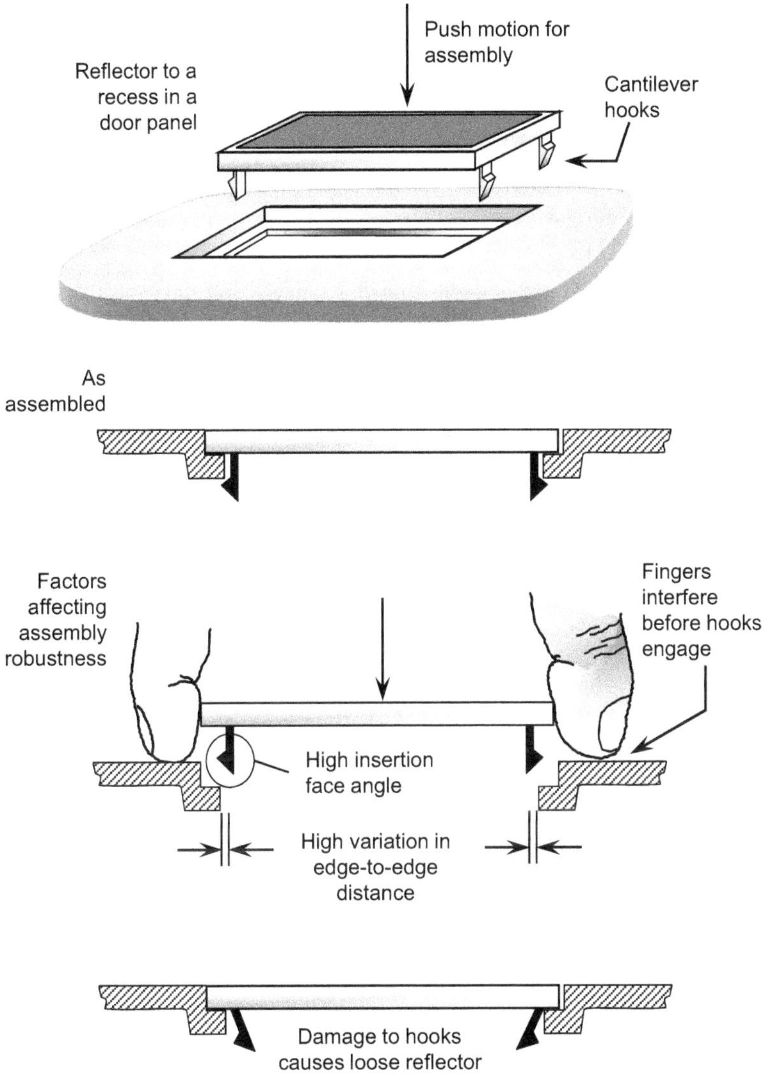

Figure 2.6 Strength vs. robustness in a reflector application

The locking features are four cantilever hooks, and when the reflector is carefully assembled, they are strong enough to hold the reflector in place. But the hooks are prone to damage during assembly, and reflectors with damaged hooks will become loose. Therefore, while the hooks themselves are strong, this design is not robust to the assembly process. The factors that reduce the interface's robustness and contribute to hook damage are:

- It is difficult for an assembler to grasp the relatively small reflector for proper placement because their fingers interfere with the panel surface before the hooks begin to engage the opening's edges. Misalignment as the reflector is pushed into the opening can damage the hooks.
- The assembly is performed blind; the reflector and the assembler's hand interfere with good visibility of the lock features' alignment with the opening.
- The insertion faces on the hooks are very steep, and as the hooks deflect, those faces become steeper. This will further increase an already high assembly force and makes a difficult assembly even more difficult.
- Variation in trimming the door panel covering at the opening causes high variation in the edge-to-edge distance across the opening.

The hooks in this example had enough *strength* to survive normal assembly deflections and to hold the reflector in place once it was engaged. But this hook retention strength was not enough. The entire system was not *robust* to the assembly process; therefore, the attachment was not reliable.

Strength is generally the ultimate goal of an attachment. But strength is a *potential*; it cannot be achieved unless the other three key requirements are met.

This discussion is limited to strength and robustness as key requirements. In addition to the issues discussed above, while the locks in this application were strong enough to hold the reflector in place when properly assembled, they were not well-designed. The application will reappear in later discussions about lock feature design and selection.

2.5 Summary

Chapter 2 described the four *key requirements* that are at the top row of the Attachment Level Construct. These are high-level characteristics shared by all reliable snap-fits. Their relationship with snap-fit attachment goals is shown in Fig. 2.7.

Note that strength appears twice. The figure illustrates how feature strength is only one component of the attachment's strength.

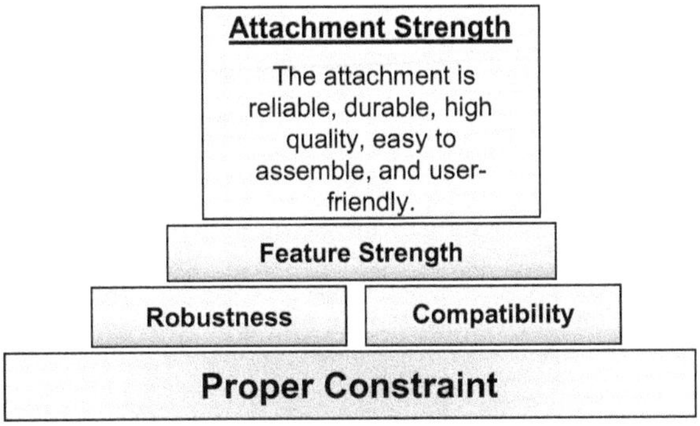

Figure 2.7 Key requirements and the attachment's goal

Important points in Chapter 2 include:

- Every snap-fit must satisfy four key requirements: *constraint*, *compatibility*, *robustness*, and *strength*.
- Attachment strength is the goal of most snap-fits and is one of the key requirements. To have reliable strength, a snap-fit must satisfy the other three key requirements.
- Constraint is the most fundamental key requirement. Proper constraint is required for success in meeting the other requirements. Many snap-fit problems can be traced to improper constraint.
- Robustness and strength are not the same thing. Strength is a potential, robustness will help ensure strength requirements are met.
- The assembly motion must be compatible with the basic shapes and constraint feature style and location.
- The assembly and disassembly motions should be the same, (tip/tip or push/push, for example) but opposite in direction. If they are the same, special care is required to ensure constraint features will not be damaged.
- Specific application requirements (durability, usage cycles, quality level, appearance, etc.) as described in a product's performance specifications cannot be efficiently or consistently met unless the key requirements are satisfied.

3 Introduction to the Snap-Fit Development Process

In Chapter 2, we described the *key requirements* of a snap-fit application. We will skip over the *elements* row in Fig. 3.1, and jump to the snap-fit *development process* shown in the bottom row.

The development process is introduced now to provide some context for the contents of Chapters 4 through 9. Then in Chapter 10, we will step through the development process in detail and discuss how the elements (row 2 of the ALC) and concepts described in Chapters 4 through 9 are used in the development process.

This is an idealized process and the realities of product development may force some modifications, but the core principles of the process will always apply.

The terms *development* and *design* are sometimes used interchangeably. In this book they have very specific meanings:

Design is a subset of the development process.

- *Development* is the entire process, from concept development through completion of a snap-fit application.
- *Design* is the development step in which feature behavior is analyzed, dimensions and tolerances are established, and detailed drawings are made. It is the fourth step in the development process.

In Section 1.6.3, five important snap-fit development *capabilities* were defined: technical knowledge, spatial reasoning, attention to detail, creativity, and communication. The development process enables the snap-fit developer to apply these capabilities.

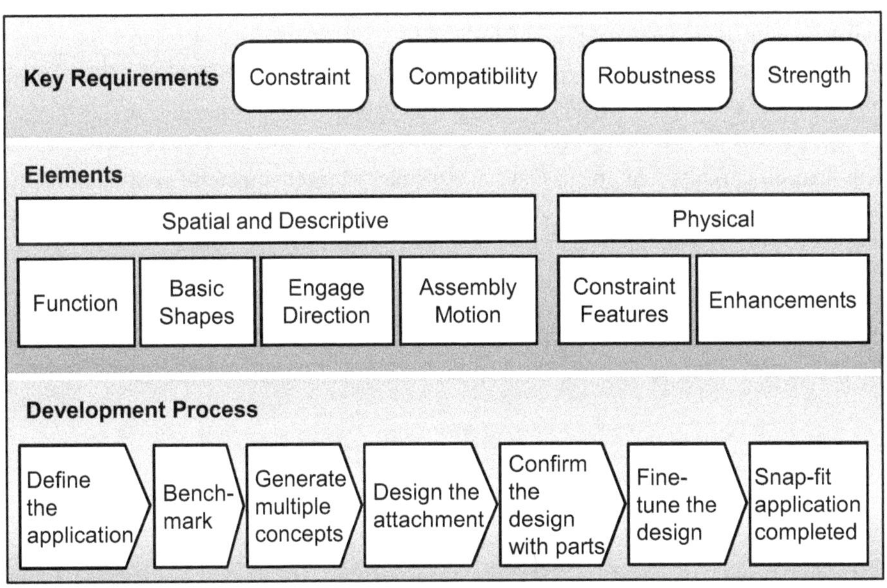

Figure 3.1 The snap-fit development process in the ALC

The development process and related decisions may appear intimidating and possibly not worth the effort. The reader should keep in mind that this is no different than adopting and learning about any new technology. Once embedded into the engineering culture, the process becomes automatic and easy. If the reader has not yet developed a snap-fit application, you are at the beginning of the learning curve and things will quickly become easier.

3.1 Concept vs. Detailed Design

Early in the development process, time is spent defining, benchmarking, and developing concepts before the actual design work begins. The reader might wonder, *why can't I just start designing my parts right away?*

It is tempting to jump quickly to part design. If an existing part is simply being modified, there may be security in knowing the attachment has been used before. But simply copying or modifying what has been done before prevents consideration of other attachment options. It can also lead to repeating mistakes if the design being copied is not very good. Jumping directly into part design also ignores the very important *concept development* step.

Much of a product's quality and cost are established during concept development.

Studies [2, 3] have shown that as much as 70 to 80% of a product's *total installed cost* is established (locked into the product) by decisions made during the concept stage. Studies have also shown that changes made later in the development process become much more expensive, and once tooling has been made or the parts are in production, the cost to make changes (fixes) is often prohibitive [4, p. 128]. Other studies note the high leverage one has over the product in the concept stage in terms of quality and the ability to implement changes [5].

In other words, the concept stage can make or break an application in terms of both total installed cost and quality. This is a basic principle of design for assembly and is true of the attachment as well as the product as a whole.

This snap-fit development process does not conflict with other general product development processes. In Fig. 3.2, we compare the snap-fit development process to a process for plastic part development described in Malloy [4, p. 130].

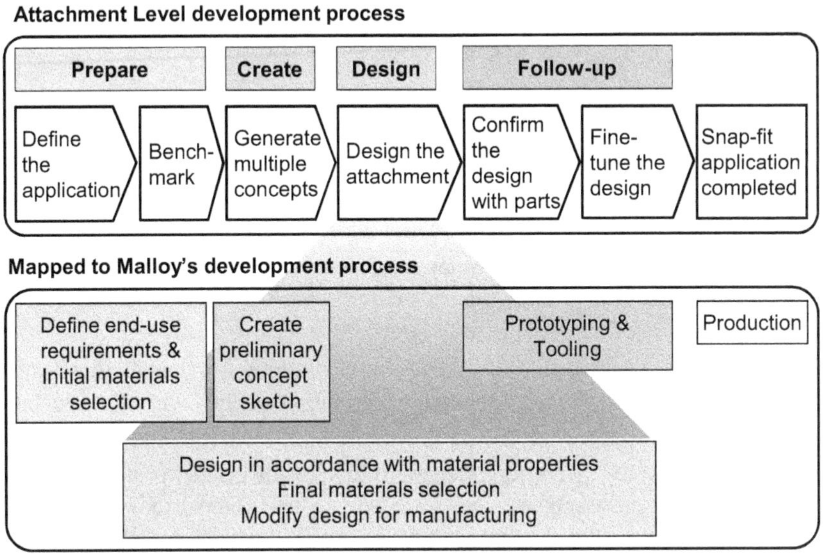

Figure 3.2 The snap-fit development process compared with another development process

3.2 The Value of Multiple Concepts

When we start with concept development, if only one concept is considered before jumping into design, we cannot be certain it is the optimal attachment for the application. Selecting from several concepts ensures a *best* concept is chosen. In the first three steps of this process, multiple attachment concepts are generated and one *preferred* concept is selected. This need not be difficult or time-consuming.

Creating a few concepts is not a lengthy or difficult part of the process.

Generating concepts requires some creativity, but creativity without understanding can be counterproductive. The best approach is a form of *controlled creativity* in which some preparation occurs.

Hands-on activities are an important part of the creative process during product development. They include making concept sketches, handling real parts during benchmarking, and sometimes making crude models of parts. These activities are critical to the spatial reasoning and creative aspects of the development process and they should not be ignored: *...the hand speaks to the brain as surely as the brain speaks to the hand* [1].

Get parts into your hands!

Once a preferred concept is selected for further development, the process proceeds through design, prototyping, testing, and if required, changes. One of the goals of this process however, is to minimize the need for these changes once a design is completed.

3.3 Step 0: Is a Snap-Fit Appropriate?

The very first step occurs before the formal process begins. A decision must be made if a snap-fit is appropriate for the application.

If snap-fits are a new experience, everyone in the engineering organization (including management) must understand a few things about snap-fits:

- The time and effort required to develop a reliable and cost effective snap-fit attachment will most likely exceed the time spent on a more traditional attachment for the same application.
- Benefits that far exceed the initial engineering costs are realized when that design is assembled thousands of times (by human operators or robots) with no tools and no loose fasteners.
- The business case for a choosing a snap-fit attachment instead of a loose fastener must comprehend the assembly's *total installed cost.* This is important because if only part piece-cost is considered, parts with snap-fits will likely have a higher piece-cost and the business case may not be made for a snap-fit.

Product decisions cannot be based on piece-cost alone.

- If a business case based on total installed cost supports a snap-fit, the next step is to decide if the effort is likely to succeed. Tables 3.1 through 3.4 list potential development issues. Some of the technical items in these tables may rule out a snap-fit, but many will simply make development more difficult or more time-consuming. *Unfavorable* answers may not prevent use of a snap-fit, but an application with many *favorable* answers is probably a better candidate.

Consider the following four tables as "thought starters" for beginning an application development discussion. A second method that focuses on the *resources* required for snap-fit application development is presented in the next section.

If an application is a candidate for a snap-fit attachment, the development process can begin.

VERY IMPORTANT!

However, even if a snap-fit attachment method is not chosen, time has not been wasted. The thought process required to understand the application's attachment issues will result in a better product regardless of the final attaching method.

In Chapter 10, we will see that the snap-fit or loose fastener decision often does not have to be made until the design step of the process.

The initial steps of this process can be applied to attachment development in general.

Table 3.1 Application Considerations

The checked response is generally more favorable to use of a snap-fit.	Yes	No	Why
Are manufacturing volumes high?	✓		Must recover higher initial costs.
Does a validation procedure exist for the application and will it test the snap-fit?	✓		End-use testing is important.
Are performance requirements available for the application?	✓		Snap-fit must meet them too.
Is the application spring-loaded? Can it fly apart during assembly or service?		✓	May cause injury, a "booby-trap."
Is sealing required in the application? Will gaskets be used?		✓	Sealing may require clamp load.
Is clamp load required in the application?		✓	Plastic snap-fits cannot, as a rule, provide reliable clamp load. Special care is required in assembly motion / feature selection and design.
Will high or sustained forces be applied to the attachment?		✓	Increases possibility of plastic creep.
Will the application experience shock or impact loading?		✓	Careful analysis & strong constraint features are needed.
Is the application subject to functional or structural loads?		✓	Careful analysis & strong constraint features are needed.
Is the mating part low mass and subject to only acceleration or deceleration forces?	✓		Constraint feature strength is less of an issue.
Is the application subject to a high frequency of service?		✓	Take this into account when setting maximum allowable strain in the lock.
If service is required, is disassembly obvious or are instructions available?	✓		Reduces chances of damage.
Is there adequate space on the parts for snap-fit features?	✓		Space for lock deflection & protrusions.
Is the mating part: Trim Bezel Panel Cover Switch	✓		These applications often have little or no functional loads.
Is either of the parts expensive?		✓	Backup attachment may be necessary. Extra care required in development.
Do federal safety, health, or other standards regulate the application?		✓	If so, thorough documentation is required.

Table 3.2 Material Considerations

The checked response is generally more favorable to use of a snap-fit.	Yes	No	Why
Is one or both of the parts made of plastic?	✓		Easier to do a snap-fit in plastic.
Do the joined materials differ signif-icantly in rate of thermal expansion?		✓	Care is needed in developing constraint.
Are the parts made of "engineering" polymers?	✓		More predictable & higher performance.
Is the application exposed to ultraviolet light?		✓	Performance degradation is possible.
Is the plastic exposed to chemicals in the environment?		✓	Performance degradation is possible.
Is high dimensional variation likely?		✓	Care is needed in developing constraint.
Is the application used in a high temperature environment?		✓	Performance & long-term degradation is possible.
Is the application used in an extreme low temperature environment?		✓	Causes brittle behavior in plastics.

Remember, snap-fits are not limited to plastic-to-plastic applications. They are also appropriate for some plastic-to-metal and metal-to-metal applications. Material properties are, of course, different for metal parts, but the fundamental snap-fit rules will still apply.

Table 3.3 Information and Data Considerations

The checked response is generally more favorable to use of a snap-fit.	Yes	No	Why
Do you have access to polymers expertise?	✓		Materials data interpretation and guidance is critical to success.
Do you have access to processing expertise?	✓		Needed guidance for process-friendly design.
Will accurate load information be available for analysis?	✓		A necessity for critical or high demand applications.
Is accurate material property data available for both parts to be joined?	✓		Needed for accurate analysis.
Will accurate dimensional data be available?	✓		Used to determine position & compliance.
Is part/base packaging known or predictable?	✓		For needed access for assembly motions & service.
Do you know the possibility of misuse or unexpected loads on the attach-ment?	✓		Needed for complete analysis of reliability.

Table 3.4 Organizational Considerations

The checked response is generally more favorable to use of a snap-fit.	Yes	No	Why
Do you have design responsibility for both the mating and base part?	✓		It is much easier if one person or one design group has responsibility for both the mating and base parts.
Does your organization have local design responsibility for both parts?	✓		Coordination and communication is much easier.
Do different companies or suppliers have responsibility for the mating part and base part?		✓	Communication and cost-sharing decisions will be difficult. Monitoring and coordination is critical.
Is one or both of the mating part and base part development teams overseas?		✓	Communication and cost-sharing decisions will be difficult. Monitoring and coordination is critical
Is there enough lead time to accommodate a possible longer design time?	✓		A longer development time for snap-fits is possible.
Does the organization understand the difference between a piece-cost penalty and savings in total installed cost?	✓		Understanding will improve management support for the effort.
Does the part supplier(s) have experience with making snap-fit applications?	✓		They should have a deep understanding of manufacturing requirements and issues.
Will the prototype supplier also be the final supplier?	✓		If so, they will learn much from prototype development.
Will the part supplier(s) participate in design meetings?	✓		They are a stakeholder in the design and can give advice during development.

The developer's attachment choice is frequently between threaded fasteners or a snap-fit. Table 3.5 lists some of the advantages and disadvantages of each. The pros and cons are generally applicable for plastic-plastic, plastic-metal, and metal-metal attachments.

Table 3.5 Snap-Fit and Threaded Fastener Technology

Snap-Fit Technology	
Pros	**Cons**
▪ Fewer parts in the product therefore in inventory. ▪ Reduced assembly time. ▪ No visible fasteners, clean appearance. ▪ Can be made nonreleasing and permanent. ▪ Can give direct feedback to operator of good assembly. ▪ No investment for power tools. ▪ Can be made tamperproof or tamper-evident.	▪ Parts are more complex and piece-cost is generally higher. ▪ Tooling cost can be higher ▪ Development costs can be higher. ▪ Close control of dimensions is required. ▪ No adjustment possible after assembly. ▪ Lock feature strength is limited by parent material strength. ▪ Hidden locks may be hard to service. ▪ Snap-fit knowledge and development capability may be limited.

Table 3.5 *(Continuation)*

Threaded Fastener Technology	
Pros	**Cons**
▪ Robust to dimensional variation. ▪ Can allow adjustment after assembly. ▪ Fastener strength is independent of joined material strength. ▪ The attachment interface is generally simple. ▪ Tooling cost can be lower. ▪ Part processing is easier. ▪ Initial development cost is usually lower. ▪ Supports low-volume productions. ▪ Part piece-cost is lower. ▪ Disassembly for service is obvious.	▪ Clamp load may crack plastic. ▪ Additional parts required in the product and in inventory. ▪ Each fastening site may require as many as three additional fasteners: screw, washer, and nut. ▪ Additional assembly labor/time. ▪ Assembly tools incur capital costs, require energy to run, and require maintenance. ▪ Visible fasteners may be undesirable on beauty' surfaces. ▪ Fasteners may strip during assembly, creating a hidden failure.

3.4 The Demand-Complexity Matrix©

Chapter 15, Section 15.4, is a detailed discussion of the Demand-Complexity Matrix©.

Every product development effort requires decisions about the resources required for successful completion. This is not unique to snap-fits, but because of possible unfamiliarity with the technology, a discussion of factors that can influence the need for these resources is important.

Engineering organizations should already have an understanding of the resources needed to develop threaded fastener attachments. A change from a threaded fastener to a snap-fit interface is a significant change from an attachment method that relies on clamp-load to one that cannot provide clamp load. The organization should understand what that shift entails and allocate resources accordingly.

Some applications are relatively easy to adapt to a snap-fit attachment. For a beginner (individual or organization), these applications are a good starting point for gaining experience and confidence. At the other end of the spectrum are applications involving some combination of interface complexity, expensive parts, and critical performance/reliability requirements. The Demand-Complexity Matrix© can help an organization evaluate a potential application by considering its own capabilities and the resources required to create a snap-fit attachment. Every organization will find its own comfort level on this subject.

A simplified matrix is shown in Fig. 3.3. Look again at the applications shown in Chapter 1, Fig. 1.2. Imagine where you would place them on this matrix. The matrix is presented in much more detail in Chapter 15, Section 15.4, where the discussion is directed to engineering managers who must make resource allocation decisions. However, all readers should, sooner or later, take a look at Chapter 15 for more insights into the role snap-fits can play in a product development organization.

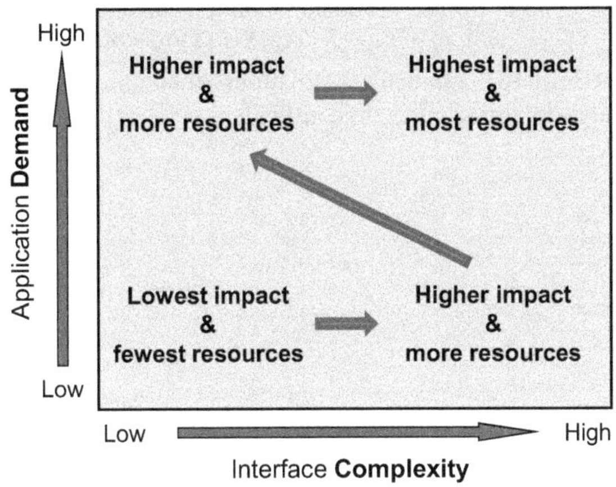

Figure 3.3 Simplified Demand-Complexity Matrix

The matrix is a continuum of application demand vs. interface complexity as well as a number of other potential factors. The total effect of the factors can be thought of as *Impact* on specified desired or undesired outcomes of the development project.

- *Demand* - Requirements for quality, reliability, and durability in the application as well as consideration of the environment in which the application must perform. For example, applications involving safety, very costly parts, and where (costly) loss of function could occur would belong higher on the demand axis.
- *Complexity* - Interface complexity, mating and base part geometry, and the parts' manufacturing requirements are part of the complexity in the matrix.
- *Resources* - Expected resources needed for successful completion of the snap-fit development process. This knowledge can help guide the decision to proceed with a snap-fit attachment. Resources considered may include in-house and supplier talent, testing, and validation.

The path from low demand and low complexity to higher demand and more complex applications is a suggested progression for individuals and organizations with limited snap-fit experience.

Resource needs are presented on the matrix as a function of complexity and demand because they are normally the most fundamental factors in resource decisions. Other factors can add more dimensions to the two-dimensional demand-complexity matrix shown here. Additional factors include:

- *Information* - Confidence in and availability of reliable information about part materials and application requirements.
- *Design capability* - The level of material and snap-fit expertise available within the design-responsible organizations.
- *Manufacturing capability* - The mold-maker and part manufacturer's expertise.
- *Communication* - The preferred scenario is when one individual, one department, or one supplier (in that order) is responsible for developing both the mating and base parts in the application. Development is more difficult when the parts to be joined are

the responsibility of different suppliers. Difficulties are compounded if the suppliers are distant from each other, and will be further compounded if they are in different countries.

- *Cost distribution* - The preferred scenario is similar to that described for communication. Design decisions will drive some costs into either the base part or the mating part. If these parts are the responsibility of different suppliers, making the best decision for attachment quality can become difficult.

3.5 Summary

This chapter introduced a development process for snap-fit applications.

Important points in Chapter 3 include:

- The design step is only one part of the entire product development process.
- A significant portion of a product's total cost and quality is determined in the concept stage of development, *before* detailed design begins.
- Part piece-cost may be higher and a snap-fit application may cost more to develop than an application with loose fasteners. Significant savings are realized when lower total installed cost is matched with high production volumes. A snap-fit may not be appropriate for low-volume products.
- The development resources required for snap-fit development are a function of many factors. Two of the most influential are *demand* and *complexity*.
- In Chapter 10, *Applying the Snap-Fit Development Process*, we will step through this process in detail and show how the elements described in the proceeding chapters are applied in the development process. As the elements are discussed in the following chapters, the reader should refer back to this figure from time to time.
- The next three chapters describe the elements in the center row of the ALC. Figure 3.4 shows where these elements are used in the development process.

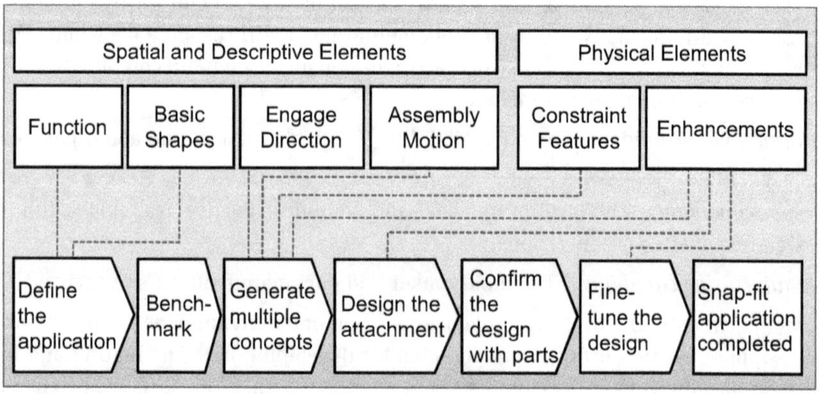

Figure 3.4 Snap-fit elements and the development process

References

[1] Wilson, Frank R., *The Hand*, Pantheon Books, New York (1998) p. 291

[2] Boothroyd, G., *Design for Manufacture and Life-Cycle Costs*, SAE Design for Manufacturability, TOPTEC Conference, Nashville, TN (1996)

[3] Porter, C.A., Knight, W.A., *DFA for Assembly Quality Prediction during Early Product Design*, Proc. 1994 Int. Forum Des. Manuf. Assem., Boothroyd Dewhurst Inc., Newport, RI (1994)

[4] Malloy, Robert A. *Plastic Part Design for Injection Molding,* Hanser Publications, Munich, Germany (1994)

[5] Ford, R.B., Barkan, P., *Beyond Parameter Design - A Methodology Addressing Product Robustness at the Concept Formation Stage*, Natl. Des. Eng. Conf., Chicago, IL (1995)

4 Descriptive Elements

Chapters 2 and 3 discussed the top and bottom rows of the ALC. The center row of the ALC contains the spatial/descriptive and physical elements used to describe and construct the snap-fit interface. Those elements and some related concepts are the subjects of Chapters 4 through 9.

Collect parts that use snap-fits. As new terms are introduced, identify them on the parts.

The first four elements in this row provide useful concepts and a consistent terminology for capturing, organizing, and describing snap-fit attachments, see Fig. 4.1. These elements facilitate:

- *Spatial reasoning,* which is very important to successful snap-fit development.
- *Knowledge transfer* between snap-fit applications. This powerful concept is discussed in Section 15.7.5, "Create and Maintain a Library of Preferred Concepts."
- *Creativity,* which is supported by two of these elements. *Engage direction* and *assembly motion* both help the snap-fit developer create fundamentally different interface options during the development process.

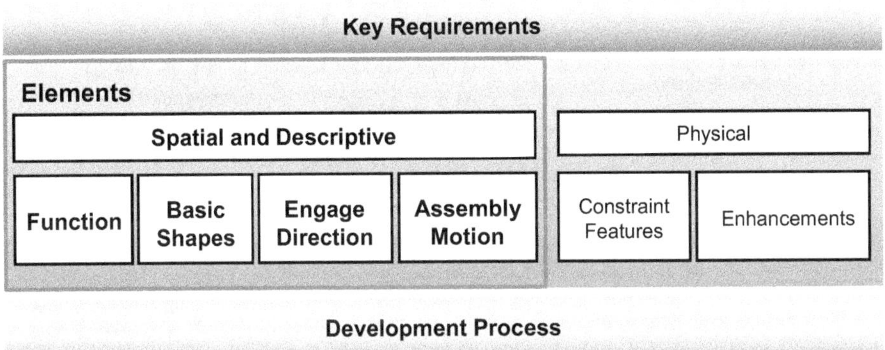

Figure 4.1 Spatial and descriptive elements in the ALC

4.1 Function

We know a snap-fit attachment holds parts together, but there is more to it than that. We'll use the concept of *function* to define the attachment's role in terms of *action, purpose,* attachment *type, retention,* and *release.*

4.1.1 Action

Action is the potential for movement designed into the attachment, see Fig. 4.2.

In *fixed* snap-fits, no relative motion between parts can occur. Once the mating and base-parts are locked together, the application is constrained in exactly twelve degrees-of-motion, (DOM). Most snap-fit attachments are fixed.

In *moveable* attachments, relative movement between the joined parts can occur. When the motion is not limited by constraint features, it is *free* motion and the application is always constrained in less than twelve DOM. When motion is regulated, as by a detent feature or a lock, it is *restricted* motion and the application is *sometimes* constrained in less than twelve DOM.

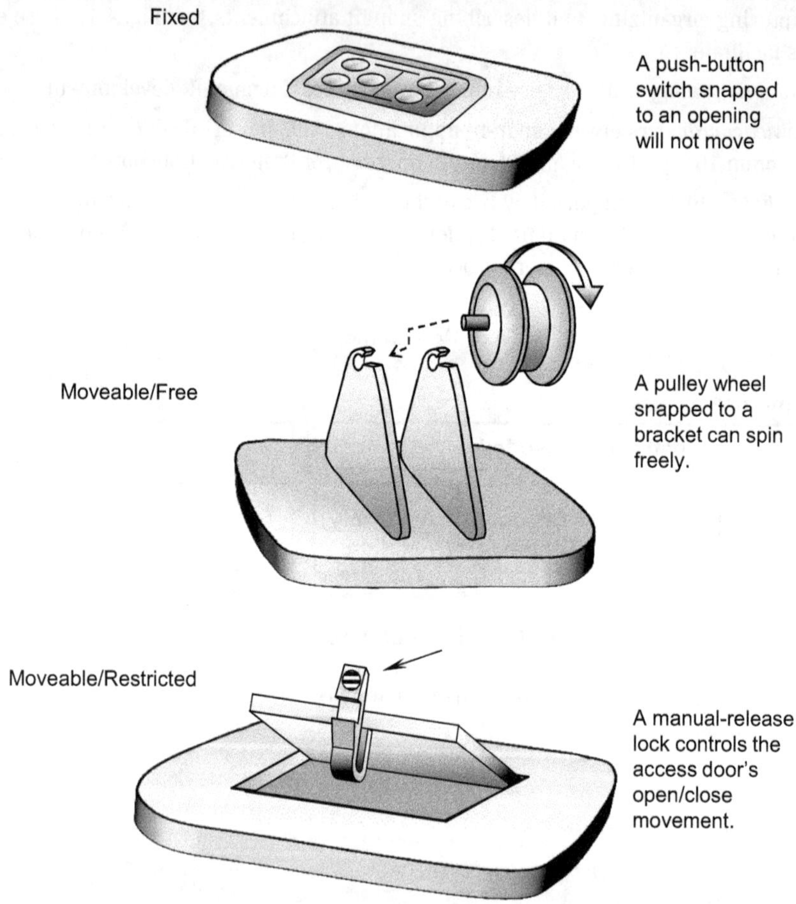

Figure 4.2 Action-fixed and moveable attachments

4.1.2 Purpose

The snap-fit may be the *final* attachment or it may be *temporary* or *utilitarian*, to be used as an aid to another (final) assembly operation.

The snap-fit is final when it is the only attaching method holding the application together for its useful life. Most snap-fits fall into this group, and all the examples shown thus far have been final attachments.

Temporary/utility snap-fits need to be reliable enough to hold the mating-part in position to the base-part only until the final attachment is made.

Temporary snap-fits can support design for assembly by allowing buildup of several parts prior to final attachment. They may reduce processing costs by allowing a more efficient and/or less expensive final attaching process to be used. For example, if parts can be held together by a temporary snap-fit, a slow-cure adhesive may be feasible as a final attachment instead of a rapid-cure adhesive requiring an oven/heating cycle.

Temporary snap-fits are also indicated in applications that require manual handling and manipulation of multiple parts, loose fasteners, and tools. They are particularly useful where hands must be free to perform other tasks and part positioning is required. Examples include extreme environments like elevated assembly operations (high construction and air-frame assembly for example) and out-of-vehicle operations in space or underwater.

Figure 4.3 illustrates one spacer in an application where spacers must be in-place at multiple sites between parts before final assembly is performed. Temporary snap-fits can hold each spacer in place.

In this example, a ring of flexible fingers (trap style locks) holds a spacer in place on a shaft until a second part is added and a final attachment is made elsewhere.

The shaft could be a bolt and the final attachment could be a nut holding the second part in place.

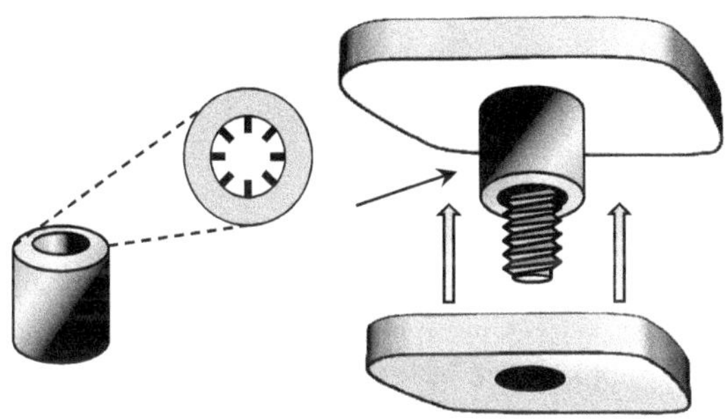

Figure 4.3 Purpose-temporary or utility attachment

4.1.3 Retention

Retention describes how the locking feature(s) are used in the attachment.

Permanent attachments, once engaged, are not intended to be separated at any time, see Fig. 4.4. Sometimes they can be released with tools or by using high effort, but damage to the lock features or to the parts themselves may result. Permanent attachments are indicated for nonserviceable applications or where evidence of product tampering is required. They may also be appropriate where an application must resist forces that could cause a nonpermanent attachment to release.

Do not confuse permanent attachments with the manual release locks described in the next section.

Unlike permanent attachments, *releasable* attachments can be separated; examples are shown in Fig. 4.5.

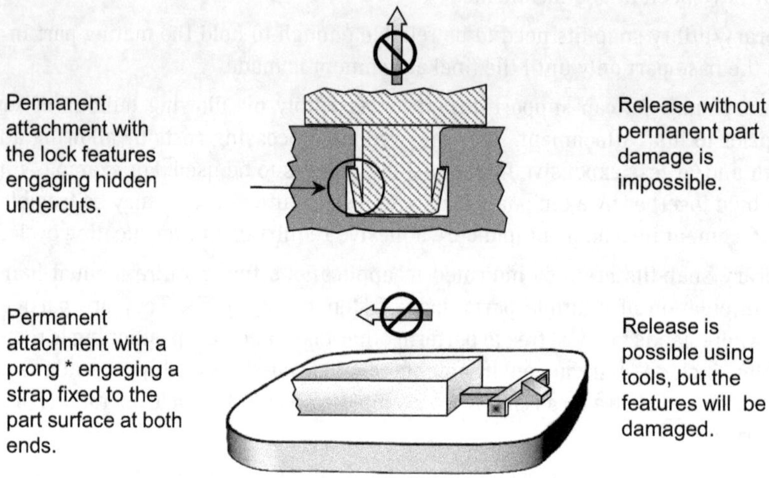

* The locking feature is the flexible strap. The feature engaging the strap may look like a cantilever hook but it is not because it does not deflect during assembly. It is a locator feature called a *prong*.

Figure 4.4 Retention-permanent and releasable attachments

4.1.4 Release

Release describes how the lock feature is deflected to allow part separation, see Fig. 4.5.

Manual-release locks require manual lock deflection for part separation. They are useful when an application must resist applied forces or when disassembly is generally not necessary or rare. Access to the lock must be provided.

Manual release locks are not necessarily a guarantee against unintended part separation. They may release under certain conditions, for example, very high separation force.

Self-releasing locks allow part separation when a predetermined force is applied to the parts. Access panels are common applications that use releasing locks. These lock features can be hidden since no direct access is required for their release.

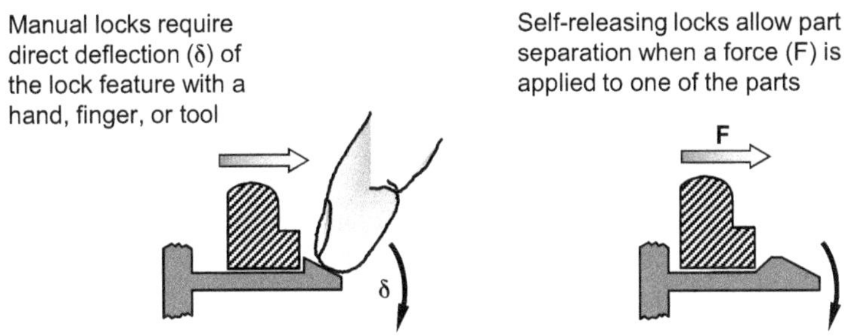

Figure 4.5 Release-manual and self-releasing locks

Figure 4.6 summarizes the *function* descriptive element.

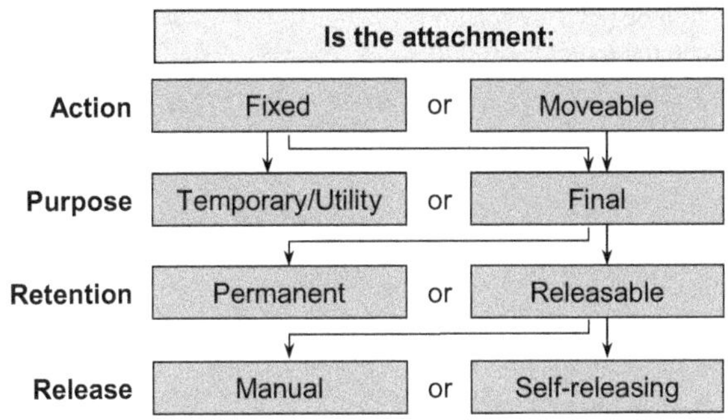

Figure 4.6 Function defines the locking requirements in the application

4.2 Basic Shapes

Basic shapes are simple geometric shapes that describe the parts being attached. Classifying parts in terms of basic shapes allows us to think of an application in generic terms. This is important because it helps the developer visualize the attachment and perform product benchmarking early in the development process. Basic shapes also facilitate the spatial reasoning needed to develop good snap-fit concepts.

4.2.1 Mating-Part and Base-Part

The two components in a typical snap-fit are the *mating-part* and the *base-part*. Recall the discussion of constraint in Chapter 2, and see Fig. 4.7.

The mating-part is typically smaller than the base-part. It is held and manipulated in the assembler's hand(s) or by a robotic end effector and moved into engagement with the base-part. The mating-part will frequently be a *solid*, *panel*, or *enclosure*.

The base-part will generally be larger and may be stationary or held in a fixture. It is usually a *solid, surface, opening,* or *cavity.*

We can usually identify the mating and base-parts by using the size and movement criteria described above. We can also use basic shapes themselves for identification because some are unique to either the mating or base-part. If these fail to distinguish the mating from the base-part, then the parts are probably so similar that an arbitrary selection can be made.

These distinctions are true most of the time. Exceptions do occur, but that does not reduce the value of having these definitions.

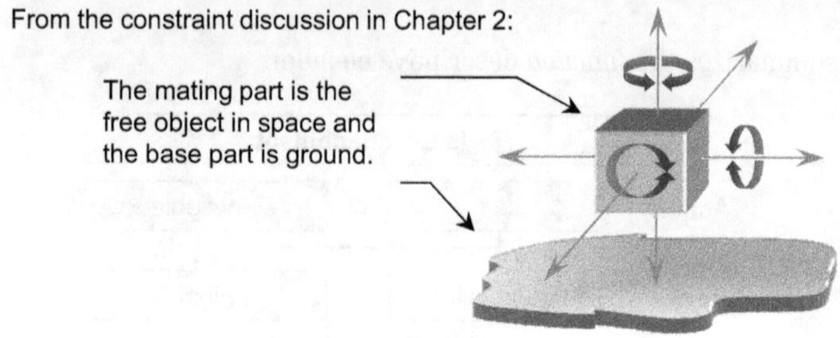

Figure 4.7 Mating-part and base-part relationship

4.2.2 Basic Shape Descriptions

Six basic shapes have been identified to describe parts in snap-fit applications. Table 4.1 summarizes their usage in mating and base-parts. Examples are shown in Fig. 4.8.

Table 4.1 Typical Basic Shape Usage

	Basic Shape					
	Solid	Panel	Enclosure	Surface	Opening	Cavity
Mating-part	Common	Common	Common	Rare	Rare	Rare
Base-part			Rare	Common	Common	Common

Usages identified as *rare* are possible, but are normally not found on that part. For example, a mating-part used as a cavity is rare because cavities are usually found on the base-part.

The solid and panel usages for the base-part are not available because a surface, opening, or cavity on or in the solid or panel normally serves as the base-part, see Fig. 4.8.

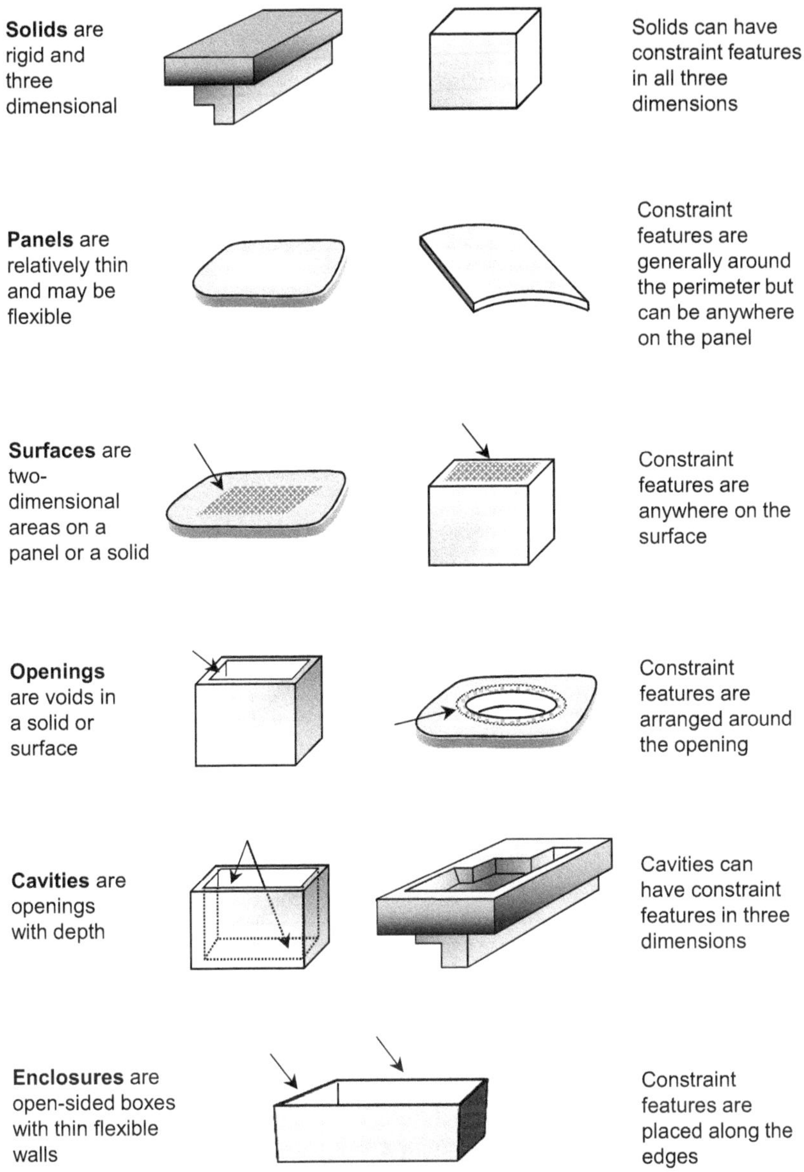

Figure 4.8 Basic shapes

4.2.3 Basic Shape Combinations

Generic part shape descriptions help in describing some important snap-fit design rules. We will also learn how basic shapes help us capture and transfer snap-fit knowledge, past experience, and lessons-learned between applications.

Table 4.2 shows how the basic shapes are distributed between mating and base-parts. The table is based on observation and part geometry.

Table 4.2 Availability of Basic Shape Combinations

		Base-part Shapes					
		SOLid (C)	PANel (C)	ENClosure (rare)	SURface (common)	OPening (common)	CAVity (common)
Mating-part Shapes	SOLid (common)	C	C	SOL-ENC Low	SOL-SUR high	SOL-OP high	SOL-CAV high
	PANel (common)	C	C	PAN-ENC Low	PAN-SUR low	PAN-OP high	PAN-CAV low
	ENClosure (common)	C	C	ENC-ENC Low	ENC-SUR high	ENC-OP low	ENC-CAV low
	SURface (rare)	C	C	SUR-ENC Low	C	X	X
	OPening (rare)	C	X	X	X	X	X
	CAVity (rare)	C	X	X	X	X	X

High - A very common basic shape combination.
Low - Less frequently observed.
C - Covered by another combination, subject to change.
X - Judged to be geometrically impossible or very rare, subject to change.

According to Table 4.2, there is a limited number of common basic shape combinations. This area is expanded and shown in Table 4.3.

The frequency of occurrence shown in Tables 4.2 and 4.3 is based on the author's study of hundreds of applications, most of them automotive. Other product applications, however, seem to be in general agreement with these observations. The frequency judgments are always subject to change as more information is gathered. Other industries may also have a different distribution within their product lines.

Very few attachments should be completely new creations.

The idea of developing frequency information like that shown in Table 4.3 is very important. A reader or a company can identify which basic shape combinations occupy most of their development effort. Design knowledge and guidelines for those combinations can be documented as best or preferred practices. This information then becomes a starting point for future product development when it is captured in a library of preferred concepts.

Snap-fit developers should begin thinking generically. For example, a new application is not just an *electronic module on a product's wall*. It is also a *solid to a surface*. This means that, regardless of the specific application, the fundamental design principles for a solid/surface snap-fit application will always apply. All the knowledge about the best way to make a solid/surface attachment then becomes available to guide development of the new attachment.

Read more about this subject in Chapter 15, Section 15.7.5.

Table 4.3 Most Common Combinations: the High Usage Area of Table 4.2

		Base-part Shapes			
		ENClosure (rare)	SURface (common)	OPening (common)	CAVity (common)
Mating-part Shapes	SOLid (common)	SOL-ENC low	SOL-SUR high	SOL-OP high	SOL-CAV high
	PANel (common)	PAN-ENC low	PAN-SUR low	PAN-OP high	PAN-CAV low
	ENClosure (common)	ENC-ENC low	ENC-SUR high	ENC-OP low	ENC-CAV low

High - A very common basic shape combination.
Low - Less frequently observed.

Some basic shape combinations are possible but not very common; Fig. 4.9 shows a few. The panel to panel example is a (now extinct) floppy disk. If you can find one, they are a good example of making an attachment between very thin parts. The panel to surface example could be an emblem or logo badge on a product.

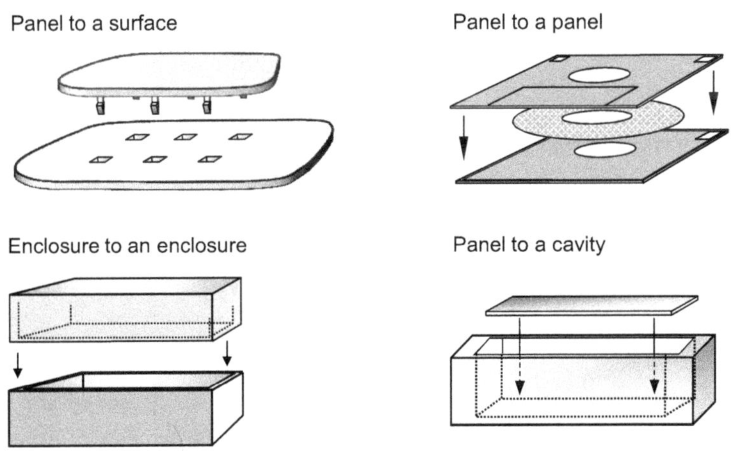

Figure 4.9 Less common basic shape combinations

Each basic shape combination can have preferred assembly motions, constraint feature arrangements, and enhancements that help ensure a good attachment. Some combinations will be preferred over others. Once the remaining elements are discussed, we will be able to summarize some desirable and undesirable characteristics for the common basic shape combinations.

4.3 Engage Direction

Engage (i.e., engage*ment*) *direction* is the third descriptive element. Engage direction is the *final* direction that the mating-part moves as lock feature(s) engage locators to form a lock pair. It is described by a directional vector defining the mating-part's movement as locking occurs, see Fig. 4.10.

Mating-part movement can occur prior the final engaging motion; those preliminary movements are not considered when we identify the engage direction.

Note that engage direction refers to *movement of the mating-part.* It is not the direction that the lock deflects for engagement.

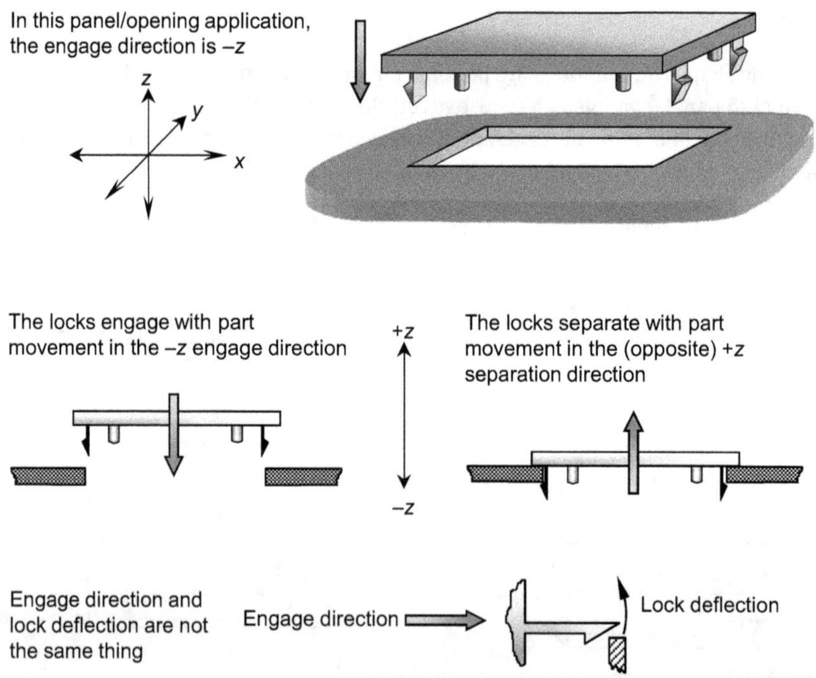

Figure 4.10 Engage direction

The relationship between separation direction and the locking features is extremely important. The locking features (lock pairs) will be required to resist any forces on the attachment that tend to separate the parts and, generally, locking features are the weak link in the attachment system. An important rule when identifying allowable engage directions is:

As we select an engage direction, we are also, by default, selecting a separation direction.

Select an engage direction so that the (opposite) separation direction is not in the same direction as any significant forces on the attachment.

This simple rule means that there should be no significant transient or long-term forces trying to release the locks and separate the parts. Figure 4.11 shows a solid to opening application with two possible engage directions. The preferred engage direction is in the $-z$ direction. Thus the separation direction is in the *opposite* direction ($+z$). Proper selection of the engage direction results in a separation direction in which applied forces are carried by the flange against the surface of the solid (both are strong locators), rather than by weak locking features.

What are significant transient and long-term forces? Those determinations should be made by the product developer with material performance input from a polymers expert. The answer will depend on the operating environment, force magnitude, cycles and duration of the expected forces, and the long and short-term properties of the material chosen for the part. Sometimes, a significant force turns out to be an unexpected force due to accidental impact or misuse.

Forces on the mating-part should be resisted by locators, not locks

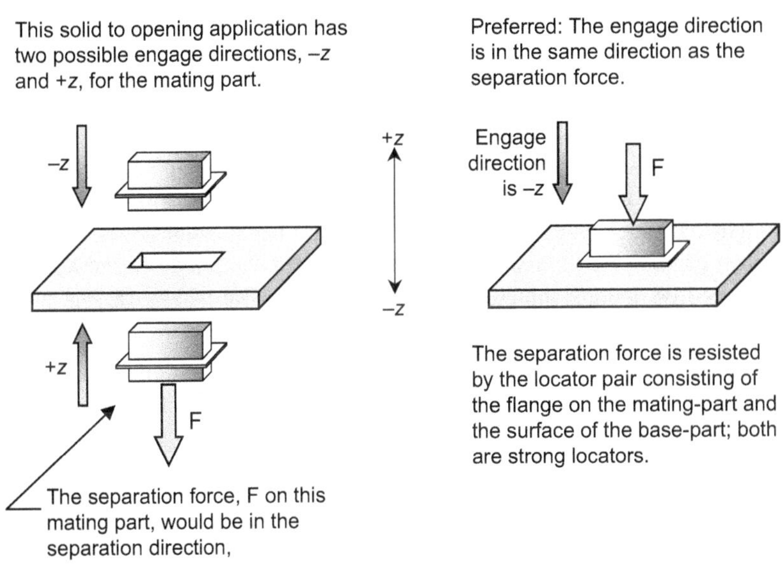

Figure 4.11 Selecting a preferred engage direction

While there may be a number of possible engage directions, the truly feasible engage directions for any particular application will usually be limited for any number of reasons, including allowable assembly motions, part shapes, ergonomic issues, packaging, and access. This means that options to avoid forces in the separation direction are often limited.

Significant force in the separation direction may be reason to consider another fastening method.

When the application is such that significant forces can occur in the separation direction, there are steps that can be taken to ensure against lock release. Be aware that simply making a lock feature nonreleasing *will not* necessarily guarantee against unintended release. This is particularly true for cantilever beam hooks.

Options for improving separation strength include:

- Selecting a different lock style. The decoupling principles described in Chapter 8 explain how and why some lock styles can have significantly more retention strength than others.
- Making the lock permanent, as defined under *function.*
- Retainers (see Section 9.3.2).

4.4 Assembly Motion

Assembly motion sounds similar to *engage direction*, but they are very different things.

Just as there are a set of generic basic shapes to describe the mating and base-parts, *assembly motion* is also described by a set of generic motions: *push*, *slide*, *tip*, *twist*, and *pivot*, see Fig. 4.12. Think of assembly motion as what a human operator must do to assemble the components. It is the *final* motion of the mating-part as it locks to the base-part.

The assembly motion can influence the attachment's strength.

Assembly motion helps the developer visualize the mating to base-part assembly process. Like basic shapes, assembly motions support generic snap-fit descriptions and spatial reasoning for snap-fit concept development. They may also have ergonomic implications in some applications where an awkward position and excessive assembly force, when combined with a certain assembly motion, can result in increased likelihood of repetitive motion injury. Most importantly, as we will learn, the assembly motion can have an indirect but significant effect on the attachment's strength.

Some assembly motions may be preferable to others depending on the basic shapes involved, application accessibility, and operator ergonomics. Table 4.4 shows some of these possibilities and provides an indication of the available motions.

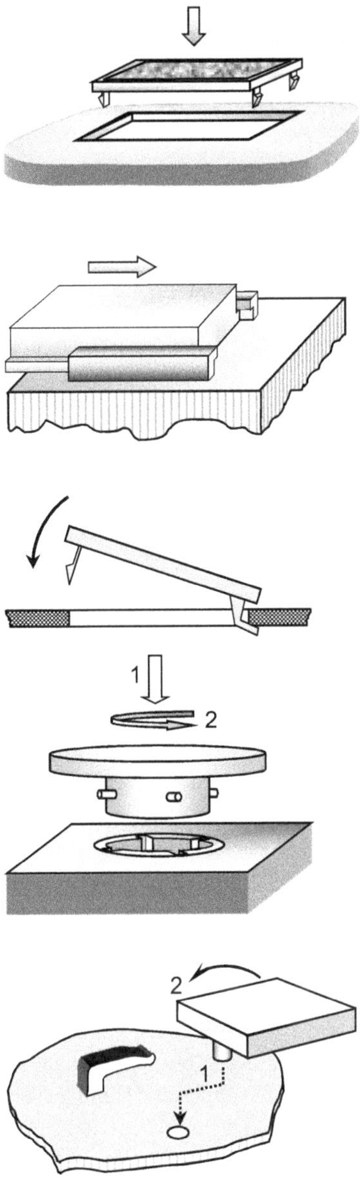

Figure 4.12 Generic assembly motions

Table 4.4 Common Basic Shape Combinations and Available Assembly Motions

		Base-part Shapes			
		Enclosure	Surface	Opening	Cavity
Mating-part Shapes	Solid	N/A	Push Slide Tip Twist Pivot	Push Tip Twist *	Slide Tip * Twist
	Panel	N/A	Push Slide Tip	Push Tip	Push Tip
	Enclosure	Push Tip	Push Tip Twist *	Push Tip Twist *	N/A

* Some availability, depending on part geometry.

4.5 Summary

Chapter 4 introduced four spatial and descriptive elements: *Function*, *Basic Shapes*, *Engage Direction*, and *Assembly Motion*. These are used to describe and classify design concepts and application behaviors in preparation for adding the physical elements to the interface

Important points in Chapter 4 include:

- During concept development, use generic descriptions of part shapes and assembly motions to help visualize the application and its behavior.
- Generic descriptions of part shapes and assembly motions allow us to capture and transfer important snap-fit knowledge, past experience, and lessons-learned on other applications to new applications.
- Regardless of the application, the fundamental snap-fit attachment design principles for a specific basic shape combination will always be true.
- Select a mating-part to base-part *engage* direction so there are no significant forces on the snap-fit attachment in the *separation* direction.
- If there are significant forces on the attachment, special attention must be paid to lock feature selection and design.

The remaining elements are physical features that are used to construct the snap-fit interface:

- Chapter 5: *Physical Elements-Locators*
- Chapter 6: *Physical Elements-Locks*
- Chapter 7: *Physical Elements-Enhancements*

5 Physical Elements: Locators

Chapter 4 introduced spatial and descriptive elements for defining a snap-fit attachment. The remaining physical elemeqents are used to build the attachment.

Locators are the first feature considered in snap-fit development.

Constraint features are the first group of physical elements (Fig. 5.1). Recall that constraint is the most fundamental of the key requirements. Not surprisingly the features that provide constraint, locators and locks, are the most important features in a snap-fit interface.

Locators, the subject of this chapter, should be the first physical features considered as we begin developing the snap-fit attachment. In a snap-fit, a locator on the mating-part will engage another locator on the base-part. Different locator combinations will result in different *locator pair* performance characteristics.

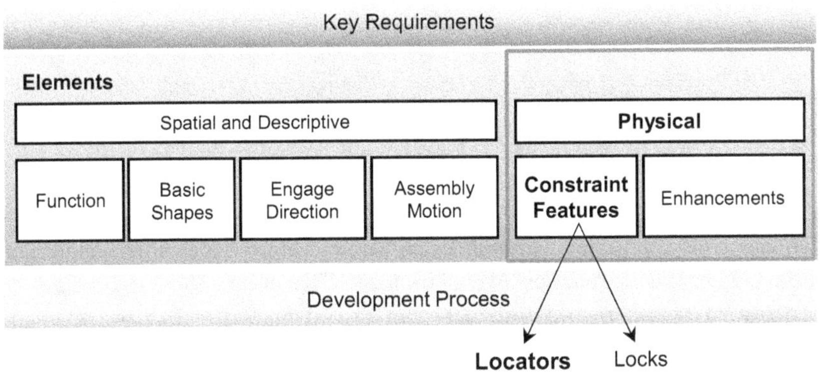

Figure 5.1 Physical elements–locators

Locators are strong and generally inflexible features. They *position* the mating-part to the base-part, and they can transmit forces across the part-to-part interface.

Locators *nest* the mating-part to the base-part.

Locators can be specific features added to the attachment strictly for constraint purposes, or they can be areas of pre-existing part features like surfaces and edges that are used as *natural locators* (Fig, 5.2).

Development issues with locators involve selecting locator styles and arranging the locators in the interface. Because locators are, by definition, strong and inflexible, analysis calculations are normally unnecessary. If they are analyzed at all, locators require only a basic analysis of behavior under simple tensile, shear, combined, or compression loading. One notable exception is when a locator is used in bending, such as a low-deflection lock feature, see Chapter 6, Section 6.3.4.

Locators are classified into one of three groups: *protrusions, voids,* and *surface*-based. Living hinges are also treated as locators because they provide constraint when present in a snap-fit interface

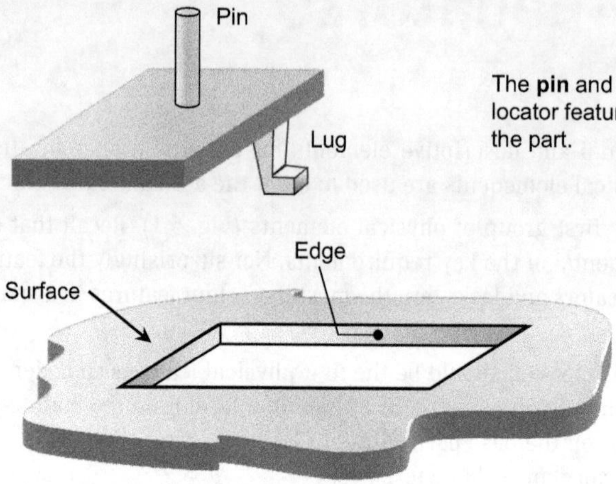

Figure 5.2 Locator feature examples

5.1 Protrusion-Based Locators

We will begin our discussion with protrusion-based locators, which are made by *adding* material to a part. They generally extend from a surface or an edge.

The idea of a variety of locators with very different attributes and the interface options they provide is more important than the specific names assigned here. The names permit discussion of general characteristics, which is important, but there will be gray areas. For example, the answer to *when does a surface become an edge?* is not as important as understanding the shared and differing attributes of surfaces and edges.

Note that, although a part feature like a surface or edge may appear in an illustration of a protrusion, when we consider constraint, the surface and edge are *separate* locators from the protrusion locator.

5.1.1 Pins

Pins will have either a constant section or a slight taper (draft angle) along the axis of symmetry. They may have round, square, or complex sections, Fig. 5.3.

Pins typically engage holes, slots, cutouts, or edges and constrain in a plane orthogonal to the pin's axis. Figure 5.20 shows constraint in a pin/hole *locator pair.* Recall that a locator will always engage another locator to create a locator pair, see Section 5.5.1.

In injection molded parts, protruding features like pins can have a truly constant section only if they are formed along the parting plane of the mold. Depending on orientation in the mold, pins are usually slightly tapered for draft angle.

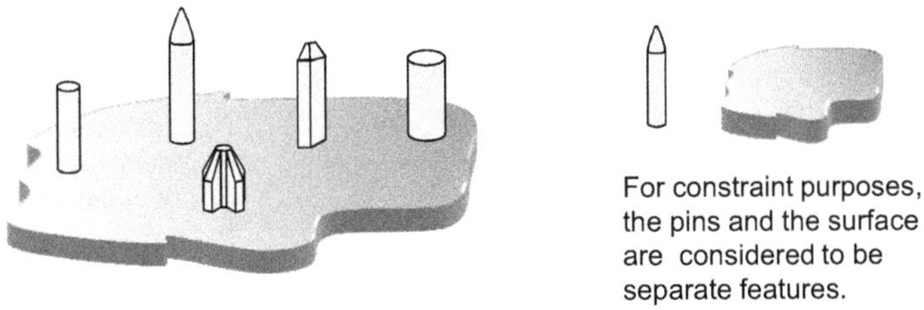

Figure 5.3 Pins

5.1.2 Prongs

Prongs are protrusions with an enlarged area at the end, see Fig. 5.4. They can engage another locator to form a locator pair, but they are more often used with trap style locks to create a very effective lock pair. In Chapter 6, see Section 6.7.1 for a discussion of lock pairs and Fig. 6.33 for application examples.

Prongs are the only locators that can effectively resist a force in pure tension. In contrast, a lug resisting a tensile force will also experience a bending component.

One of the prongs in the illustration below *appears* to be a cantilever hook lock. However, this lock shape can also be a prong when used in an application where it is not allowed to deflect. See Fig. 4.4 in Chapter 4 for one example; more examples will follow in the lock discussion in Chapter 6. This demonstrates how features are defined by *what they do* and not by how they look, which is not uncommon with snap-fits.

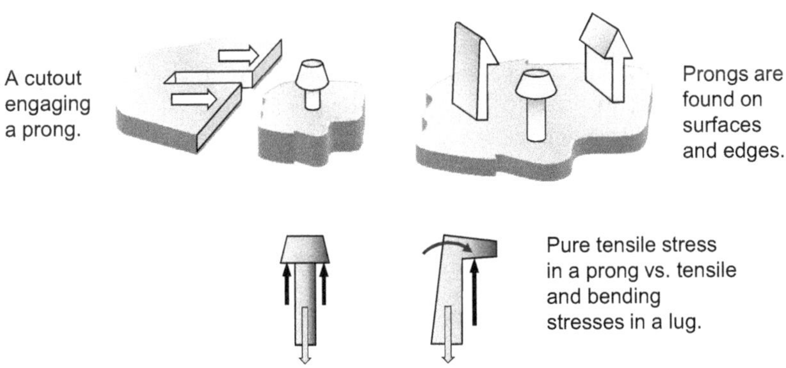

Figure 5.4 Prongs

5.1.3 Tabs

Tabs are protrusions with parallel or slightly tapered sides. They are thin relative to their width and height and normally engage an edge or a slot, see Fig. 5.5.

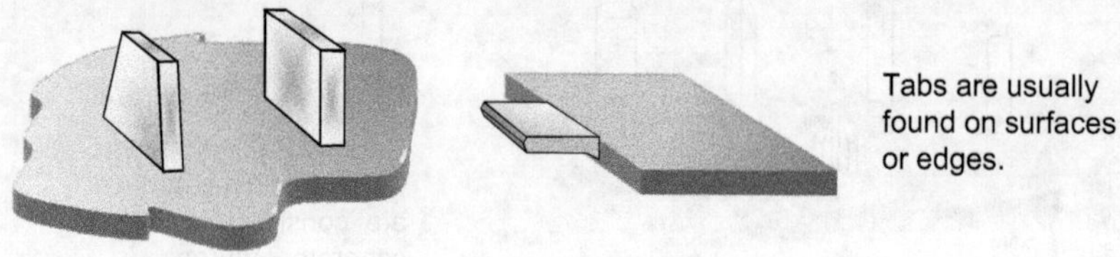

Figure 5.5 Tabs

5.1.4 Lugs

Lugs are characterized by an L-section, see Fig. 5.6, but there are numerous variations on that basic shape. A lug will typically engage an edge.
Lugs can sometimes also be used as *low-deflection* locks, particularly when a slide, twist, or pivot assembly motion is used.

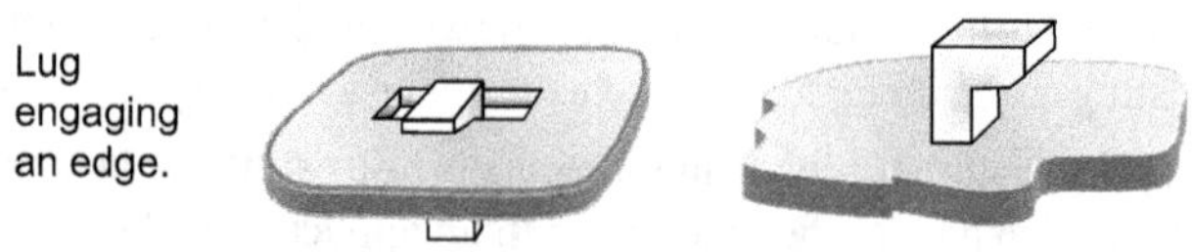

Figure 5.6 Lugs

5.1.5 Tracks

Tracks are formed when two separate lugs facing toward or away from each other, or joined as a T-section, are extended to create a strong locator that allows parts to engage with a slide assembly motion, see Fig. 5.7. Tracks can remove multiple DOM and can be very effective locators.

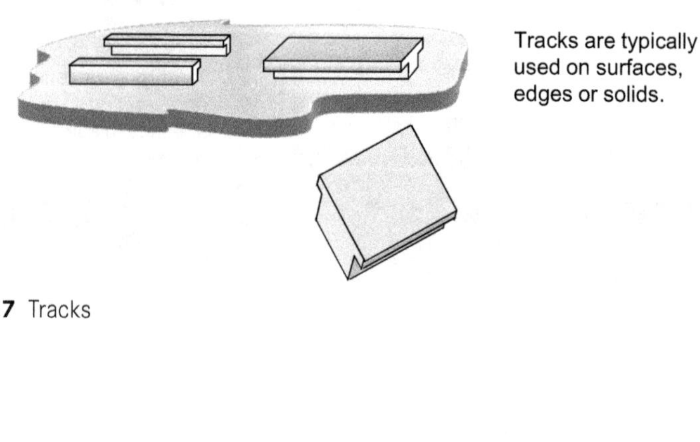

Figure 5.7 Tracks

5.1.6 Cones

Cones are a variation of the pin locator, where a section at the base is significantly larger than a section toward the end. This allows cones to engage holes and provide locating in the axial direction as well as in lateral directions (Fig. 5.8).

Cones may have a round or a square section. The square section cones are only intended for engagement with a square hole. Cones with round sections are preferred over those with square sections because they are easier to mold, and when engaged with a hole, rotational alignment is not an issue.

Features with thick sections, like cones and wedges or large-diameter pins, require special attention and may be restricted as to location. Thick sections tend to create sink-marks on the opposite surface and internal stresses in a part. Good mold design practice requires that thick sections be cored out. This means that any appearance requirement for the opposite surface can limit their use.

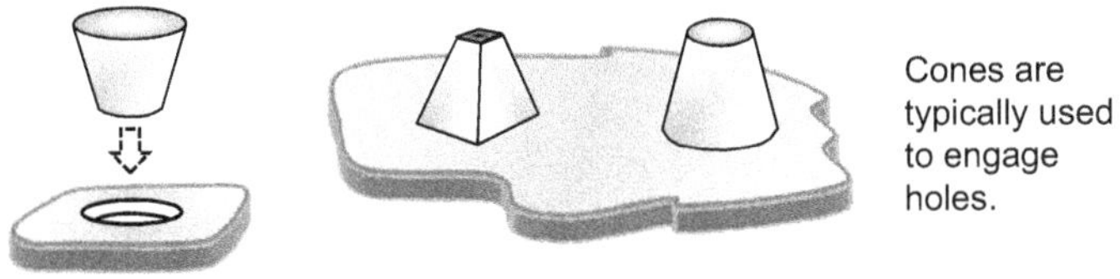

Figure 5.8 Cones

5.1.7 Wedges

Wedges are a variation of tab locators in which the base has a much greater area than a section toward the end. The greater thickness at the base makes them much stronger than a tab, see Fig. 5.9.

Wedges engage slots, and like cones, they provide constraint along the axis of the taper as well as in lateral directions. Wedges are not axisymmetric, they have a long and a short axis.

Alignment of the long axis of the wedge with the long axis of the mating slot adds a dimensional control requirement that does not exist when round cones engage round holes.

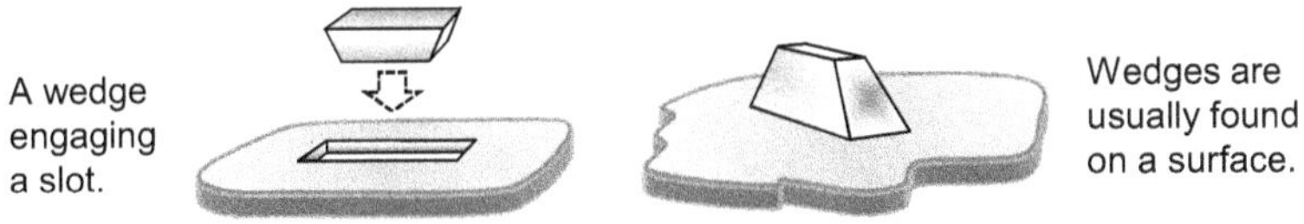

Figure 5.9 Wedges

5.1.8 Catches

Catches are ramp-like locators and are usually found on surfaces and edges. They engage with other edges or surfaces, see Fig. 5.10. Catches are also frequently used as a retaining member on cantilever beams and certain other lock styles.

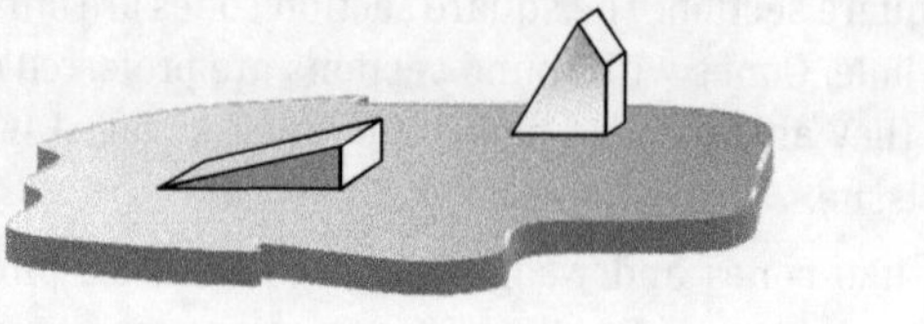

Figure 5.10 Catches

5.2 Surface-Based Locators

Surface-based locators, while not necessarily flat, are essentially two-dimensional areas with length and width, but no depth. Contact with other locators occurs on the surface.

Note that the term *surface* is also used as the name for a basic shape so there is a possibility of confusion. To clarify: a base-part can have the basic shape *surface*. An *area* of that surface can also be described as a *surface locator*. The (basic shape) surface could also carry other locators, such as protrusions or holes.

Surface-based locators are typically pre-existing areas or features of a part.

5.2.1 Surfaces

A surface that is used as a locator will have some width and length, but it may not be perfectly flat (Fig. 5.11). A surface locator may be a *natural locator* where the part surface, as designed, serves as the locator. Or the surface can be modified by adding a specific area, called a *land*, to allow for local dimensional control.

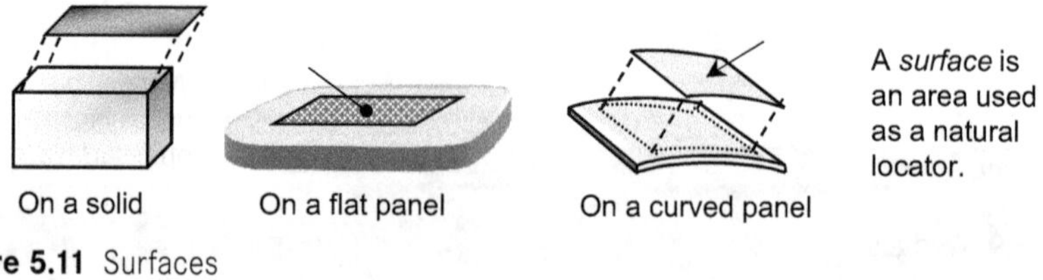

Figure 5.11 Surfaces

5.2.2 Edges

Edges are relatively narrow surfaces (Fig. 5.12). An edge is generally on a part wall or on a rib, and like a surface, it can be either a natural locator or modified with a land.

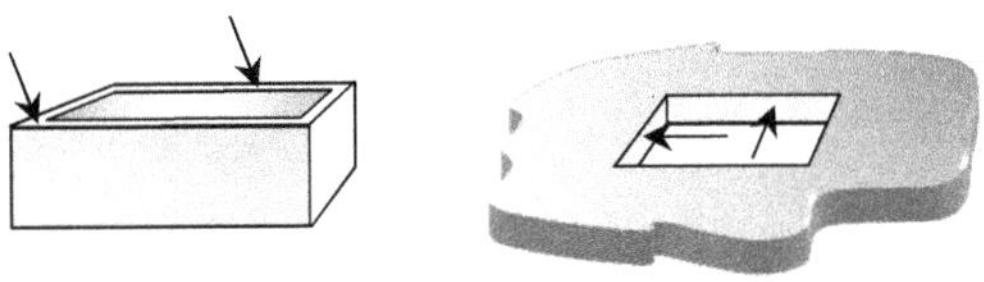

Figure 5.12 Edges

5.2.3 Lands

Surfaces and edges used as natural locators may be difficult (costly) to fine-tune because major changes to the mold may be required. Adding *lands* to surfaces or edges at strategic locations for fine-tuning purposes can make mold adjustments easier and less costly. While they are typically raised areas, see Fig. 5.13, lands can also be recesses in a surface or cutouts on an edge.

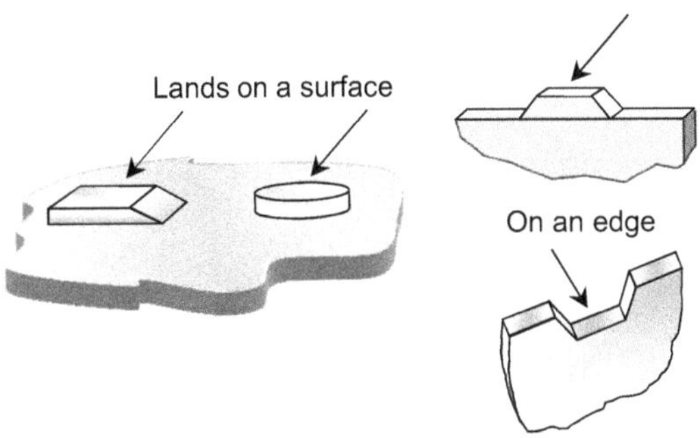

Figure 5.13 Lands on a surface and an edge

■ 5.3 Void-Based Locators

Void-based locators are created by removing material from a part. All constraint in these locators occurs at the boundary of the void.

As was the case with protrusions, illustrations of voids must show then in place on some other feature, often a surface. That feature is a separate locator and, for constraint purposes, is not considered a part of the void.

5.3.1 Holes

A hole is an opening in a part wall or surface. Typically, holes are round, square, or rectangular (Fig. 5.14). Holes constrain in 4 or 5 DOM depending on whether the mating locator is a pin or a cone. Constraint is continuous around the perimeter. As with the protrusion locator pins and cones, round holes are preferred over square or rectangular holes for dimensional robustness.

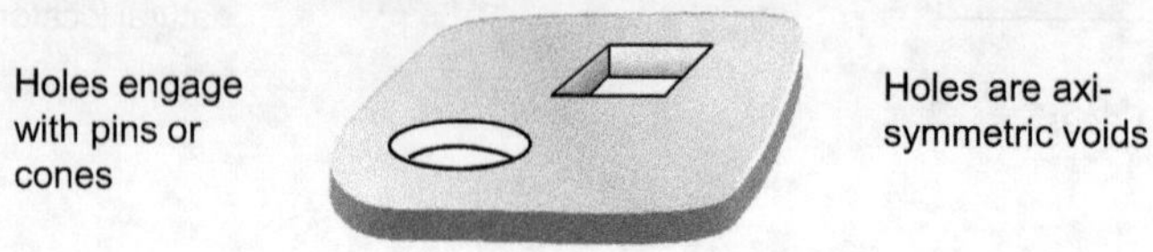

Figure 5.14 Holes

5.3.2 Slots

A slot is an elongated hole, see Fig. 5.15. Elongation removes contact (constraint capability) along the long axis of the slot. A slot constrains in at least 2 and possibly 3 DOM.

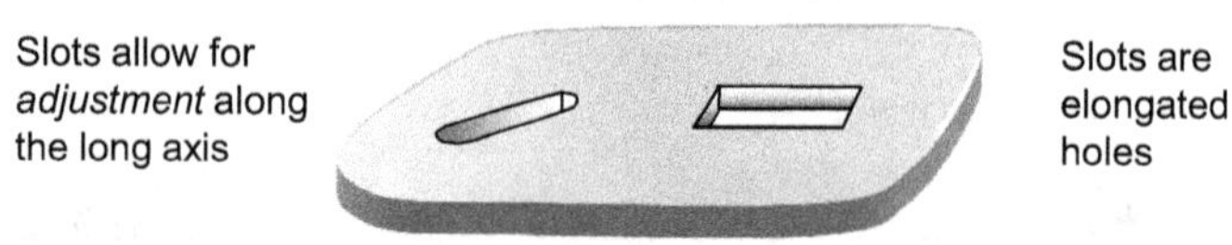

Figure 5.15 Slots

5.3.3 Cutouts

Cutouts are a hybrid of the hole and edge locators. A cutout has three active or useful edges rather than one, see Fig. 5.16. Like a hole, the cutout provides some lateral constraint capability, and like an edge, it provides more accessibility for assembly.

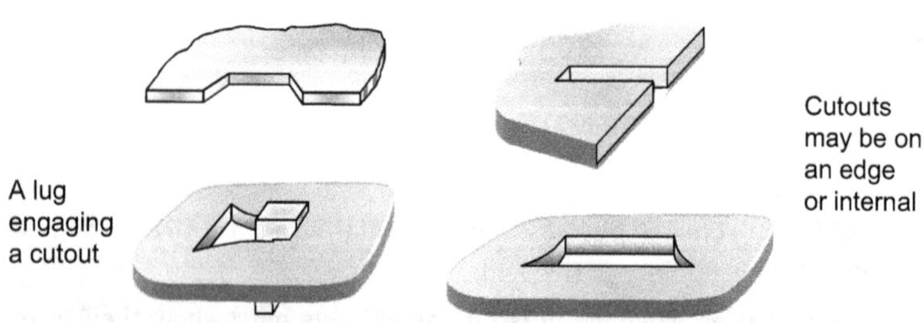

Figure 5.16 Cutouts

5.4 Living Hinges

A living hinge is a relatively thin connective section between two parts, see Fig. 5.17.

Like a moveable application with opposing collinear pin/hole locator pairs, a living hinge allows rotational movement of one part relative to the other, while removing 10 DOM (see Section 4.1.1 in Chapter 4). Living hinges are often found on lids and covers.

Living hinges can also serve as manufacturing/assembly aids. A living hinge can allow material to flow between two areas so a whole part can be produced in one mold. The part is completed when its two halves, connected by the living hinge, are folded and snapped together.

Living hinges act as the first engaged locator pair in an interface and, like more conventional locators, provide part-to-part positioning and strength.

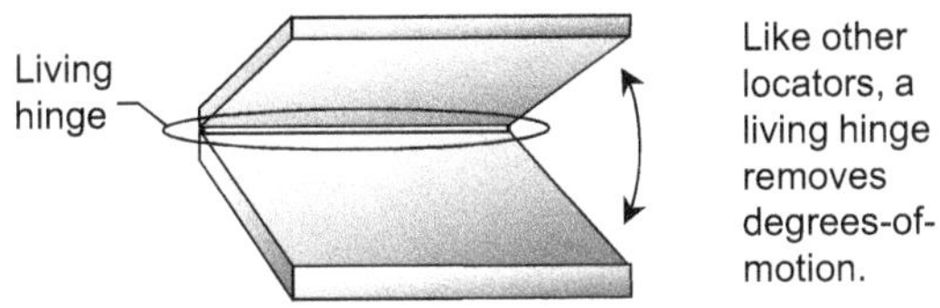

Figure 5.17 Living hinge

5.5 Using Locators

This section explains how the various locator types described above are used in snap-fit attachments.

5.5.1 Locator Pairs

A locator by itself cannot provide constraint. A *locator pair* is a locator on the mating-part engaging another locator on the base-part (Fig. 5.18).

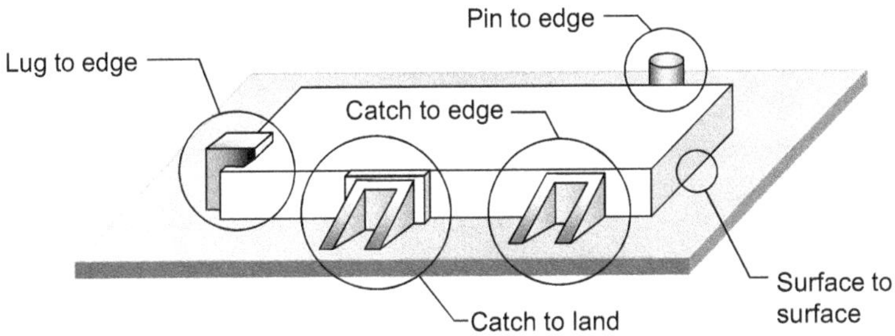

Figure 5.18 Examples of locator pairs

Each locator pair in an attachment provides mating to base-part positioning, and sometimes the locator pair will also transmit forces across the interface. These capabilities can be said to occur along a *line-of-action* which is coincident with the position and force vectors. It is convenient to use lines-of-action when discussing relationships between locator pairs in an attachment.

In an attachment, a locator pair may act alone to constrain in a certain direction(s) or multiple locator pairs may act together. Catch and edge locator pairs in Fig. 5.19 show how lines-of-action are used to describe the relation between multiple locator pairs.

The term *line-of-action* will appear in later discussions. In all discussions and illustrations, the convention for showing locator pair position and strength vectors is that locator(s) on the *mating-part* are being acted on by locator(s) on the *base-part* as if a force trying to move the mating part is being resisted by the locators on the base part.

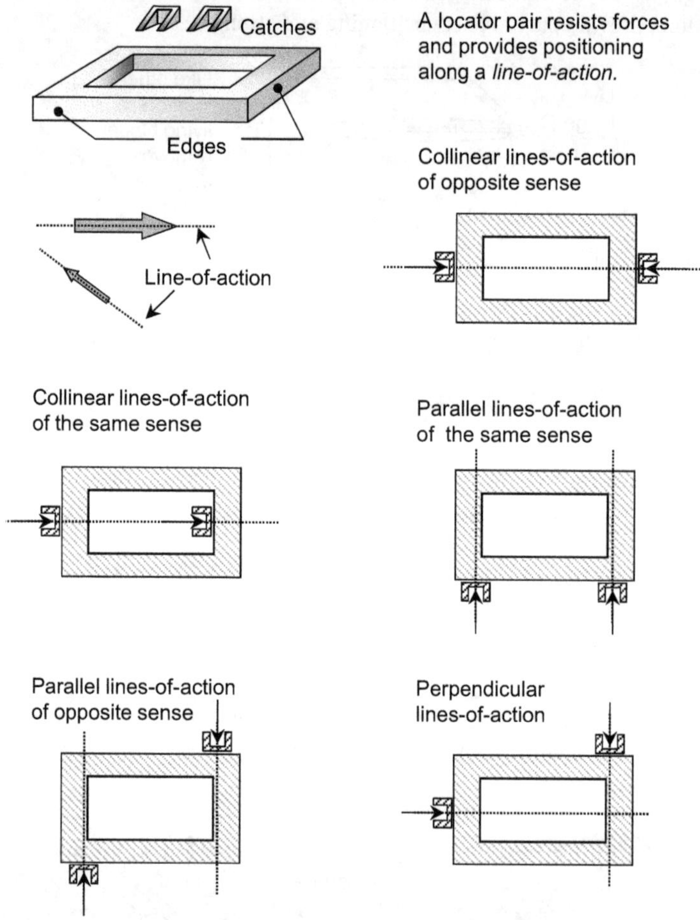

Figure 5.19 Examples of locator pair interactions

Figure 5.20 shows how forces that are applied to a mating-part would be resisted by various locator pairs. A locator pair may act alone (the lug/edge pair), or locator pairs may act together (parallel lines-of-action). In this example, two locator pairs with parallel lines-of-action resist a force in the −Y direction. One of those pairs also acts with a different pair as a couple to resist rotation, and it may need to be stronger than the other catch/edge pairs for that reason.

Think about two locator pairs with coincident lines-of-action of the same sense. We have not discussed them here, but we will talk about them in Chapter 8. Think about how they could affect constraint? Would that effect be good or bad?

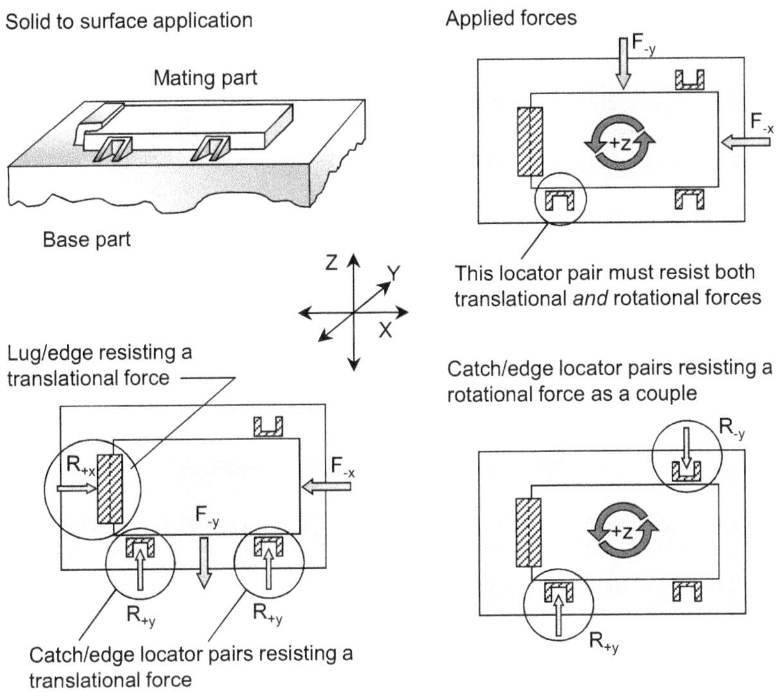

Figure 5.20 Locator pairs and applied forces

5.5.2 Providing Constraint

Figure 5.21 shows the DOM removed by a variety of constraint pairs. In these examples, the protrusion-based locators are on the *mating-part*, identified as *MP*, and the void-based locators are on the base-part. The DOM removed is always shown with respect to the mating-part.

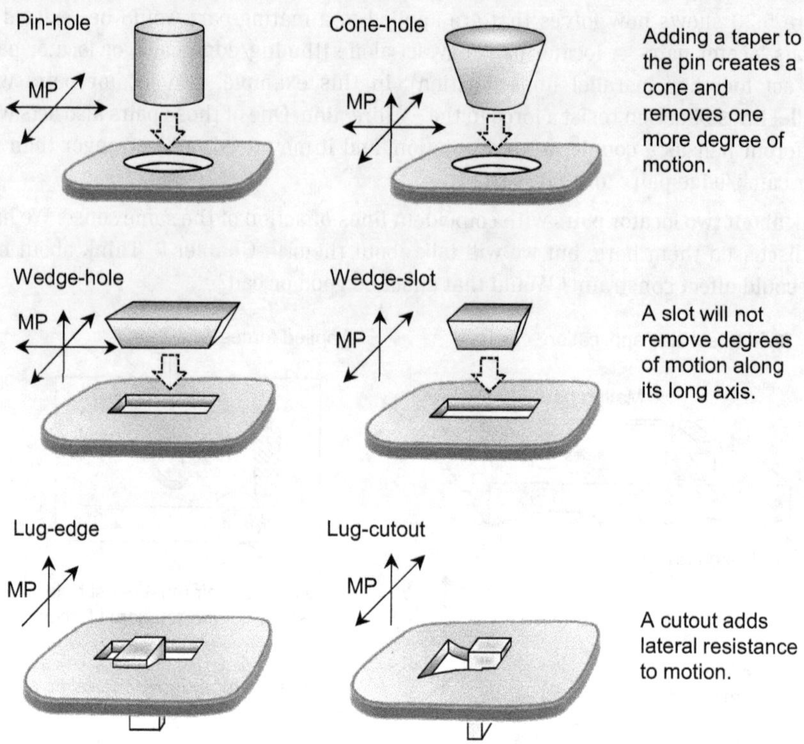

Figure 5.21 Locator pairs and degrees-of-motion

Notice how identical or similar locators can have different constraint behavior and even different names depending on the other locator member of the pair. A pin/hole locator pair differs from a cone/hole pair in the DOM removed. A wedge/hole differs from a wedge/slot pair. A rectangular opening provides an edge in a lug-edge pair and that edge differs from a cutout in the lug/cutout pair. The same rectangular opening appears as three different kinds of locator (hole, slot, and edge). In other words, a locator feature is not defined until both members of the locator pair are known.

The constraint characteristics of the protrusion-based locators are defined independently of the surface or edge from which they extend.

Think about locator pairs that remove only 1 DOM and pairs that remove multiple DOM. What might be the advantages or disadvantages of each?

5.5.3 Assembly Motion and Strength

Because locators are strong relative to locks, the more degrees-of-motion (DOM) that can be removed with locator pairs in a snap-fit, the stronger the attachment. This is an extremely important point. The obvious follow-up question is *how can I get more locator features and fewer locks into an attachment?* The answer begins with the spatial element *assembly motion* introduced in Chapter 4, Section 4.4.

An extremely important snap-fit principle is that *attachment strength is influenced by assembly motion*. Recall the importance of decisions made during concept development.

This is really important!

Figure 5.22 illustrates how the assembly motion can affect locator and lock selection and placement in the interface. The tip assembly motion allows use of a strong locator pair (lug/edge) at the high mass area of the mating-part for improved resistance to separation. The push assembly motion requires hooks everywhere and does not permit the use of stronger lugs.

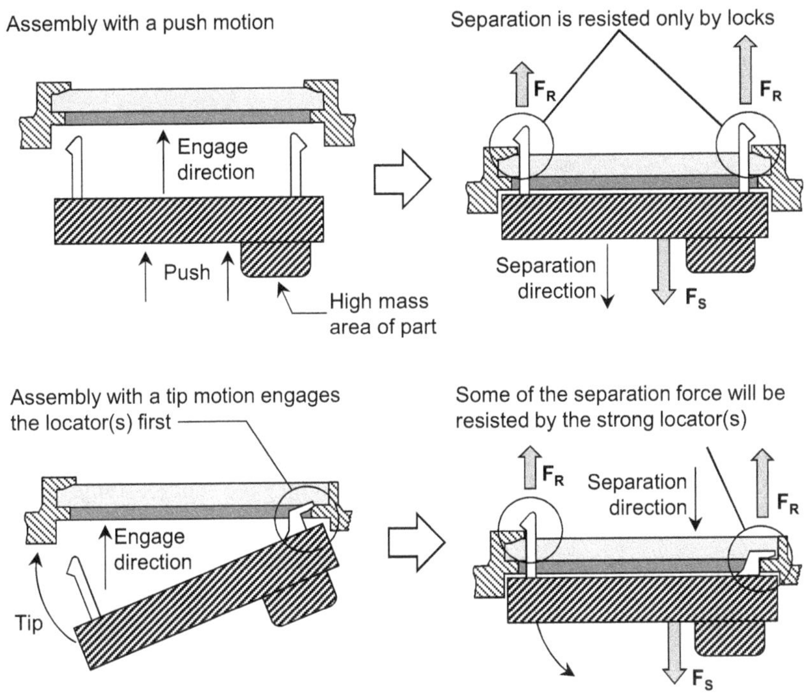

Figure 5.22 Assembly motion and attachment strength

Table 5.1 shows the capability of the five generic assembly motions to maximize the DOM removed by locators and minimize the DOM removed by locks. The potential for DOM removal by locators is highest with the slide, twist and pivot motions, next highest for a tip motion and lowest for the push assembly motion.

While the slide, twist, and pivot motions may remove the most DOM, their usage can be limited by the mating and base-part shapes. The tip motion is generally a more versatile option because of its adaptability to the more common basic shape combinations and it also removes many DOM.

A tip can often replace a push assembly motion.

The push assembly motion (a popular and frequently used assembly motion) should be avoided when any separation forces are acting on an application because the push motion requires the most DOM to be removed by the lock features. The exception is when the only external force on the mating-part is in the assembly direction and that force is resisted only by locator pairs.

Table 5.1 Assembly Motions and Maximizing the DOM Removed by Locators

		Best Case		Worst Case		
		Maximum possible DOM removed by all locators	Remaining DOM to be removed by locks	Minimum possible DOM removed by all locators	Remaining DOM to be removed by locks	Usage
Assembly Motion	Slide	11	1	10	2	**Limited by basic shapes**
	Twist	11	1	10	2	
	Pivot	11	1	10	2	
	Tip	10	2	10	2	**High adaptability**
	Push	7	5	7	5	
	DOM TOTAL	12 DOM		12 DOM		

5.5.4 Fine-Tuning

Plastic parts should be designed to permit easy modifications to the mold. To get and maintain dimensional accuracy, molds require fine-tuning for initial setup as well as adjustment for ongoing wear.

Protrusions and voids are tunable locators. In other words, adjustments to their sizes, attributes, and locations are relatively easy because they are additions to pre-existing part features (Fig. 5.2).

Surface-based locators are often naturally occurring part features. Areas on surfaces and edges can be used as locators but their role as primary part geometries can make them difficult or costly to fine-tune.

Identifying a special area on a surface-based locator to accept the dimensional changes required for fine-tuning avoids the need to modify an entire part feature. That is the purpose of *land* locators introduced in Section 5.2.3.

A land is a *specified area* that provides a locating surface and also allows easy local dimensional control for fine-tuning. Figure 5.13 shows lands as raised areas on a surface and as a raised or recessed area on an edge. Figure 5.23 shows a solid to opening application where lands allow local fine-tuning. The additional clearance around the part perimeter also makes assembly easier.

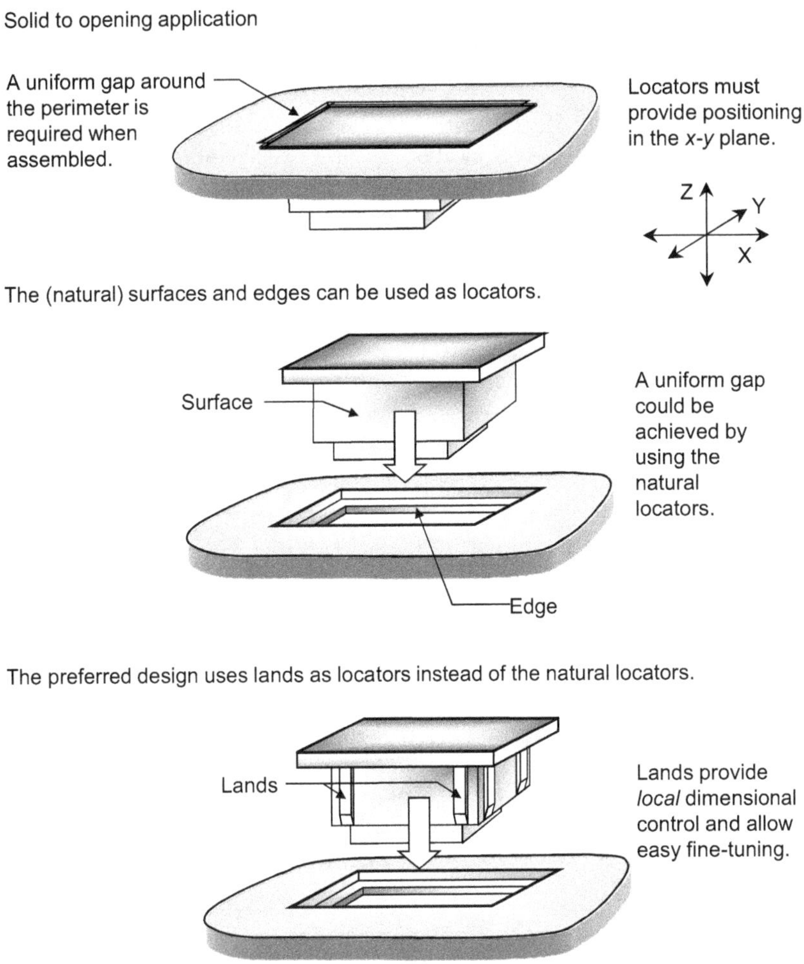

Figure 5.23 Using lands for fine-tuning and ease of assembly

5.5.5 Dimensional Robustness

Three aspects of dimensional robustness are the *positioning* of locator pairs, the *compliance* between locator pairs, and the use of *datum points* in the interface.

5.5.5.1 Positioning

Some locator pairs in an application may be *position-critical* because they control important positioning or alignment behavior of the parts. For this reason, they will be potential sites for fine-tuning the attachment. Keep this in mind as these sites are identified and use caution if two natural locators make up a position-critical pair where fine-tuning will likely be necessary. Position-critical locator pairs should be placed with their line(s)-of-action as close as possible to the site where alignment is required.

When two locator pairs have parallel lines-of-action, they will interact to affect part location, see Fig. 5.24. In this example, the lines-of-action are of the same sense but the interaction also occurs when they are of opposite sense (a couple), as discussed in Section 5.5.7 and as shown in Fig. 5.27.

Point a represents the mating-part location on the x-axis. Point a's position along the x-axis is less sensitive to the y-axis positions of catches 1 and 2 when the catches are as far apart as possible. For fine-tuning purposes, a tunable land could be added to the edge at one of the two catch/edge locator pairs or the catches themselves could be tuned.

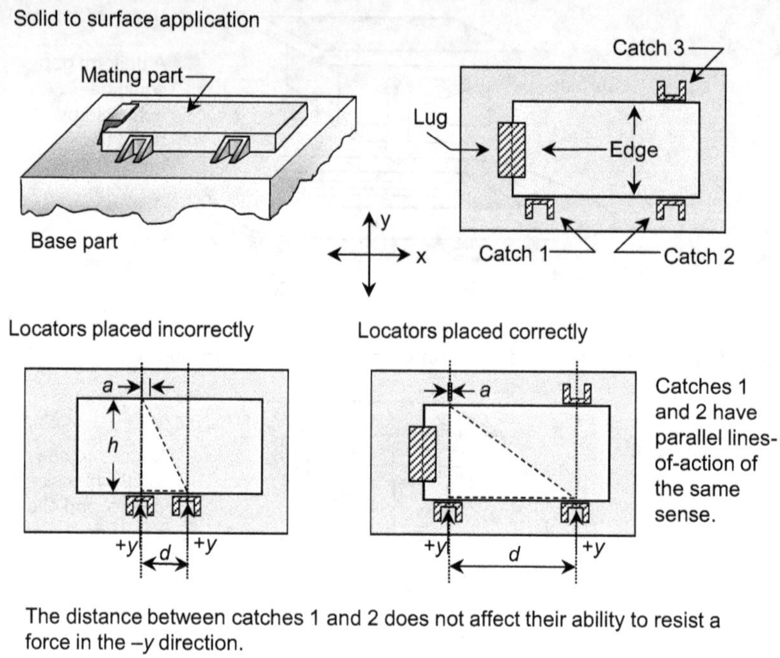

Figure 5.24 Feature positioning and dimensional robustness

We'll ignore the arc traveled by point a in this example because all distances are very small. Point a's position along the x-axis is a function of the ratio h/d:

$$\Delta a = \frac{h}{d}(\Delta y) \tag{5.1}$$

Where: Δa is the *x-axis* variation in the position of point a

h is the *y-axis* distance between locator pairs 1 and 2 and point a

d is the *x-axis* distance between the locator pairs' parallel lines-of-action

Δy is the *y-axis* tolerance between locator pairs 1 and 2

Table 5.2 Locator spacing effects from Fig. 5.24 and Eq. 5.1

Locator Spacing	h/d	Δy	Δa
Close	2.5	±0.1	±0.25
Far	0.67	±0.1	±0.067

Obviously the smaller variation in point a's position is preferred.

Because point a moves on the x-axis, any locator pair constraining along the x-axis will be sensitive to this dimensional change. In this example, positioning at the lug/edge locator pair is affected.

Unlike dimensional robustness, the strength to resist translational movement is not affected by the distance between locator pairs with parallel lines-of-action, but dimensional considerations aside, a wide spacing between locator pairs will also improve part-to-part stability. Stability is related to part rotation where locator pairs act as a couple. This is discussed in a later section.

5.5.5.2 Compliance

Compliance is treated as an enhancement and is discussed in detail in Chapter 9, Section 9.3.3. For now, it is sufficient to define compliance as a strategy for managing dimensional variation and over-constraint when constraint pairs interact, see Fig. 5.25. Designing compliance into the snap-fit interface allows for the use of normal tolerances rather than costlier fine tolerances to maintain close, rattle-free constraint between parts.

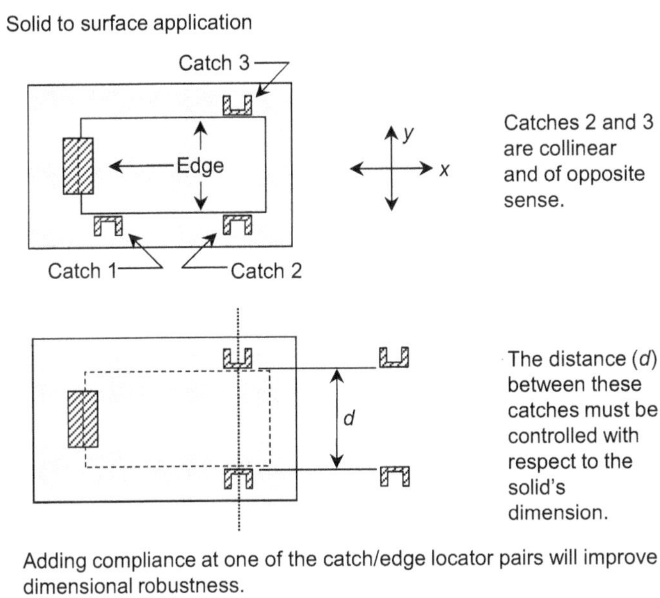

Figure 5.25 Dimensional robustness and compliance

5.5.5.3 Datum Points

A datum point should be selected in each plane at a position-critical locator pair. A locator pair constraining along multiple axes can provide the datum for all locator pairs along each of those axes. A pin/hole locator pair, for example, can constrain along both *x*- and *y*-axes so it would be an effective datum for all other locator pairs in that plane.

Some applications may not have critical positioning or alignment requirements and this requirement can be relaxed.

An understanding of Geometric Dimensioning and Tolerancing (GD&T) is important when making datum and dimensioning decisions for snap-fit interfaces.

5.5.6 Constraint Efficiency

Efficiency means minimizing the number of constraint features in the interface. This reduces cost and complexity of the part and the mold and it improves dimensional robustness because opposing force/position vectors are contained *within* one locator pair rather than across a (possibly significant) distance between locator pairs.

Locator pairs that remove multiple DOM are preferred over pairs that remove only 1 DOM, Fig. 5.26.

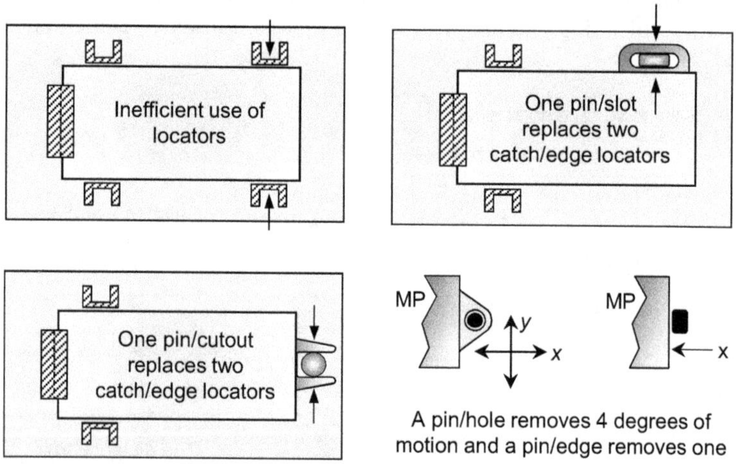

Figure 5.26 Improving constraint efficiency

The efficiency of common locator pairs is summarized in Table 5.3. A good starting point for locator pair selection is to identify the most efficient locator pair possible given the assembly motion being considered. Use that pair as the datum point for all other locator pairs in that plane with collinear lines-of-action to that first pair.

See Chapter 8 for a full discussion of constraint.

Table 5.3 Constraint Efficiency in Common Locator Pairs

Locator Pair	DOM Removed	Locator Pair	DOM Removed
Track/track	10	Lug/edge	2
Living hinge*	10	Tab/slot	2
Cone/hole	5	Pin/edge	1
Lug/cutout	4	Tab/edge	1
Pin/hole	4	Catch/surface	1
Wedge/slot	3	Catch/edge	1
Catch/cutout	3	Surface/surface	1
Prong/cutout	3	Surface/edge	1
Pin/slot	2	Edge/edge	1

Surfaces and edges may be either natural locators or lands
* Living hinge is a special case locator

5.5.7 Mechanical Advantage and Stability

For rotational constraint, two locator pairs will act as a couple to resist rotational motion.

For maximum mechanical advantage (strength), those locator pairs should be placed with their (parallel) lines-of-action as far apart from each other as possible, see Fig. 5.27.

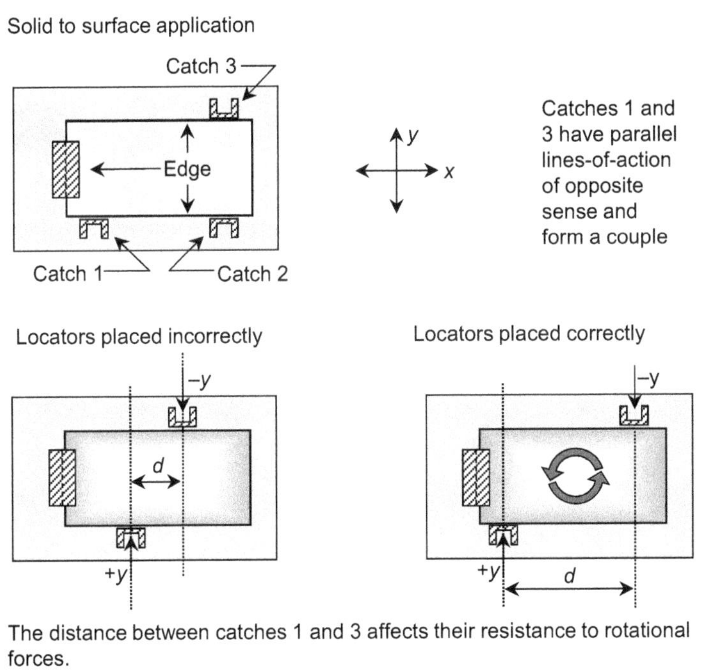

Figure 5.27 Locator pair placement for rotational strength

Also see the recent discussion in Section 5.5.5.1.

In all cases of linear or rotational forces, maximum distance between locator pairs with parallel lines-of-action is also recommended for part stability and dimensional robustness.

These rules for mechanical advantage, stability, and dimensional robustness also apply to lock pairs.

5.5.8 Ease of Assembly

Locks can be damaged during assembly if the mating and base-parts are misaligned. *Guide* enhancements, discussed in Chapter 9, Section 9.1.1, prevent lock damage by ensuring proper part-to-part alignment during the assembly process. Often, locator features can also function as guide features to eliminate extra interface features. The first locator feature(s) to make contact with the base-part should be considered for the guide function.

In the push assembly motion example shown in Fig. 5.28, locator pins are extended far enough to engage the edges of the opening before the cantilever hook locks make contact with the rim of the opening. In the tip example, the lug/edge locator pair provides initial positioning and guidance at one end and the pin ensures alignment at the other end.

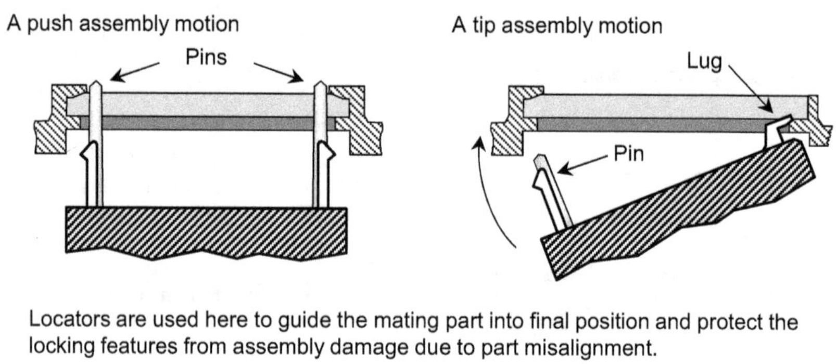

Figure 5.28 Locators used as guides for ease of assembly

5.6 Summary

This chapter introduced locators as strong and inflexible constraint features. They provide mating to base-part positioning and strength across the interface to prevent relative motion due to applied forces. Lock features, described in the next chapter, provide final constraint (locking) of the mating-part to the base-part to prevent separation.

General information about constraint feature design:

- Specify a radius on all interior and exterior corners of constraint features. This applies to the feature intersection with the parent material and to all corners within the feature itself.
- Size each feature according to the demands placed on that feature, see Figure 5.20 and the related discussion. In many designs, there seems to be a temptation to make similar locator features identical to each other without regard to strength requirements.
- Constraint features that do not require die-action will be preferred over those that do. But, do not sacrifice reliability just to avoid die-action.
- Some clamp load is possible in certain circumstances, but do not expect to get any significant or long-term clamp load in most snap-fit attachments.
- Because plastic materials tend to creep, avoid long-term or sustained forces on locators unless these forces are low, and long-term performance is verified by analysis and end-use testing.

Important points in Chapter 5 include:

- Snap-fit reliability depends on using constraint features to establish and maintain a line-to-line fit between the mating and base-parts with minimal residual force in the system. This begins with the proper use of locators.
- Constraint features prevent movement between and separation of the joined parts by removing degrees-of-motion (DOM). They are the necessary and sufficient conditions for a snap-fit attachment.
- Locator pairs should be the first constraint features added when developing the snap-fit interface.
- The first locator pairs considered should be the one(s) making initial contact during assembly. This locator pair(s) should also be used to provide a guide (enhancement) function.
- Locators will be the strongest features in the constraint system. For the most reliable attachment, design the interface to maximize the DOM removed by locators and minimize the DOM removed by locks.
- A locator pair is a locator feature on one part engaging a locator feature on the other part.
- Some locator pairs can constrain in as many as 5 DOM, others in only 1 DOM.
- A living hinge in snap-fit attachments is treated as a locator pair.
- Locator pair selection and arrangement in an application is a function of the assembly motion. Therefore, the assembly motion will determine the potential strength of a snap-fit attachment.
- The potential for the DOM removed by locators is highest with the tip, slide, twist, and pivot motions, so they are generally preferred over the push motion. Of these preferred motions, tip is usually the most practical choice.
- Locator features are based on protrusions (material addition), voids (material removal), or surfaces (existing material).

- Locators, particularly lugs, can be used as low-deflection locks, particularly when a slide, twist, or pivot assembly motion is present.
- Surface-based locators may be natural locators or, for fine-tuning and dimension-al-control, they may be lands added to a natural locator. A land may be a protrusion or a recess.
- Where two natural locators make up a position-critical pair, and fine-tuning may be necessary, use a tunable locator as one member of the pair.
- Maximize part stability and minimize sensitivity to dimensional variation by placing locator pairs constraining the same rotation or translational movement as far apart from each other as possible.
- Maximize mechanical advantage against rotational forces by placing locator pairs acting as a couple so their (parallel) lines-of-action are as far apart as possible. Locator efficiency is a measure of how many DOM are removed by a locator pair. More efficient pairs are preferred.

6 Physical Elements: Locks

Chapter 5 introduced locators as strong and inflexible constraint features that position the mating-part to the base-part. The second constraint feature group is *locks*, see Fig. 6.1.

Don't even THINK about lock features before locators have been identified!

Lock features complete the snap-fit attachment by holding parts together in their *located* positions. From the snap-fit definition in Chapter 1:

Locks are flexible features that must deflect for engagement with locators on a second part, followed by return of the lock feature toward its original position to create interference and latch the parts together.

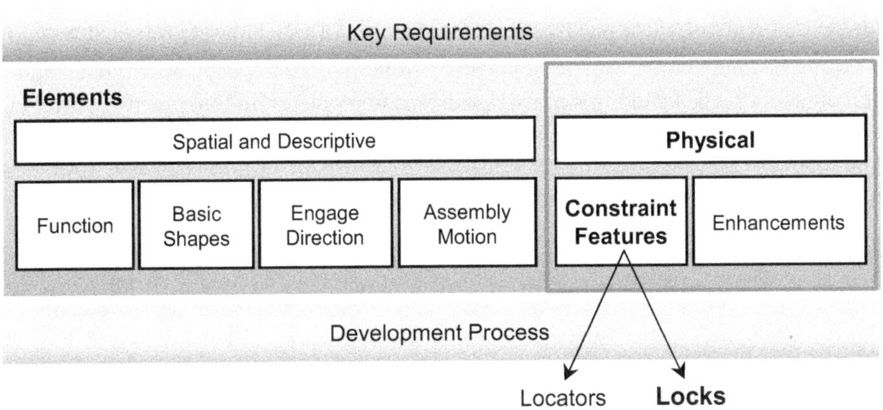

Figure 6.1 Lock features in the ALC

Locks present a design dilemma: They must be relatively weak and flexible so they can deflect for assembly. *But, they must also be strong for retention*. Thus, in every snap-fit application, lock features must satisfy conflicting requirements.

Locks require much more attention than locators.

How well a lock satisfies these conflicting requirements depends on how well it separates or *decouples* its assembly and separation behavior. Chapter 7 will discuss decoupling and lock strength in detail.

One way to quantify decoupling effectiveness is *lock efficiency*, the ratio of lock separation strength to assembly force [1].

$$E_L = \frac{F_R}{F_A} \qquad (6.1)$$

Where: E_L is lock efficiency.

F_R is retention strength.

F_A is maximum assembly force.

Think of separation strength as the attachment's value and assembly force as a cost. The higher the lock efficiency ratio, the better.

Although lock efficiency can be calculated for a lock design, its primary value lies in the concept itself. In this book, we'll refer to lock efficiency when discussing performance differences between lock styles.

Timing of lock feature selection is important. While it may be common practice in snap-fit development to make an early lock style selection, that is not the best approach. Follow the recommended snap-fit development process in the ALC because lock feature performance depends on first understanding constraint in the interface and on the wise use of locators.

Consider the assembly process: The mating-part is first brought into a pre-locked position with the base-part using locators, and then the locks are engaged. Chapter 10 explains why we should make our development decisions about these constraint features in the same order.

There are exceptions, but a lock pair is typically a lock feature on one part engaging a locator feature on the other part. Unlike locator pairs, which can remove multiple DOM (degrees-of-motion), a lock pair(s) should remove only 1 DOM in the separation direction.

Warning: Do not simply copy a lock feature from another application without understanding its true behavior: this practice will lead to problems. In the copied application, materials may be different and sometimes subtle differences in part geometry and enhancement features may be affecting feature behavior in unrecognized ways. For example, a cantilever hook with a very low L/T ratio may extend from a wall in an application where significant wall deflection contributes to lock deflection for assembly and release. When such a lock is copied to an application where all deflection must occur only in the lock, the lock is likely to fail.

It is also common practice to design one lock feature (using CAD) and then copy it to all other locations in the interface where a lock is needed. This practice ignores the possibility of very different lock strength requirements at other locations, see Section 5.5.3 and Fig. 5.22 for examples.

6.1 Lock Deflection and Separation Behavior

Lock features have two purposes: *deflection*, which allows for part assembly and separation, and *retention*, which is resistance to (unintended) separation. Deflection and retention are accomplished by different areas of the lock: a deflecting member and a retaining member, see Fig 6.2.

There can be confusion between the terms *separation*, *lock release*, and *retention*. We will use 'separation' as a general term to refer to all lock behavior during separation movement. If we are being more specific and emphasizing lock feature resistance to *unintended* separation, we'll use *retention. Lock release* may also be used in certain contexts.

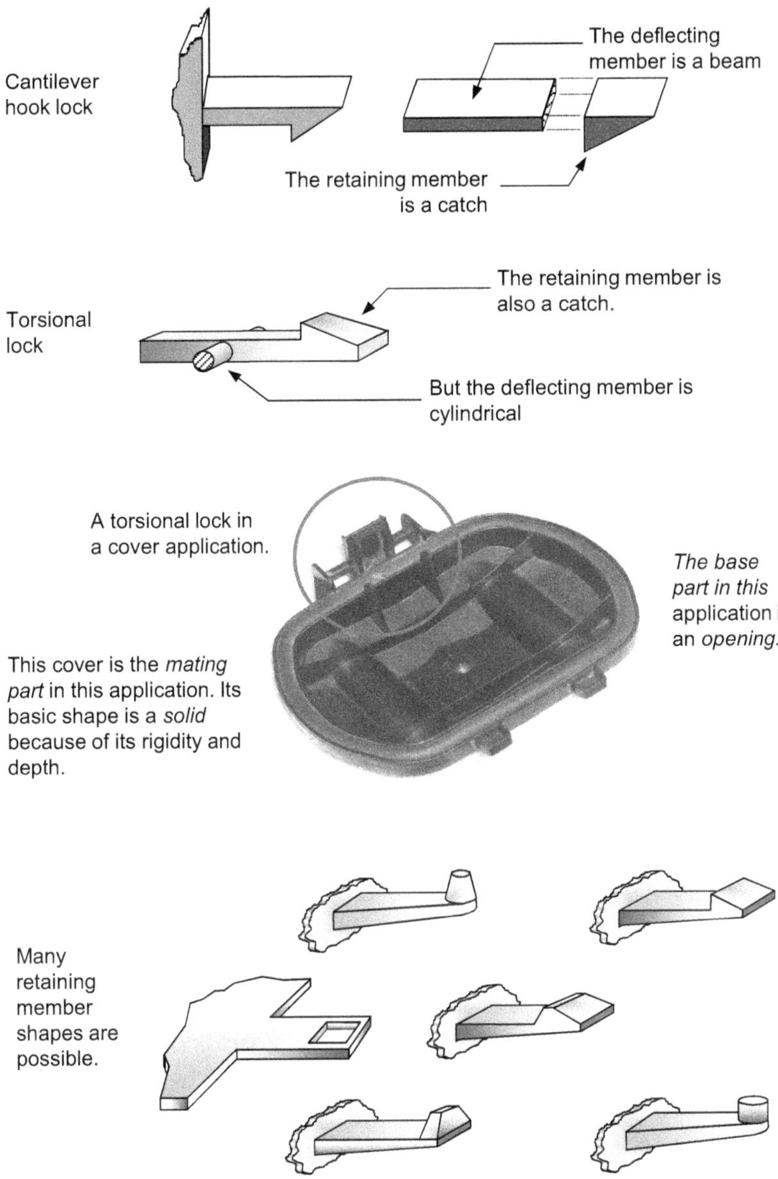

Figure 6.2 Deflecting and retaining members of a lock feature

6.2 Lock Styles

Locks are identified by their deflecting member. Locks based on cantilever beam deflection include hooks, loops, and traps. Other lock styles are torsional, annular, and planar, see Fig. 6.3.

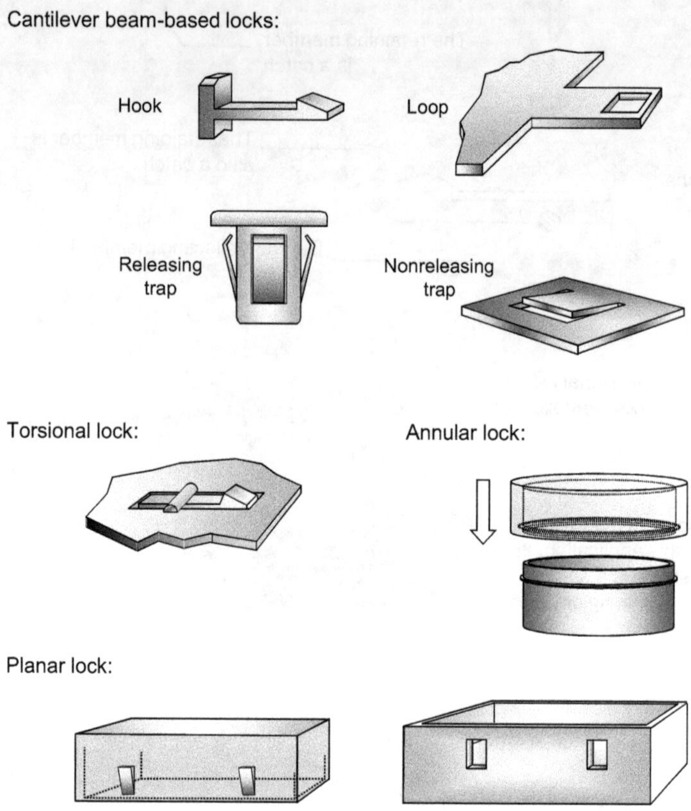

Figure 6.3 Lock styles

Assembly and separation behaviors are much different for each of these lock styles. This makes each style better suited for some applications and worse for others. Keep in mind, however, that for many applications, the lock style is selected primarily to satisfy other requirements such as available space, part geometry, molding requirements, or access for assembly. The most feasible lock style may not always be the best one for performance. But the variety of lock features allows for many options and solutions.

6.3 Cantilever Beam Locks

Locks based on cantilever beams are by far the most common lock style. Because they are so common, we will spend more time on them than on the other styles. For the same reason, they are sometimes used by default throughout this book when a lock is needed to complete an example or an illustration. In a real application, another lock style may be preferred.

However, despite their frequency in product applications, as well as in illustrations, examples in this book, and in the literature, never forget that *cantilever hooks*, one of the beam-based lock styles, are the *least preferred* lock for many applications. This topic is discussed later in this chapter as well as in other parts of the book.

In cantilever locks, the deflecting member is a beam. The most common beam shapes have a rectangular section and may be straight or tapered in length or width or both.

The retaining member is selected independently of the deflecting member.

Analysis of beam behavior for assembly and separation is based on classic bending equations for a cantilever beam fixed at one end. Exceptions are a beam fixed at both ends and the nonreleasing trap style lock which is analyzed as a column in compression. The purpose of analysis is to determine the beam's bending force and maximum strain. Beam bending force is then used in assembly and separation behavior calculations. These results determine the lock's final dimensions.

Common cantilever lock configurations use beams similar to those in Fig. 6.4 and have a rectangular section. Other sections are possible. Beams having a gently curved section are sometimes used in a circular arrangement of locks. Sometimes this circular arrangement of cantilever hooks is incorrectly called an annular lock, see Section 6.6 for a discussion of annular locks.

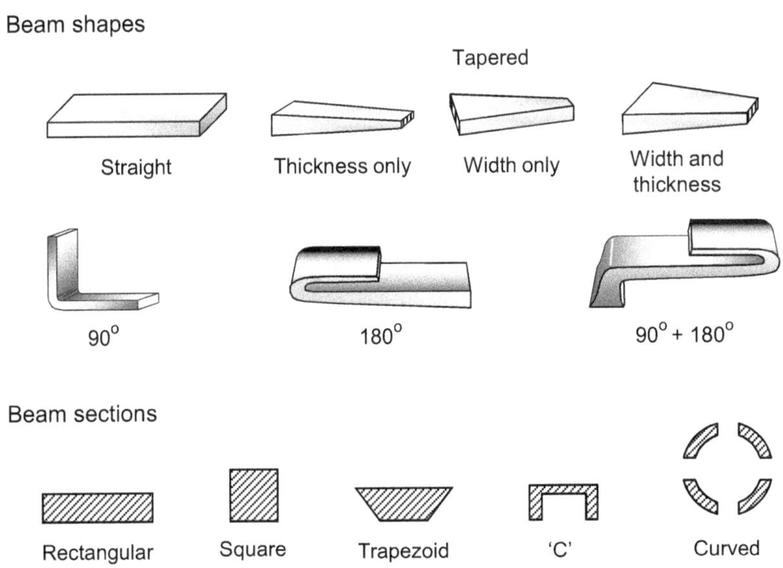

Figure 6.4 Cantilever beam deflecting members

Common lock orientations relative to a part are shown in Fig. 6.5. Note the interchangeability of the catch and the rectangular opening as retaining members on some of the beams. Again, as with locators, the lock feature is considered a separate feature from the surface or edge on which it is mounted.

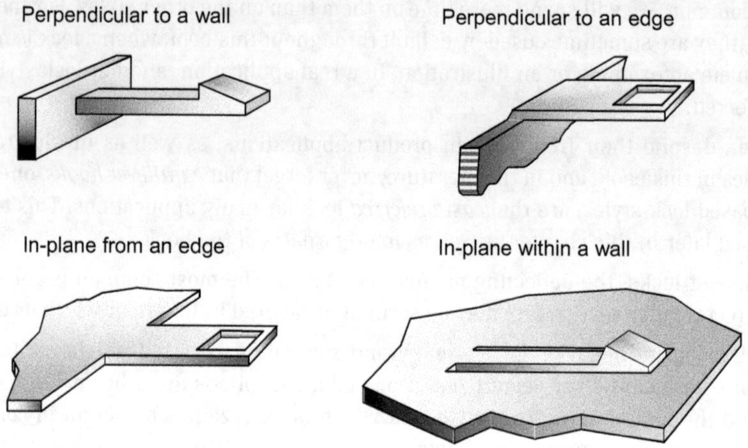

Figure 6.5 Common beam orientation to local part geometry

Lock features should be expected to constrain in the separation direction only, see Fig. 6.6.

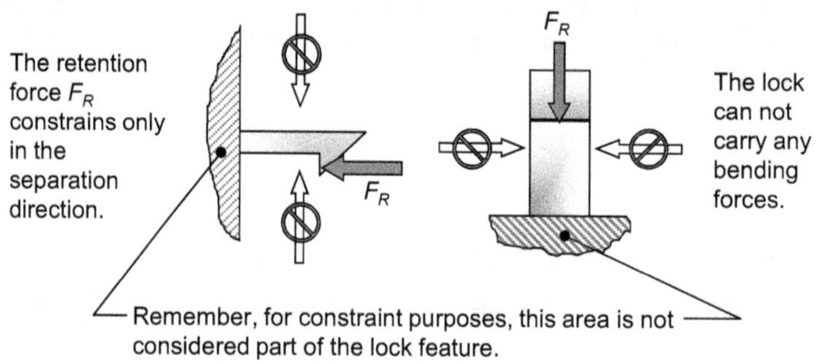

Figure 6.6 Locks should contribute to constraint only in the separation direction

In other words, all the lock should do is hold the mating-part in position.

One of the most common mistakes made in snap-fit design is to use the lock to react against forces other than just the force in the separation direction. This causes an under-constraint condition. Always ensure that locator features are present to carry these other forces.

It is also preferred that lock features carry no *significant* forces in the separation direction. This is because most locks tend to be relatively weak in that direction. It will be up to the developer to determine whether separation forces exist and are significant or not.

In reality, although we try to avoid it, an application may require a lock(s) to resist (sometimes significant) separation forces. As we will see, there are some cantilever beam-based lock styles that can be quite strong in retention and are capable of resisting separation forces.

Figure 6.7 illustrates some very bad cantilever hook lock designs. All these mistakes have been found repeatedly on beam-based lock features in many different applications. Because they are mistakes associated with the deflecting, and not the retaining member, they can occur on any cantilever beam-based lock. Later in this section, we'll describe a common scenario illustrating how one of these bad designs can occur.

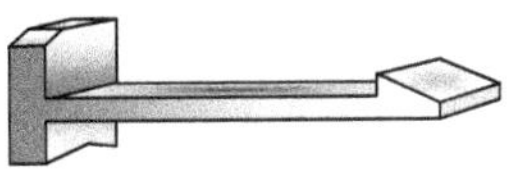

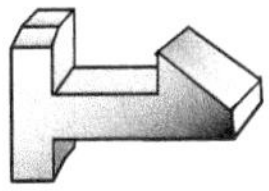

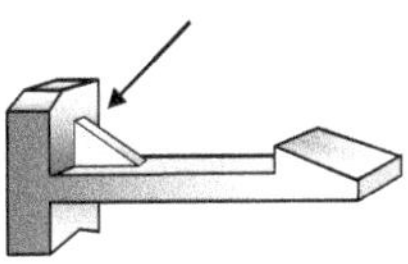

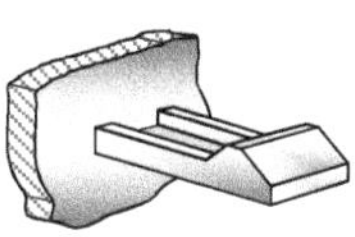

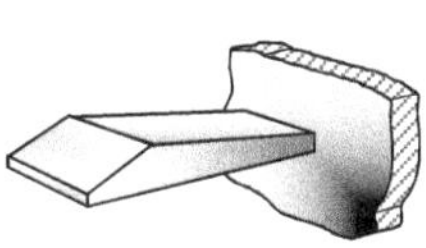

Figure 6.7 Examples of bad cantilever lock designs

In the author's experience, the original mistakes in these designs are made due to a lack of snap-fit knowledge. But the poor design is often carried forward far too long in the development process. The reason for this is failure to admit or to recognize that a cantilever hook style lock was a bad choice in the first place. By this point, molds may have been made and it may seem that the design is locked in and that it is too late to change.

Of course, the best option is to avoid making a bad lock feature choice in the first place. Figure 6.36 (at the end of this chapter) shows how, in the author's experience, selecting the wrong lock feature style (usually this is a cantilever hook) is the single biggest cause of snap-fit problems.

If problems do occur, the first solution should be to redesign the lock area with minimal mold changes. Because the lock is part of a lock pair, changes to the lock area may also include changes to the locater area on the other part. This is another reason why resistance to making late changes can be so strong. An example of how minimal changes can sometimes be made to improve lock performance is given in the next section.

Chapter 14, "Diagnosing Snap-Fit Problems," also lists possible lock feature changes to minimize redesign.

6.3.1 Hooks

One of the most important lessons to be learned is when not to use hooks.

Cantilever hook locks are by far the single most common snap-fit lock style. They are relatively easy to understand, analyze, design and manufacture as an integral attachment. Their popularity can lead to the perception that cantilever hooks represent snap-fit technology. (See Section 1.2 in Chapter 1 and Section 15.3 in Chapter 15 to read about the *battery cover syndrome.*) One bad experience with a cantilever hook can cause a developer or even an entire company to avoid using snap-fit technology at all.

Hooks have their place as an attachment option, but they should not necessarily be the default lock feature selection. The reader will learn about cantilever hook limitations and about methods to improve hook retention performance when needed.

Figure 6.8 identifies the major features of cantilever hooks.

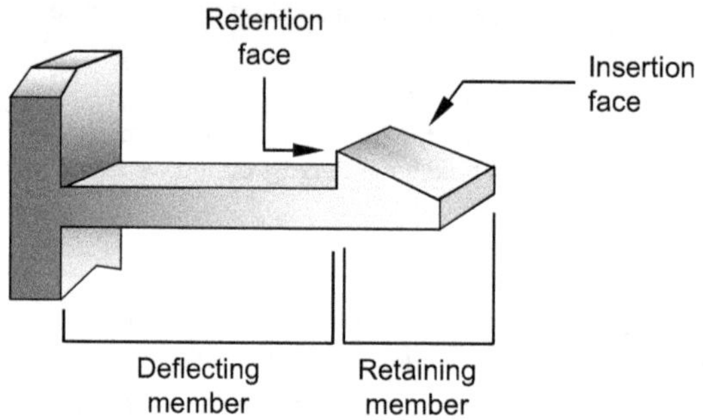

Figure 6.8 The basic hook

Figure 6.9 describes an all too common hook development scenario. The author has seen this particular poor hook design so many times and in so many different products that it deserves special attention. It has been found, broken, on very expensive copying machines, a high (?) quality home vacuum cleaner (our own), and multiple other low and high cost products.

Adding a rib shortens the beam's length.

While they do have their place in lock design, the presence of a supporting rib is often a good clue that the original design had issues which lead to a series of attempted fixes. Often the first fix tried is adding a rib to support the beam. Often, this does not work as expected. If a cantilever hook requires a rib, then a different lock style is probably indicated.

Several illustrations in this book show parts with ribs added to cantilever hooks with a low *L/T* ratio (short hooks). These hooks were not the proper lock choice for the original design.

This scenario begins with a common error: *I've seen snap-fits, therefore, I can design them.*

Figure 6.9 Common (bad) hook development scenario

A variety of cantilever hooks are shown in Fig. 6.10. While caution is recommended, when used in an appropriate application and the lock pair is properly executed, they are viable lock options.

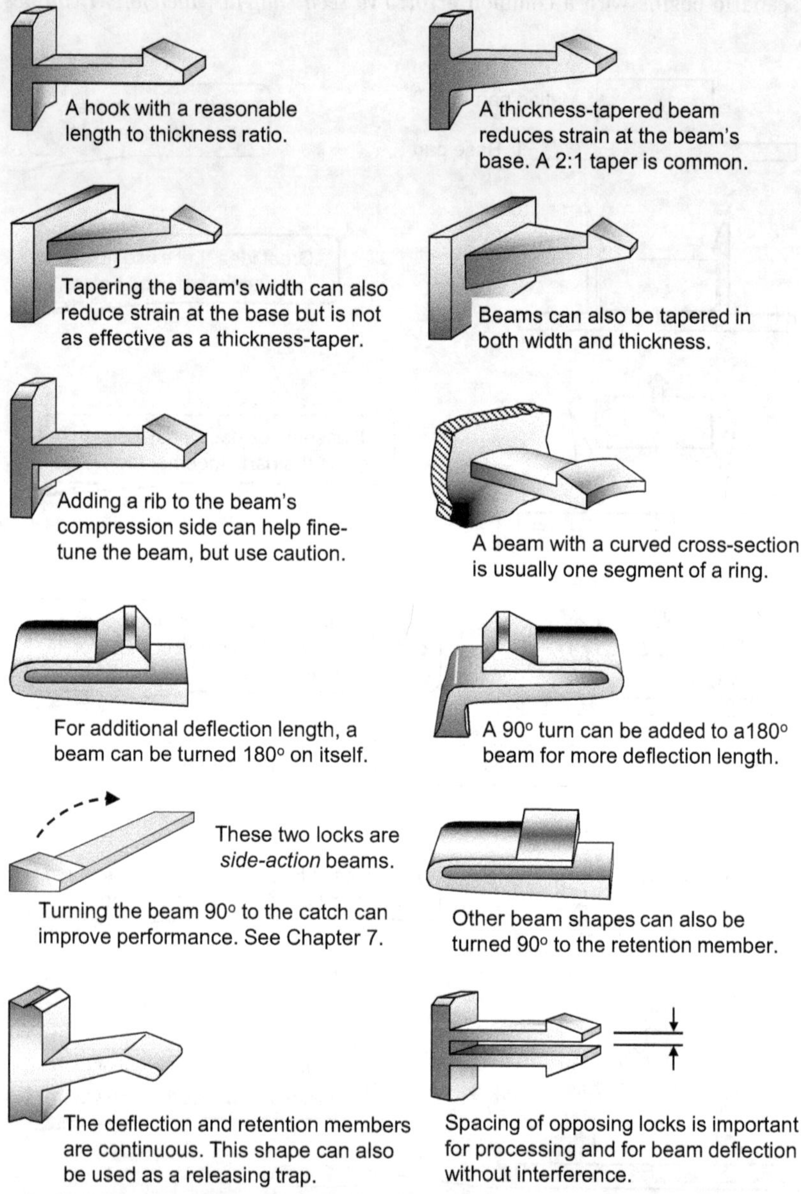

Figure 6.10 Hook examples

6.3.1.1 Hook Assembly Behavior

With a cantilever hook and a flat insertion face on the retaining member, the insertion face angle increases as the hook deflects during engagement, see Fig. 6.11. This causes the hook's force-deflection signature to increase geometrically. Most of the time, this is not a problem. However, the resulting high force can sometimes cause difficult assembly and as a result, provide poor feedback to the assembler.

If this is the case, adding a profile to the insertion face can make the (assembly) force-deflection signature more operator-friendly by reducing the maximum assembly force and changing the lock's assembly *feel*.

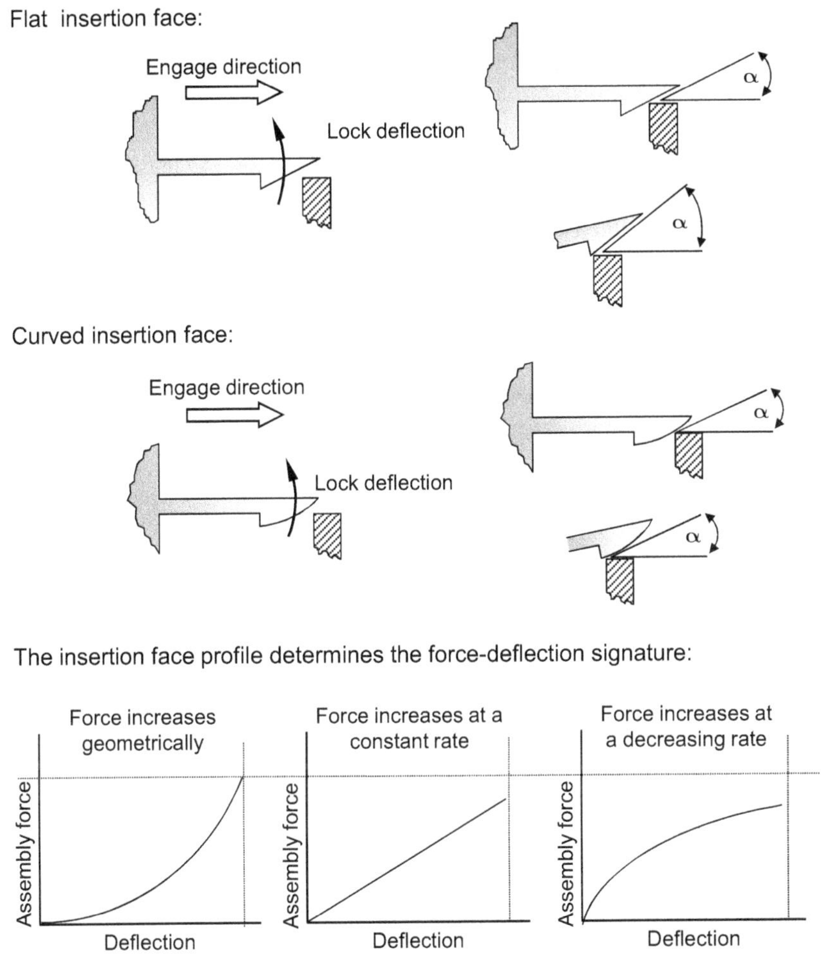

Figure 6.11 Hook assembly behavior

The profile's contour can be adjusted to provide a constant rate of change or a decreasing rate in the force-deflection signature. The latter will give an over-center feel when the lock is engaged. Calculations of profile shape are included in Chapter 13, Section 13.6.1.3.

In an application where locks are *releasing* or the *action* is controlled-moveable, the product user may be operating the lock frequently and the force-deflection signature can create a perception of either high or low quality. In this case, the insertion face

Recall Chapter 4, Section 4.1.

should always have a profile to improve user feel and perceived quality. The over-center signature is preferred. High assembly forces repeated many times in a moveable application may also cause eventual damage to one or both members of the lock pair. An insertion face profile can minimize the chances of long-term damage by reducing the maximum assembly force.

6.3.1.2 Hook Separation Behavior

The retaining member for cantilever hook locks is a catch (a locator feature) at the end of the beam. Although the retention face angle is a factor, the hook's separation resistance is ultimately based on the beam's deflection force. An exception we'll discuss is a hook with a retention face angle greater than 90°.

One weakness of the hook is that when separation force is applied to the lock, the reaction force is not along the beam's neutral axis. As shown in Fig. 6.12, there is an offset (d) and the hook is destined to bend. Of course, bending is the beam's weakest direction for strength and resistance to deflection.

Unlike assembly, where the insertion face angle increases with deflection, when the hook deflects for separation, the retention face angle *decreases*. This leads to lower separation force than expected unless this change in the retention face angle is accounted for in the analysis. As with the assembly force calculations discussed above, most published lock feature calculations do not recognize this, sometimes significant, effect on the hook's retention strength.

Even nonreleasing hooks with a retention face angle at or near 90° will release under a sufficiently high force, see Fig. 6.12. When a nonreleasing hook does release under load, the typical pattern of release begins with beam-end rotation/distortion at the retaining member. This causes a reduction in the retention face angle which then enables additional slippage along the retention face and finally, release due to beam bending. The shear or tensile hook failure modes shown in Fig. 6.12 are very unlikely unless the hook end is restrained from rotating.

When a retention face angle greater than 90° is used on both the hook and the mating feature, then high strengths are possible. This very strong hook lock is frequently found on strap closure buckles used with products like luggage, backpacks, book bags, and laptop computer cases, see 6.12 and Fig. 9.13 in Chapter 9.

A retention face angle greater than 90° requires clearance or compliance in the system to allow the lock face to move past the engagement point and then return. For this reason, it is not practical in many applications.

It is important to differentiate between the kinds of forces to which a lock might be subjected. Separation forces may be low, which is good, but if they are continuous and long-term, they may result in plastic creep and part separation. Forces may be high but transient, and in a properly designed application, have no effect. In an application with poorly designed locks, the same transient forces may cause unintended separation. It is these transient forces we are concerned with here. Of course, the terms *high*, *low*, *long-term*, and *short-term* are relative and depend entirely on the mechanical properties of the plastic(s) in any given application.

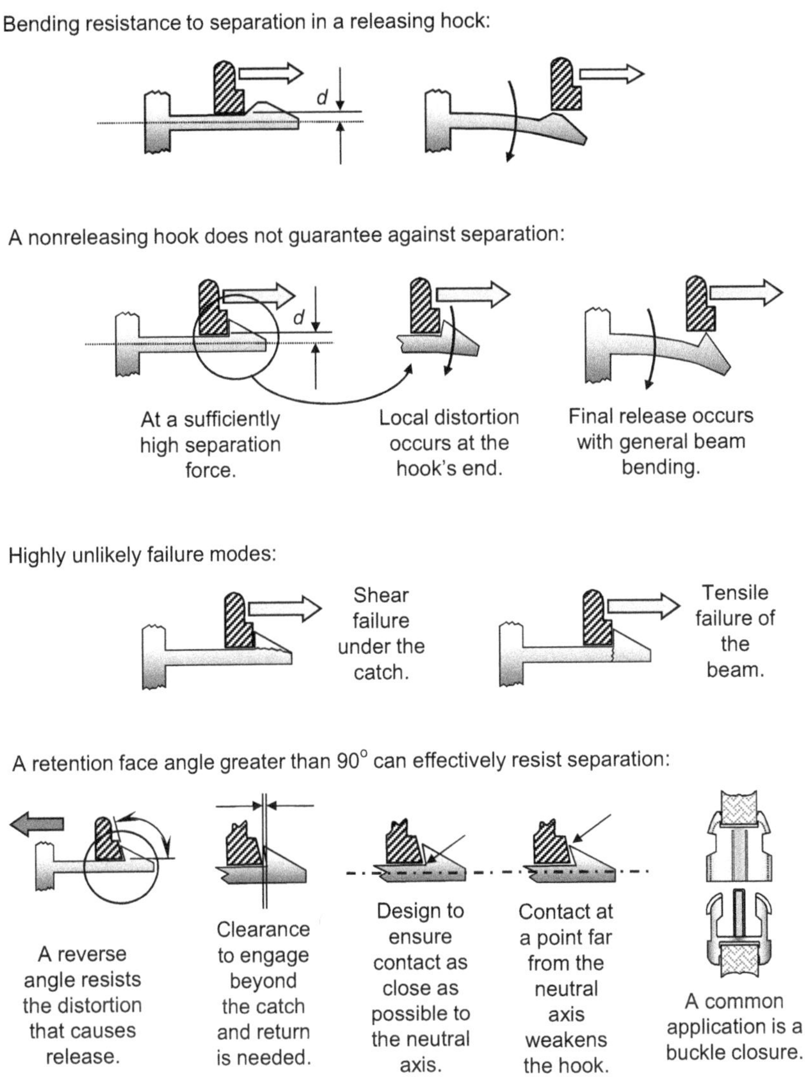

Figure 6.12 Hook separation behavior

What happens during application of a sudden transient force to any lock? The energy is either absorbed by the locking system or the lock feature breaks or releases. The goal is to absorb the energy before separation and without permanent lock damage.

As with adjustments to the insertion face profile for hook assembly, separation performance can sometimes be improved by adding a profile to the hook's retention face, see Fig. 6.13. The profile compensates for the reduction in retention face angle during deflection and ensures that the instantaneous angle remains constant. This allows the lock to absorb more energy before releasing, as shown in the force-deflection signature. When comparing force-deflection signatures, picture the separation energy absorbed as being proportional to the area under the curve. The signature can be modified for maximum effectiveness by adjusting the profile.

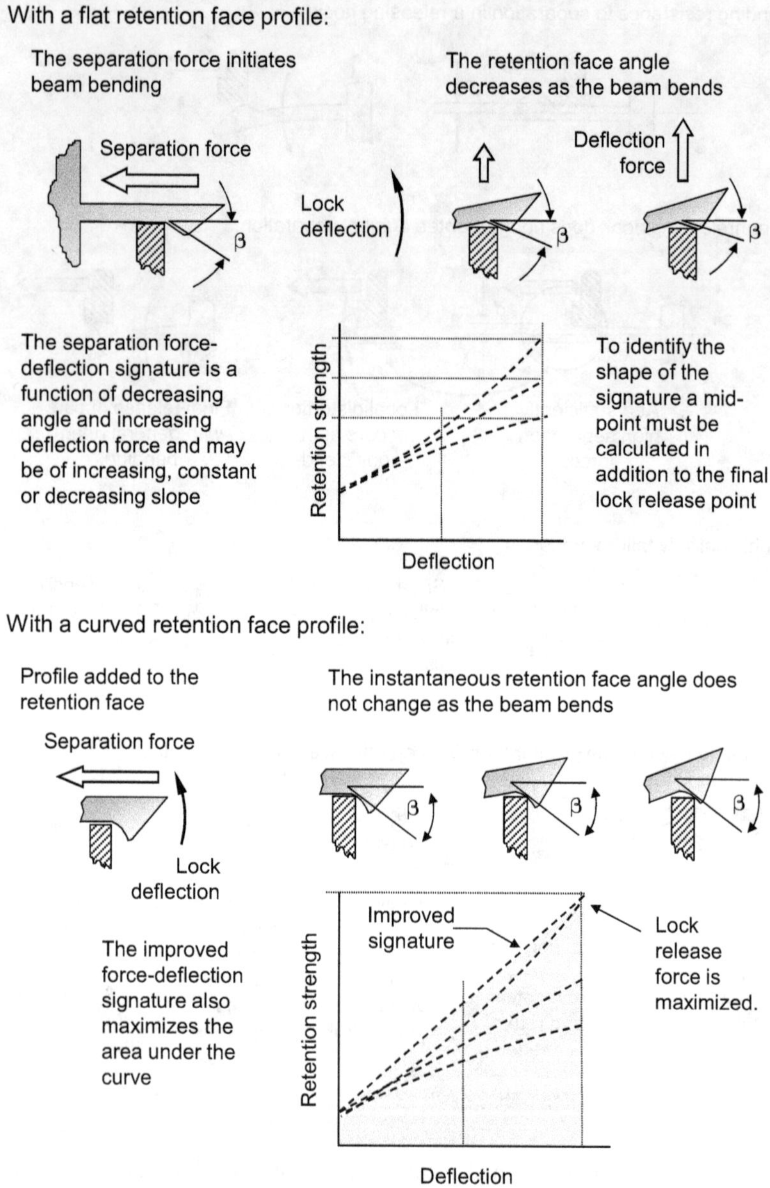

Figure 6.13 Hook retention face profile

Retention face profile calculations are discussed in Chapter 13, Section 13.6.2.3. Adjustments to the retention face profile are a relatively subtle change and can be effective when forces are of very short duration, as might occur in a drop test. The principle of energy absorption can also be applied using some features as springs to add elasticity to the system. This is covered in Chapter 9, Section 9.3.3, in the discussion of compliance enhancements.

There are also other ways to make cantilever hooks stronger.

One effective method is shown in Fig. 6.10 where, in two of the locks, the beam is rotated 90° to the retaining member's assembly direction. (We refer to these as *side-action* locks.) This principle is also mentioned in Section 6.7.2 of this chapter and will be discussed in detail in Chapter 7.

This change makes the hook's engage direction perpendicular to the long axis of the beam and allows the beam to bend along its thin section for low assembly forces and low strain, yet resist separation across the thicker section of the beam's width.

With side-action locks, engagement is 90° to the beam's long axis.

Using a side-action design can permit use of a cantilever hook or loop style lock where part clearances or mold design constraints prevent use of a more conventional lock. They are also effective in short grip-length applications.

Another possibility is to replace the catch of the cantilever hook with a hole to create a cantilever loop which, as we will see, can be stronger than a hook.

Yet another way to improve hook performance is to add retention enhancements as described next.

6.3.1.3 Hooks and Retainers

Figure 6.14 shows an example of a *retainer* feature added to a nonreleasing hook, which, in this application, must resist a separation force. The retainer prevents distortion at the hook's end that can lead to beam bending and allow the hook to release. A disadvantage of the retainer is that it will increase the lock's assembly force. Retainers are *enhancement* features and are discussed in Chapter 9, Section 9.3.2.

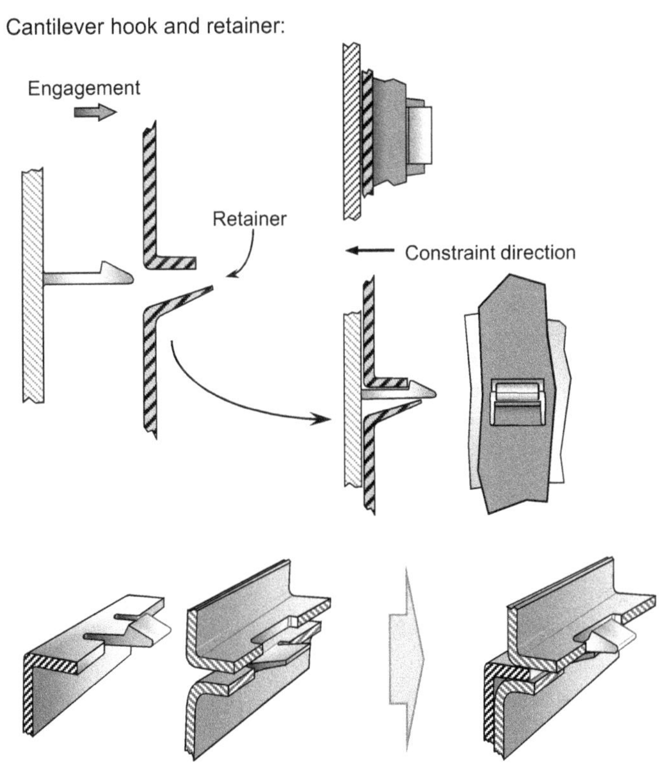

Figure 6.14 Improving hook retention strength with a retainer

6.3.1.4 Hooks and Prongs

A lock feature may not always be what it appears to be. Features are defined by their behavior, not by how they look.

In Fig. 6.14, we saw how a relatively weak hook/edge lock pair can be strengthened by adding a retainer. But the lock is still a hook with the hook style's inherent limitations.

Figure 6.15 shows how that hook/retainer lock arrangement can be transformed into a *prong* locator. The hook becomes a prong by the addition of a supporting wall to prevent deflection. The original edge locator is redesigned as a trap that now becomes the deflecting member of the new, and much stronger, trap/prong lock pair. Note how these changes require minimal redesign to the area.

A feature that looks like a hook is now a locator!

For an even stronger attachment, the prong can be further modified and the supporting wall can also become a trap to create a double-trap arrangement. Again, redesign of the area is minimized. The retention face(s) on the prong can be made releasing by adding an appropriate angle, or nonreleasing by using a steeper angle, up to 90°, also see Fig. 6.24.

More than once, the author was able make these kinds of change relatively late in a product's development process in applications where hooks specified in the original design would not have been effective.

A 'weak' cantilever hook/edge lock pair evolves into a very strong prong/trap:

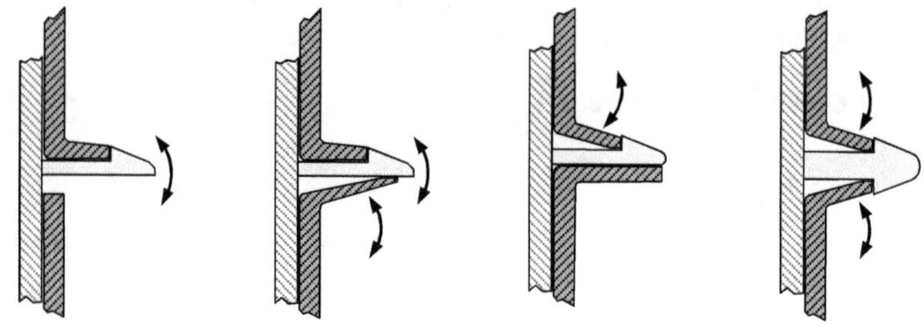

Cantilever *hook* lock and an *edge* locator: a very common lock pair configuration.

A *retainer* enhancement is added for additional retention strength.

When supported against deflection, the hook becomes a *prong* locator that engages a nonreleasing *trap* lock for a very strong attachment.

The prong can be further modified to engage two nonreleasing traps for more strength.

The cross-hatched areas in these pictures could also be (flexible) sheet metal in a plastic/metal snap-fit application.

Figure 6.15 Minimal changes create a strong trap/prong lock pair

This concludes discussion of cantilever hook locks. They are the most frequently used lock style and deserve attention, but are often *not* the best choice for a locking feature.

Many general principles for beam-based lock behavior were introduced here for the cantilever hook. Some will also apply to the deflecting members of the other beam-based lock styles.

6.3.2 Loops

A cantilever beam that uses a cross-bar at the end of an opening as the retaining member is inherently stronger than the cantilever hook. This lock style is called a *loop* [1]. It behaves like a loop of rope that has no strength in bending but is extremely strong in tension. When space for lock deflection is limited, loops will require less clearance for deflection while delivering equivalent or better retention performance than a hook. Basic loop styles are shown in Fig. 6.16.

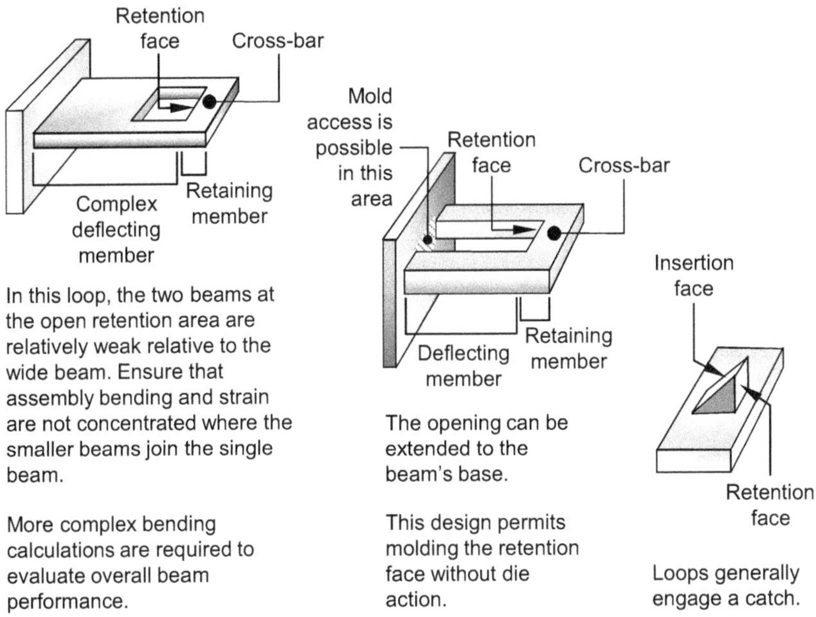

Figure 6.16 The basic loop

Good plastic part design recommends the use of radii on all internal and external corners. Because of the potential for loops to resist separation forces, specification of proper radii, especially at the inside corners of loop features, is critical. Under load, sharp internal corners will be stress risers where failure can occur.

As a rule, the loop style in which the opening extends to the mounting surface is preferred because mold access for forming the cross-bar is available. Because the beams extend from the surface to the cross-bar, beam behavior and calculations are simplified and beam deflection is not concentrated at the small opening.

However, there may be cases where knit line issues require that a large cross-sectional beam extend farther out towards the cross-bar to carry a higher volume of melt before it enters the two narrow beams and the cross-bar.

All the cantilever hook configurations shown previously in Fig. 6.10 can be converted to loops by replacing the catch retaining member with a rectangular opening. A few examples are shown in Fig. 6.17.

Just as with hooks, the beams can be rotated 90° to the retaining member's engagement direction.

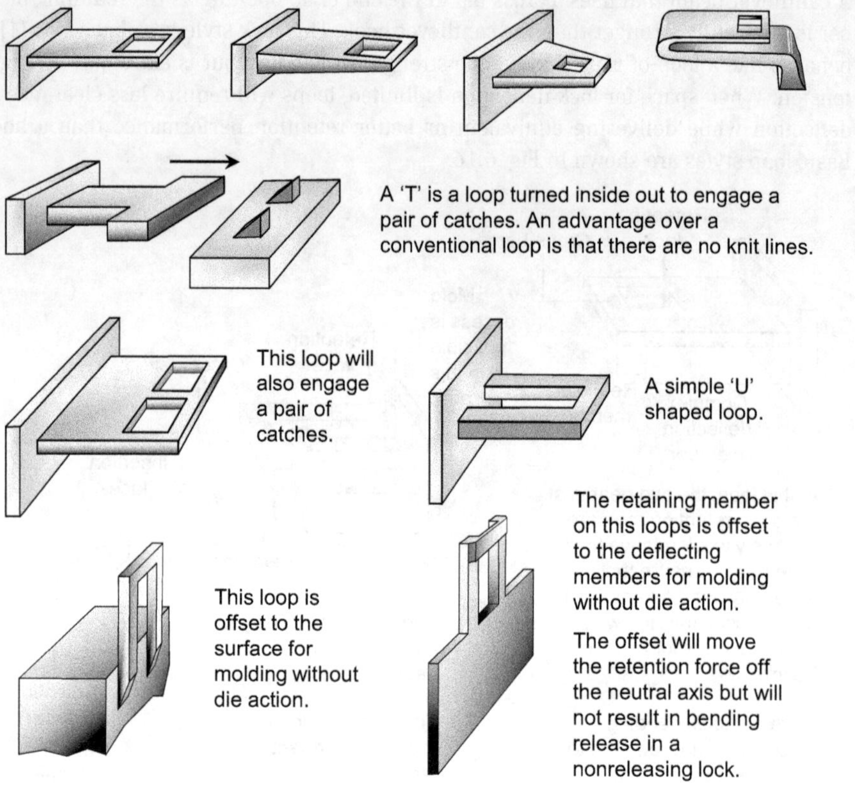

Figure 6.17 Loop examples

6.3.2.1 Loop Assembly Behavior

Loops typically engage a catch as the locator member of the lock pair, see Fig. 6.18. Because the catch does not deflect during assembly, changes in the insertion face angle are not an issue when calculating assembly force. However, a profile can be added to the catch's insertion face to adjust the force-deflection signature.

The nonlinear assembly signatures are preferred for perceived quality when the snap-fit is to be operated by the consumer.

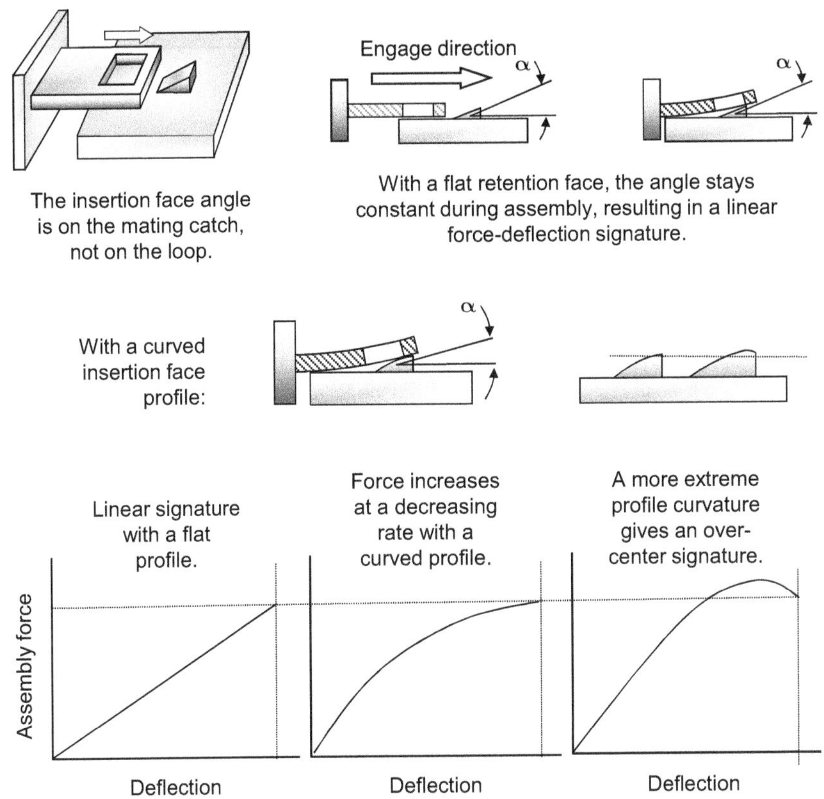

Figure 6.18 Loop assembly behavior

6.3.2.2 Loop Separation Behavior

As with assembly behavior, loop separation behavior is determined by the multiple beams and by the locator in the lock pair, usually a catch, see Fig. 6.19. The loop and catch lock pair can be very strong, and unlike a nonreleasing hook style lock, will resist unintended release under high or sudden impact loads.

When a loop is nonreleasing, the reaction force is coincident with the beam's neutral axis and no bending can occur due to a separation force. Instead, retention strength is determined by feature geometry and the tensile and shear strengths of the materials in the lock pair. This means the loop style lock can always provide better retention than a comparable hook, beams being stronger in tension and shear than in bending.

When the loop is a releasing lock, the reaction force is no longer along the neutral-axis, but it will be very close, offset by one-half of the beam thickness. With a flat retention face, as with loop assembly, the force deflection signature will be linear because the retention face angle on the catch does not change. The force-deflection signature can be tuned by adding a profile to the catch's retention face.

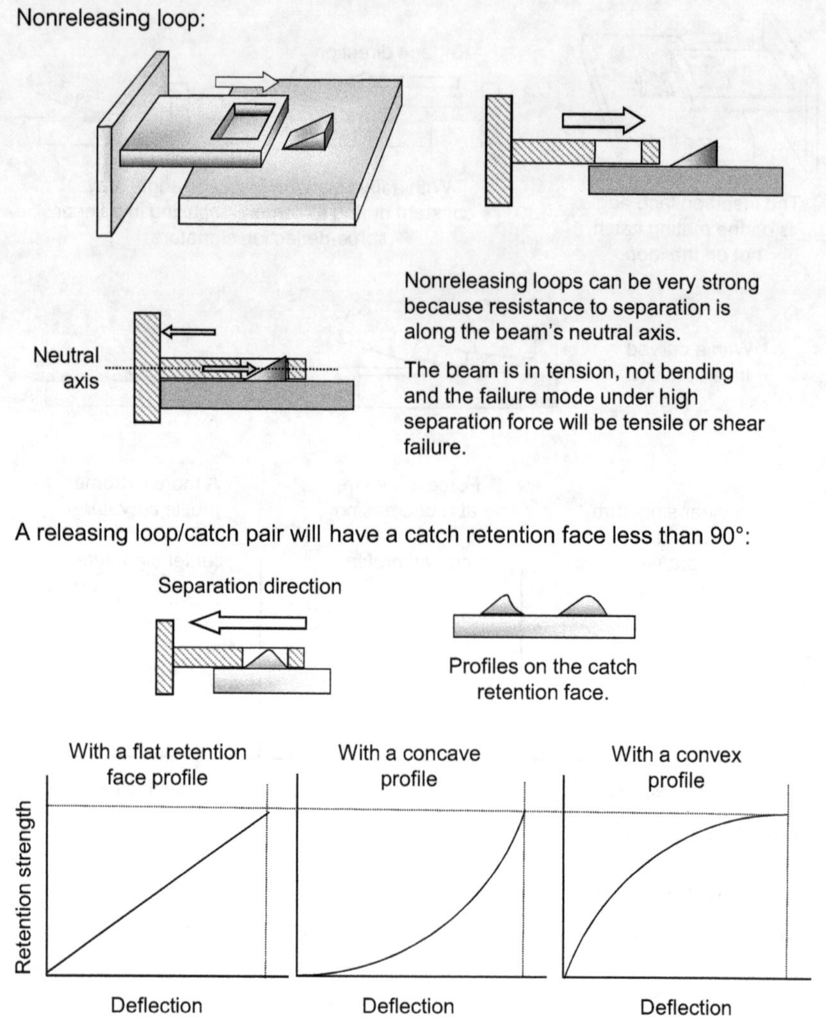

Figure 6.19 Loop separation behavior

6.3.2.3 Loops and Knit Lines

A unique issue with loop style locks is the likelihood of a knit line forming in the loop during the molding process. Knit lines occur where two melt fronts meet as the material flows through the mold, see Fig. 6.20. The magnitude of strength reduction depends on the melt temperature at the fronts and the ability of their surface layers to merge. Test data shows the strength reduction at a knit line may be as much as a 65 % depending on the material and absence or presence of a filler [2].

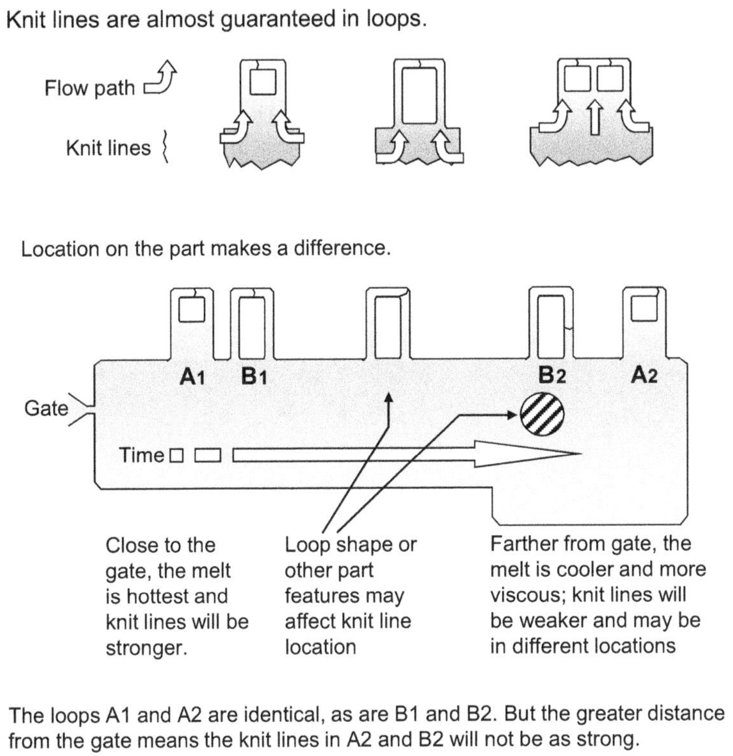

Figure 6.20 Knit lines in loops

Strength reduction is most dramatic in filled materials. The fibers do not flow across the knit line so it consists of only the polymer material. In the tests cited above, unfilled polypropylene showed a 14% reduction and 30% glass-filled polypropylene showed a 66% strength reduction. The unfilled and 40% glass-filled nylon 66 test results were 3% and 48% respectively. (These results occurred under specific test conditions and should not be considered design data.)

Refer to plastic part and mold design literature for more information on knit lines.

Loops with identical shapes but located in different areas of the same part can have different levels of knit line strength and the knit lines can occur at different locations. This is due to local flow characteristics and because the melt temperature at a given point depends on its distance from the gate and cooling effects of the mold along the flow path.

Beall [2] recommends adding a drawing note indicating *No knit line in this area* as a precaution at *any* highly loaded area of a part. This is a good idea, but the loop's shape can make avoiding a knit line impossible. In general, designers should accept that knit lines will occur in the loop and compensate for the expected loss of strength. Possible solutions are shown in Fig. 6.21. The easiest solution is to simply over-design the loops in question to compensate for knit line weakness. Mold flow analysis and prototype part testing can identify likely knit line location(s) and evaluate a solution's effectiveness.

It is worth repeating that it is good injection-molded part design practice to specify fillets and radii on all corners, both internal and external. This is especially critical in loops, where a sharp corner in the opening can be a weak site due to molded-in stresses and also becomes a stress riser under applied load. They may not always be shown in the illustrations, but always specify radii at all internal and external corners of a loop.

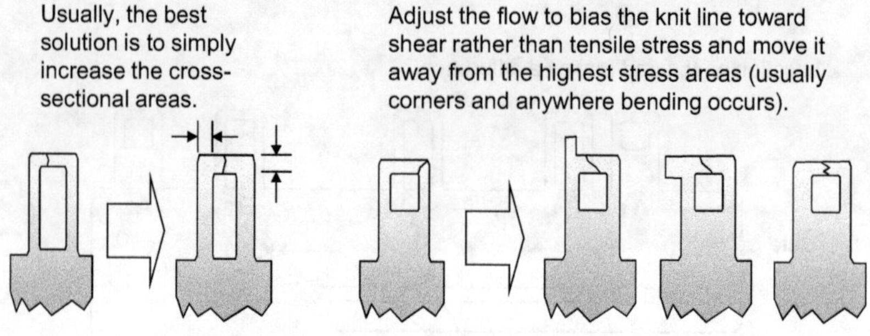

Figure 6.21 Possible solutions to knit line weakness

6.3.3 Traps

Traps are the simplest cantilever beam-based lock but they have the highest strength potential. Trap shape is typically limited to variations of the straight cantilever beam. In releasing traps, the retention face is formed by a bend in the beam. In nonreleasing traps, the retaining member is the end of the beam itself.

However, traps can also be made with a catch on the end of the beam, just like a hook. The fundamental difference between any trap and hook style lock is the direction of assembly and separation movements with respect to the trap's fixed end or base.

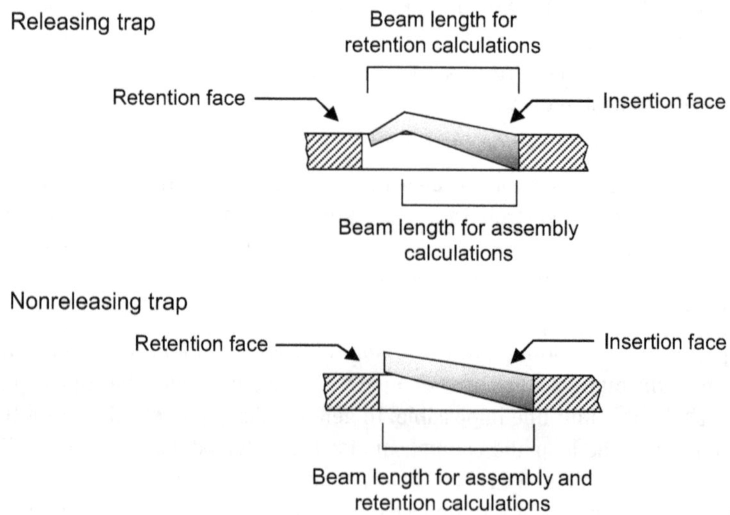

Figure 6.22 The basic trap

Releasing and nonreleasing trap *assembly* behavior is based on cantilever beam mechanics. Releasing trap separation behavior is also based on the cantilever beam. However, nonreleasing trap *retention* behavior is based on column mechanics because the beam is loaded in compression. As we will see, this is a very important and useful distinction.

Traps differ from other beam-based locks in both insertion and separation behavior, see Fig. 6.23. Cantilever hooks and loops engage with the mating feature moving *toward* the fixed end of the beam and resist separation with the mating feature moving *away* from the fixed end of the beam.

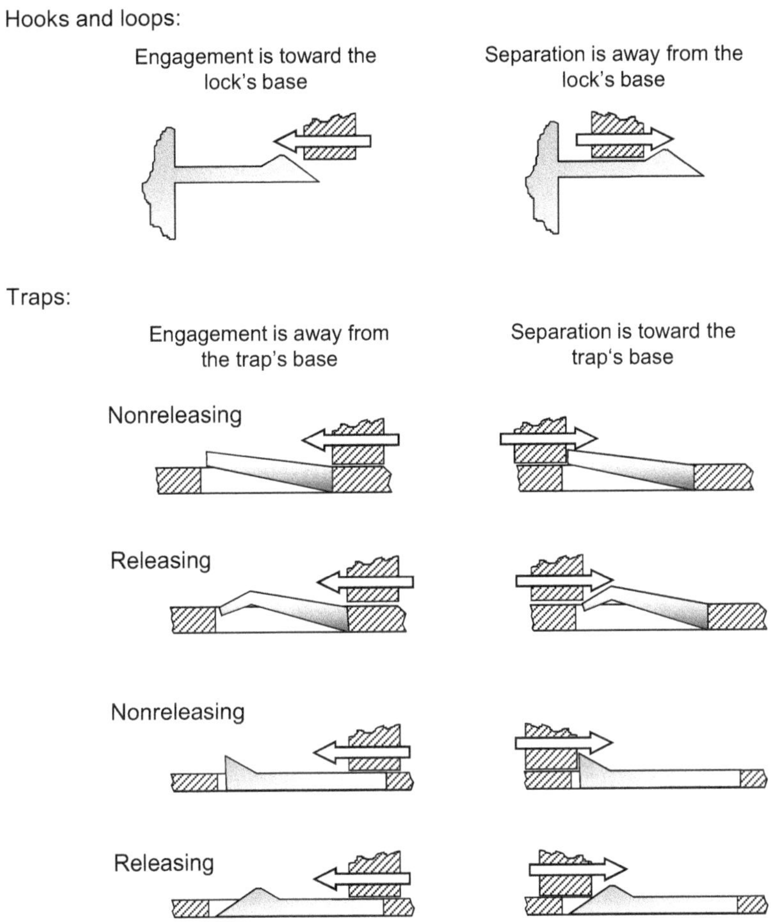

Figure 6.23 Differences between hook/loop and trap assembly and separation

Traps are just the opposite, engaging with the mating feature moving away from the lock base and resisting separation with the mating feature moving toward the lock base. These differences result in significant assembly and separation performance differences between these lock styles.

Nonreleasing beams consisting of simply a straight beam will resist separation acting primarily as a column in compression, although there may be a small bending component.

Nonreleasing traps with a catch on the end will resist separation with both compression and bending components since the separation force is applied off the neutral axis.

Traps are a preferred locking feature.

Figure 6.24 illustrates some common trap configurations. Note their versatility, for example, how the cantilever beams of a trap can be in-plane with or perpendicular to a wall, and how they can hang from a post, tab, or frame. Traps should be considered as a lock feature option in all cases, particularly when lock reliability and retention strength is essential.

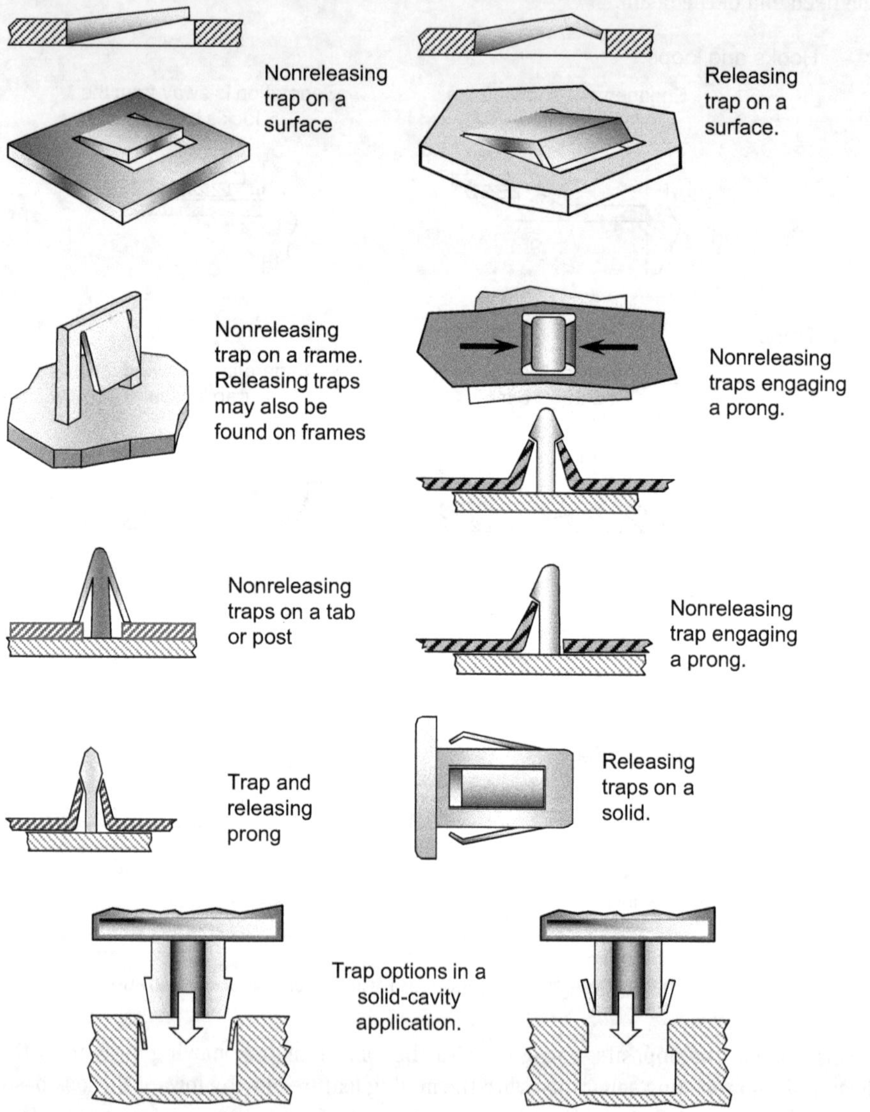

Figure 6.24 Trap examples

6.3.3.1 Trap Assembly Behavior

Sometimes force-deflection signatures for trap engagement are not easy to predict. With both releasing and nonreleasing traps, the insertion face angle decreases as the trap is engaged but, depending on the trap design, the initial deflection force *moment arm* can be very short. This may result in a high initial assembly force that decreases until engagement is complete. Or the assembly force may start out low and increase, see Fig. 6.25.

In general traps are assembly-friendly attachments because they tend to result in much lower assembly forces for the separation strength they provide (a high lock efficiency ratio, Eq. 6.1) and can be designed for an over-center action for improved assembly feel and feedback.

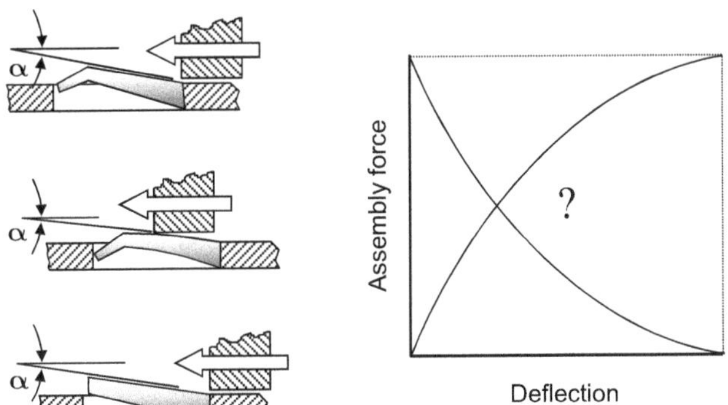

Figure 6.25 Trap assembly behavior

6.3.3.2 Trap Separation Behavior

Traps can be extremely strong and are ideal as *nonreleasing locks* in applications where parts are not intended for separation. If separation is intended, access to the trap for manual release must be provided.

Traps will be releasing or nonreleasing depending on the beam shape. Like the cantilever hook, a releasing trap resists separation through beam bending. Also like the hook, release behavior is a function of the angle and shape of the retention face and the coefficient of friction between the mating surfaces. There is one notable exception - unlike the hook, as separation occurs, the trap's retention face angle becomes *steeper* resulting in a higher separation force, Fig. 6.26.

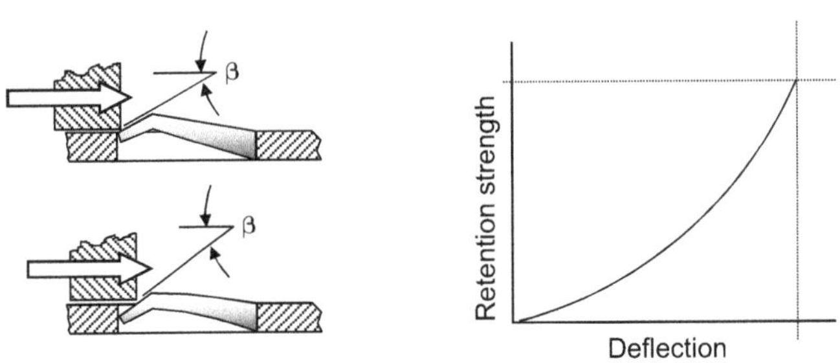

Figure 6.26 Releasing trap retention behavior

This can be an advantage in resisting unintended separation but a disadvantage when manual separation is required. This behavior can be significant and must be factored into separation performance calculations.

Nonreleasing traps generally resist separation not through bending but through compression and can be very strong, see Fig. 6.27. However, recall from Fig. 6.23 that some nonreleasing trap configurations could have a bending component in addition to compression. A nonreleasing beam will fail in compression (i. e., buckling) and trying to force part separation will most likely damage the lock. If applications using nonreleasing traps are to be serviced, provision must be made to allow access for trap deflection.

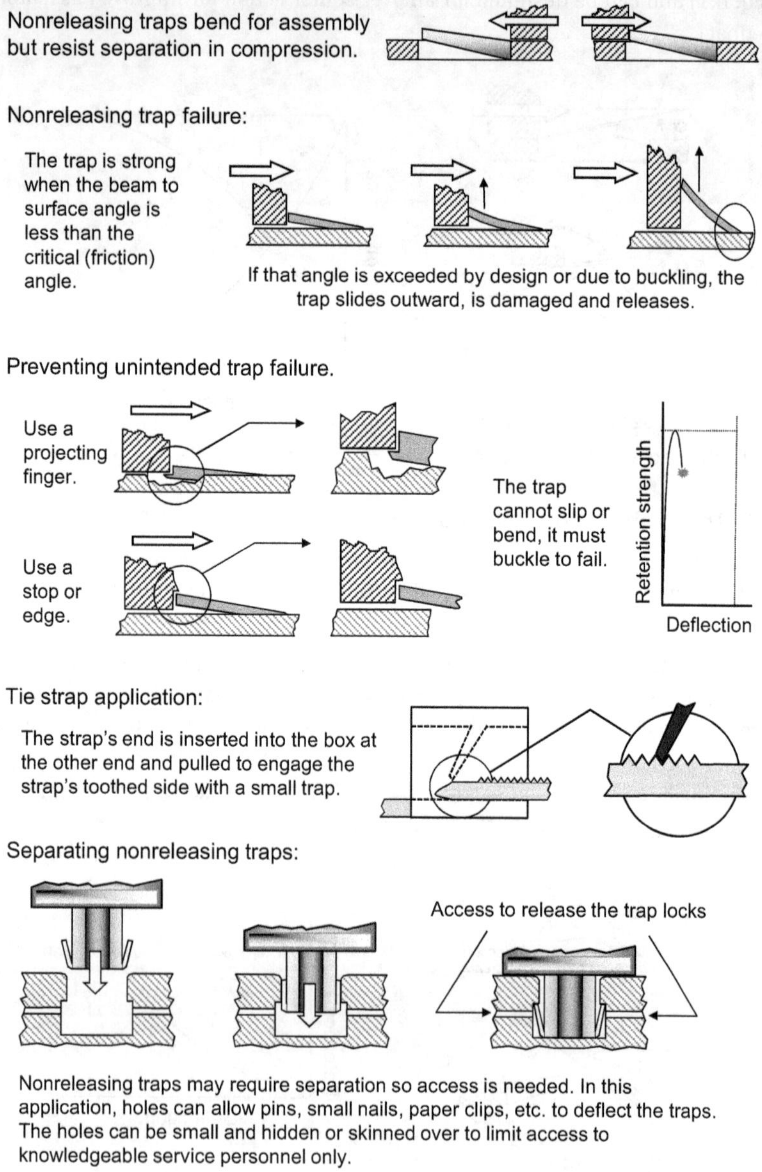

Figure 6.27 Nonreleasing trap retention behavior

A good example of the trap lock's strength can be found in common plastic tie-straps used to bundle wires and cables. Often these are small pieces with very small and thin beams acting as the trap, but their separation strength is extremely high.

Because of their strength, traps are also used on luggage buckle applications as an alternative to a hook with a retention face angle greater than 90°. (Such a hook application is shown in Fig. 6.12.)

A nonreleasing trap must ensure beam compression without buckling and protect against beam slippage and damage due to separation forces. An exception is when the application is to be tamper-proof or tamper-evident and permanent damage to the lock is desirable or acceptable.

Fish-hooking is a term describing the behavior of a nonreleasing trap where beam movement is not restricted. As separation force is applied, slippage at the beam end may occur immediately if friction is insufficient, or once initial buckling occurs, the beam will slip on the mating surface. In either situation, resistance to the separation force disappears and permanent damage is likely. A protruding finger at the end of the lock beam or a stop feature added to the mating surface can ensure against slippage and ensure the only possible failure mode is beam buckling.

The author was once involved with an application that used nonreleasing traps with protruding fingers. For no apparent reason, the application started failing. Investigation revealed that the supplier, not recognizing the functional importance of the tiny protruding finger and thinking that it was mold flash, changed the mold to remove the finger. This is a good example of a failure to understand the snap-fit.

A hidden trap can also be added to a product strictly to provide evidence of tampering and not necessarily as a primary lock feature.

6.3.4 Low Deflection Lugs

We recognize lock features by their deflection and tend to think of that deflection in terms of relatively large movements. However, lug locators can sometimes be used as low deflection locks.

The requirements for these applications are an assembly motion that involves sliding (the slide, twist, and pivot motions) and no applied forces in the separation direction. Figure 6.28 shows some examples. Note that the permissible assembly motions are those that remove 11 DOM, leaving only 1 DOM to be removed by the lock.

In all these cases, the lock/locator feature deflects over a small interference feature then returns toward its relaxed state. The principles of constraint - beam deflection, tolerances, and strength that apply to all lock pairs still apply.

Allowance for deflection and managing strain such as clearance, radii, and taper must be designed into the lug.

We will see in Chapter 9, Section 9.3.3, how the use of lug deflection as a compliance enhancement can help maintain a line-to-line fit and prevent movement in the snap-fit interface.

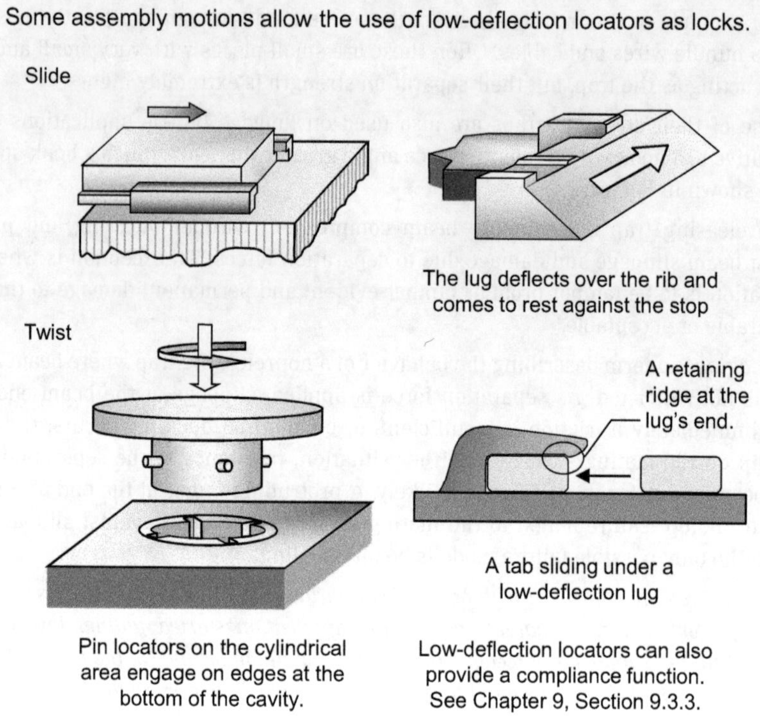

Figure 6.28 Low deflection locators as lock features

6.3.5 Other Cantilever Beam Locks

This is a catch-all section showing unique beam-based locks that do not fit into the three cantilever beam categories we have discussed.

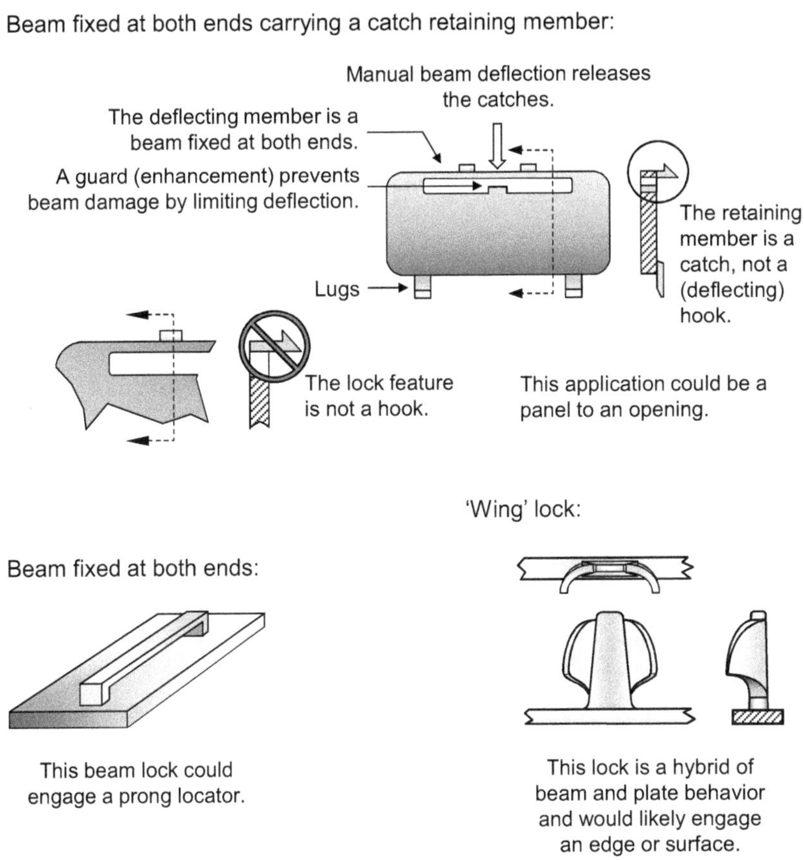

Figure 6.29 Other cantilever beam-based locks

6.4 Planar Locks

Planar locks involve one or two deflecting walls, usually with an edge and a catch on the walls. They engage through plate deflection and retain through shear and compression strength and plate mechanics.

Planar locks are so named because they are found on walls or surfaces (i. e., planes), see Fig. 6.30. These walls are thin relative to their length and width so the engagement and separation behavior of planar locks is described by plate mechanics. A planar lock pair is often a catch on one part and an edge on the other part.

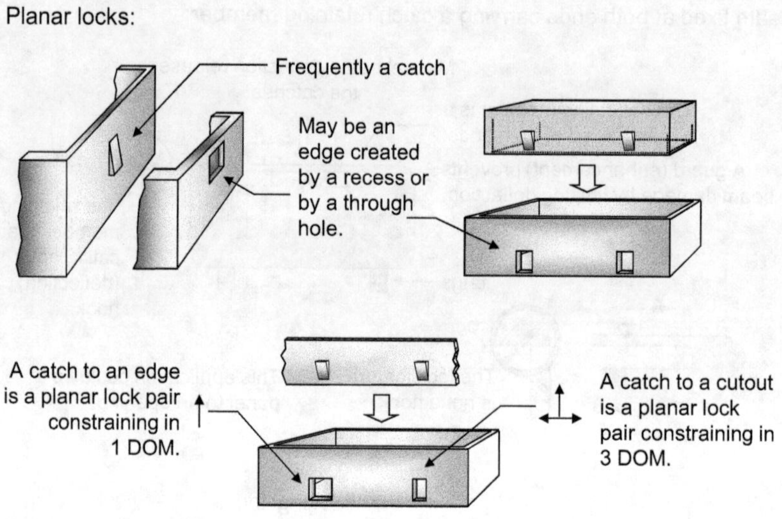

Figure 6.30 Planar locks

These locks can be made relatively strong, but because at least one member of the lock pair must sit on a surface, the reaction force will always be off the neutral axis. This creates the potential for wall distortion under high applied force and part separation. The weakness of a thin wall may require local support/stiffness in the form of ribs or additional wall thickness in the area of the lock.

An additional consideration with these locks is that both mating features can be on deflecting walls. This means both walls will deflect for engagement, reducing assembly forces and strains; but they can also deflect for separation, weakening the attachment. This is a case of a *lock/lock* pair rather than the more common lock/locator lock pair.

The principles of insertion and retention face profiles and their effects on force-deflection signatures are the same as for the cantilever hook. However, the more extreme deflections of cantilever beams are not likely to be found in a wall.

Because a wall is likely to be strong in two axes, a planar lock can be made to constrain in 3 DOM. In Fig. 6.30, the catch-cutout pair constrains in 3 DOM while the other constraint pair is a catch-edge. Note that a second catch-cutout pair at the left side location in the example would not be appropriate because it would create an over-constraint condition with the first.

6.5 Torsional Locks

Torsional locks rely primarily on torsional behavior for assembly deflection and separation resistance. There may be some bending in the system as well because the retaining member often rests on a (possibly flexible) beam attached to the torsional members.

As shown in Fig. 6.31, the torsional member is not necessarily round, although a round section is preferred. Torsional locks are relatively uncommon but are useful as an alternative to the cantilever style lock when clearances or access make hook location for disassembly difficult. For example, in an application where a hook must be flush with a panel and must be manually releasable (a nonreleasing lock), the see-saw action of the torsional lock allows release from the blind side of the retaining member.

Aside from the torsional deflecting member, the assembly and separation behaviors of these locks are similar to the cantilever beam style lock or to the trap, relying on the direction of mating feature engagement and release relative to the torsion member.

We are defining torsional locks as locks where the deflection mechanism is primarily torsion and can be analyzed as such. There are some locks where questions can arise as to their identification. In these locks, the installation deflection may be a combination of torsional shear, bending and plate deflection. Separation may involve torsional shear, plate deflection and either bending or compression. In these cases, evaluation of assembly and separation behavior will depend on which one of the deflecting members dominates, and a thorough understanding of the interactions. A finite element analysis to evaluate these combined effects may be required.

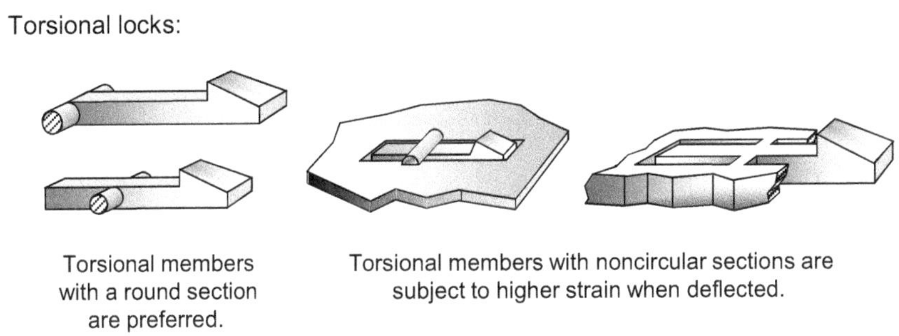

Figure 6.31 Torsional locks

6.6 Annular Locks

Annular locks use interference between concentric ridges and grooves on mating internal and external walls or surfaces and rely on radial elasticity for assembly and separation strength. An annular lock may be thought of as an extended catch wrapped around a cylinder or other shape and an extended edge wrapped around a mating shape, see Fig. 6.32.

Annular locks can be strong (permanent or nonreleasing) or they can be releasing. A snap on cap on a ball-point or felt-tip pen is a common releasing annular snap, plastic food containers are another.

Sometimes, a circular arrangement of hooks or traps is called an *annular lock* in the literature. This arrangement is not an annular lock because it requires analysis of beam bending to understand assembly and separation performance. True annular locks require analysis based on hoop behavior not on beam bending.

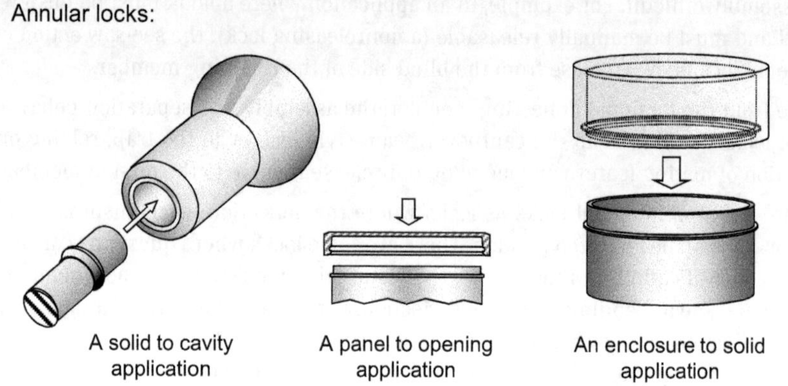

Figure 6.32 Annular locks

6.7 Using Locks

6.7.1 Lock Pairs

As with locators, a lock feature on one part requires an engaging feature on the other part. This is a *lock pair* and it is typically a combination of a lock and a locator. However, we did see an exception in the above discussion of planar locks. While lock/lock combinations are feasible, in this book the lock is always treated as engaging a locator feature as the other member of the lock pair.

The locking member of a lock pair may be on either the mating-part or on the base-part. Sometimes this is an economic/risk decision where it may be wise to put the lock on the smaller and less expensive part because locks, by their nature, are generally more susceptible to damage. Other times, lock placement is dictated by part shapes, engage direction or assembly motion. Ideally, but not always possible, locks should be placed on the part with the best material properties for lock performance.

Recall the discussion in Chapter 5 about locator pairs, constraint, and degrees-of-motion (DOM). A similar discussion is not required in this chapter because, quite simply, each lock pair should be designed to remove motion only in the separation direction. However, many of the principles of locator usage discussed in Chapter 5, Section 5.5 also apply to the use of lock pairs in the interface.

Figure 6.33 shows a number of lock pairs with locators identified in bold font. All but the wing and two-part locks are cantilever beam locks. The *two-part* application is a Level 4 decoupling lock; decoupling is discussed in the next chapter. The wing lock's deflection behavior involves both beam bending and plate behavior.

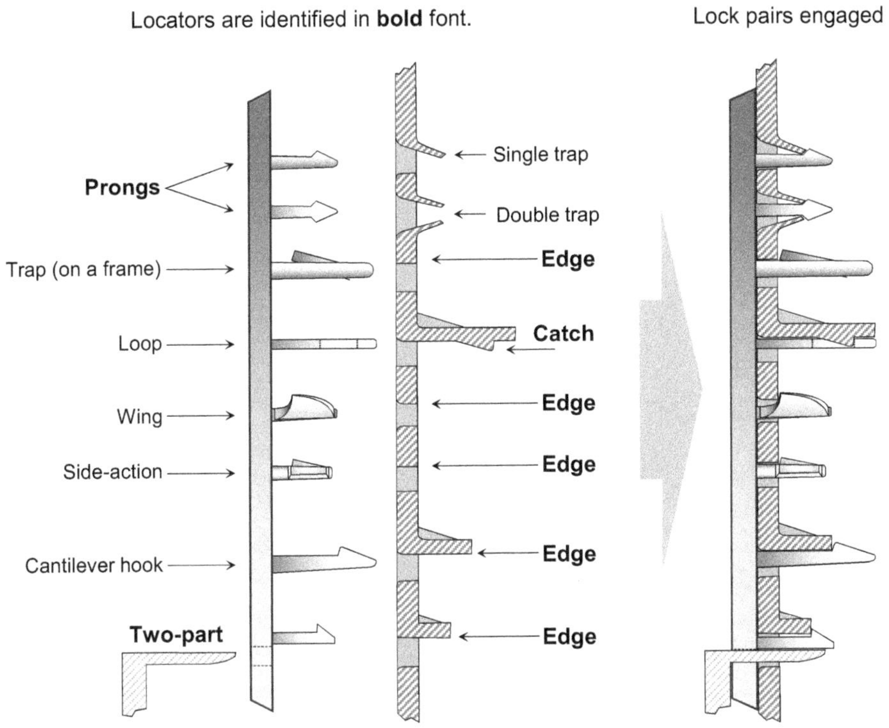

Figure 6.33 Lock pair examples with cantilever beam based locks

6.7.2 Short Grip-Length and Low-Clearance Applications

Sometimes a snap-fit application requires attaching one part to another relatively thin part. At other times, clearance for protruding constraint features is minimal. Sometimes, both situations exist in the same application. In all these cases, selecting a proper length for a cantilever lock feature can become a major design challenge. A general rule of thumb for any unplated cantilever beam-based lock is that the length-to-thickness ratio must be a minimum of 5:1 with 10:1 preferred. If a part is thin or if clearance limitations force the use of a very short hook, meeting the minimum recommended ratio of 5:1 may be impossible.

The author's experience with this situation is that, many times, the developer does not understand the options available. They will force the use of a cantilever hook lock by making the hook as short as the application requires and then hope for the best.

Forcing a cantilever hook into a short grip-length application is a mistake.

Sometimes, the resulting beam length-to-thickness ratio is as low as 3:1 or even 2:1. These short hooks are highly susceptible to over-strain and permanent damage during assembly, see Figs. 6.7 and 6.9. Once this damage occurs, the hooks cannot reliably maintain the attachment. The result is loose parts that may rattle or fall out of position.

The solution is to find a way for the beam to have an acceptable length/thickness ratio for assembly/separation deflection in these short grip-length applications. Some of these solutions will also allow for short lock features to accommodate low part-to-part clearance.

Level 4 decoupling, discussed in Chapter 7, Section 7.5, can also provide solutions to short grip-length and low-clearance applications. See Figs. 7.8 and 7.9 where lugs provide a very strong attachment in short grip-length applications. The lugs' low profile can also work well in applications with low clearance for the lock features.

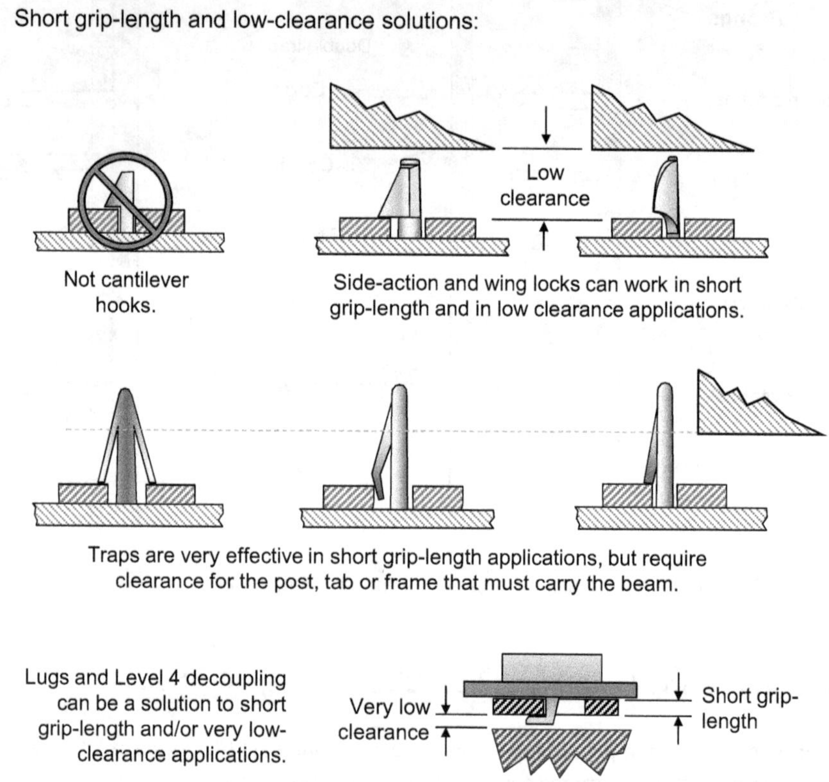

Figure 6.34 Solutions to special applications

6.7.3 High Demand Applications

High demand applications are those applications where reliable and strong lock features are absolutely critical. Examples would include applications where life or safety is at stake. Others could be applications where lock failure would be extremely costly, as in a very expensive part or a part where failure would lead to high-volume product recalls or production delays. It is up to the product development team to identify an application's place on the demand continuum.

The demand-complexity matrix, introduced in Chapter 3, Section 3.4, and discussed in detail in Chapter 15, Section 15.4, can be a starting point for thinking about the variables and resources required for these decisions.

As a rule, cantilever hooks are not recommended in high demand applications but, sometimes they can be made to work. See Figs. 6.14 and 6.15 where possibilities for strengthening or changing hooks are shown. Strong loops are possible, but nonreleasing traps are one option because of their great compression strength. Another option is the use of level 4 decoupling, to be discussed in Chapter 7.

Selecting assembly motions that maximize the DOM removed by locators is also recommended.

The next chapter will discuss lock strength in more detail and the reader will acquire a deeper understanding of the characteristics of strong and reliable locks.

6.7.4 Tamper Resistant Applications

These are applications where the manufacturer does not want anybody (or maybe an unqualified individual) to open the attachment. Internal (hidden) traps are very effective at preventing any access. In applications where access by only a qualified individual is allowed, access may be concealed.

See the nonreleasing trap example in Fig. 6.27. In that example, no access could be provided and the attachment would be permanent. Or small access holes could be provided and even skinned over when the part is molded and their location known only to qualified service technicians. Access might require drilling out the skinned-over area.

A relatively inexpensive part like a bezel can carry a tamper-proof or fragile tamper-evident feature to protect access to a critical area of the product. The bezel's removal and evident damage to the feature would indicate unauthorized access. See the discussion of Level 4 decoupling in Chapter 7, Section 7.5.

6.7.5 The Case against Cantilever Hooks

There has already been discussion about the limitations of cantilever hooks, and there will be more in Chapters 7 and 13. Here, we'll introduce a list of requirements for using cantilever hooks. This list is (intentionally) very limiting.

By now, the reader may feel that the author is not a big fan of cantilever hooks.

Use cantilever hooks *only* when all these conditions are met:

- The lock's deflecting member length can be made at least 5x the engagement deflection (if unplated), 10x if plated.
- No separation forces will be applied to the hook.
- The assembly force can be kept within ergonomic limits.
- The insertion face angle is $\leq 30°$ to the beam's neutral axis.
- Unintended release of the attachment will not have serious consequences.

Figure 6.35 illustrates, in a general way, the design limitations of cantilever hooks as compared to other beam-based lock options. The primary issue is the hook's inherent inability to effectively decouple assembly performance from separation performance. This results in a relatively small solution set or design space within which a cantilever hook can be made to work reliably.

When cantilever hooks can be made to work, the hook's limitations can still become an issue if the part material or geometry change or the application's retention requirements increase. The cantilever hook is not very robust to design changes.

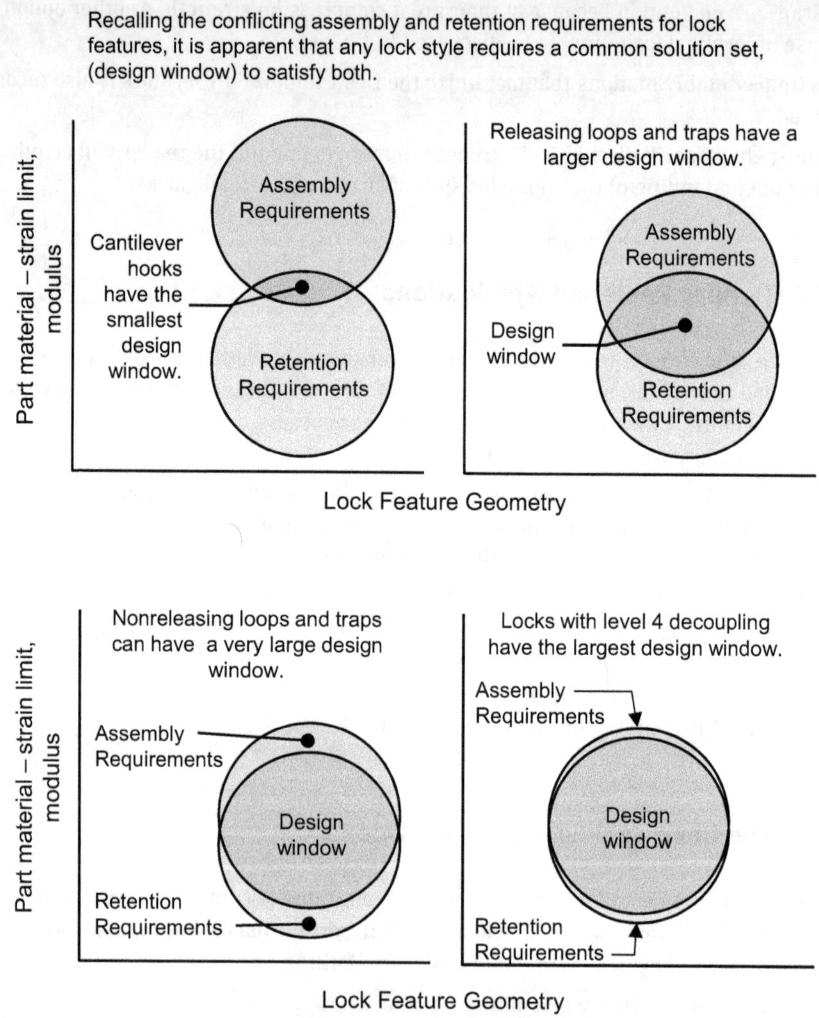

Figure 6.35 For many applications, cantilever hook locks are not the best choice

6.8 Summary

This chapter described lock features which, along with locators, are the physical features that provide constraint in the snap-fit interface.

Understanding lock behavior and making an informed decision about lock style is the single most important factor in good snap-fit development.

Figure 6.36 shows that (in the author's experience) the highest frequency of snap-fit product issues out of 150 reviewed applications originated with poor lock feature selection. The second most frequent issues were constraint related, and the third was lock feature strength.

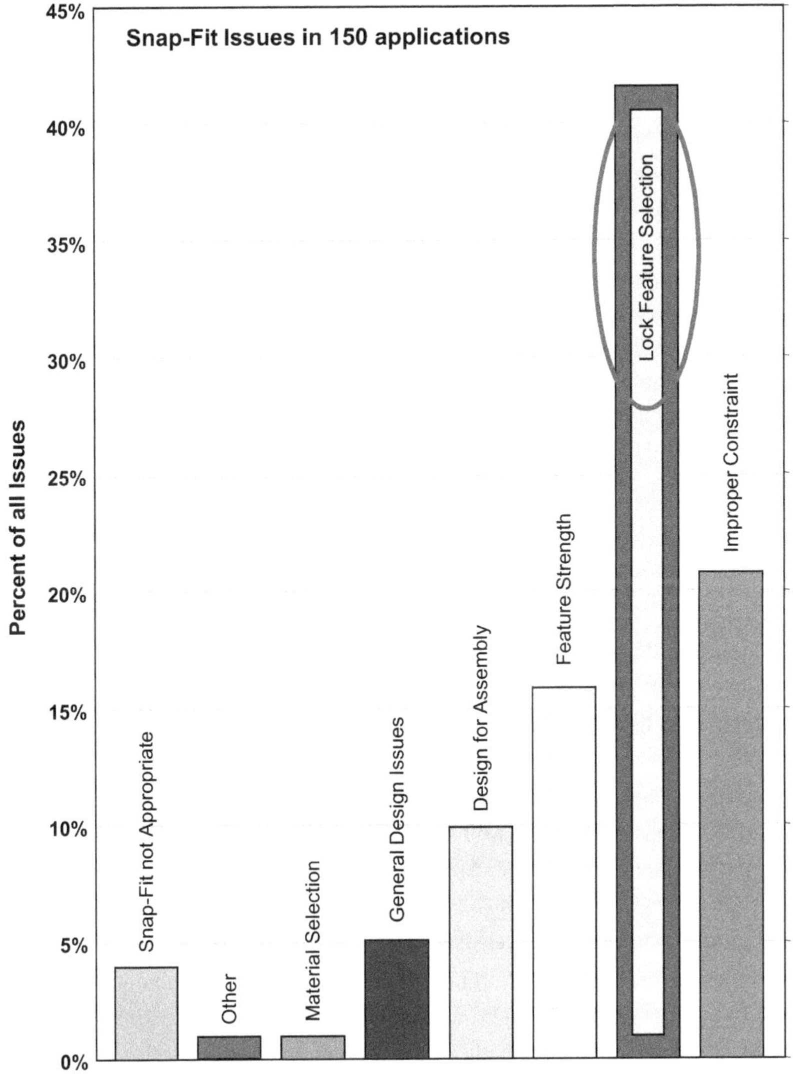

Figure 6.36 Lock selection and snap-fit application issues

General information about constraint feature design:

Yes, you saw a similar list at the end of the chapter about locator features.

- Specify a radius on all interior and exterior corners of constraint features. This applies to the feature intersection with the parent material and to all corners within the feature itself.
- Size each feature according to the demands placed on that feature, see Fig. 5.20 in Chapter 5 and the related discussion. In many designs, there seems to be a tendency to make the constraint features identical to each other without regard to their individual performance requirements.
- Constraint features that do not require die-action will be preferred over those that do. But, do not sacrifice reliability just to avoid die-action.
- Locks should not carry long-term or sustained forces. If they do, the choice of material is critical and long-term performance must be verified by analysis and end-use testing. Plan the development program with reference to the concepts of the demand-complexity matrix.

Important points in Chapter 6 include:

- Snap-fit reliability depends on using constraint features to establish and maintain a line-to-line fit between the mating and base-parts. Do not expect to get any significant or long-term clamp load in a snap-fit attachment.
- Constraint features prevent movement between and separation of the joined parts by removing degrees-of-motion (DOM). They are the *necessary and sufficient* conditions for a snap-fit attachment.
- The fundamental problem in snap-fit attachment design is that locks must be weak to deflect for assembly yet strong enough to prevent part separation. Some lock styles are better than others in resolving the assembly vs. separation paradox (i.e., decoupling).
- A lock pair will typically be a lock feature on one part engaging a locator feature on the other part.
- Locks fall into four groups depending on the deflecting member: cantilever beams, planar, torsional, and annular. The cantilever beam group is subdivided into hooks, loops, and traps.
- A good rule of thumb for establishing a minimum beam length in beam-based locks is that the beam length to thickness ratio should be at least 5 : 1 for unplated locks and 10 : 1 for plated locks.
- The cantilever hook is the most commonly used locking feature, yet many times it is not the best choice for the application.
- A supporting rib on an existing cantilever hook lock is often a sign that a hook was not the best choice for the application.
- Even nonreleasing hooks will release under sufficiently high forces.
- The cantilever hook style is inherently poor in resisting separation forces, but hooks with a retention face greater than 90° can be an exception to this rule.
- Loop style locks will be particularly susceptible to sharp interior corners at the retaining member.

- Compensate for knit line weakness in loop style locks.
- Nonreleasing trap locks must be protected from manual over-deflection and damage.
- Locators can be used as low-deflection locks, particularly when a slide, twist, or pivot assembly motion is present.
- Grip length is the distance from the retention face of a locking feature to the opposite (reacting) locator. A short grip length will rule out use of cantilever hooks.
- A loop or trap lock used with a tip assembly motion is a highly effective snap-fit attachment concept and should (almost) always be considered as a design option.
- A profile added to the insertion and/or retention faces in a lock pair can improve assembly and separation performance.
- Locks are generally the weak link in the constraint system. For the most reliable attachment, design the interface to minimize DOM removed by locks and maximize DOM removed by locators. This begins by selecting an assembly motion that will maximize the use of locators.
- Ideally, lock pairs should constrain in only one DOM – the separation direction. Designing for a lock to constrain in additional degrees-of-motion will leave the attachment under-constrained.
- The rules for mechanical advantage and dimensional robustness that were introduced and explained in Chapter 5 for locator features also apply to lock pairs.

References

[1] Luscher, Anthony F., *Design and Analysis of Snap-fit Features*, from the Integral Attachment Program at the Ohio State University (1999)

[2] Reiff, Dave, Motorola Inc., Fort Lauderdale, FL, *Integral Fastener Design*, Plast. Des. Forum 16, No. 6, Sept./Oct. (1991) pp 59–63; Described loops as a unique lock feature.

[3] Beall, Glenn L., *Plastic Part Design for Economical Injection Molding*, Libertyville, IL (1998): Test data reproduced from LNP Cloud, McDowell & Gerakaris, Plast. Technol., Aug. (1976)

7 Lock Strength and Decoupling

We tend to think of snap-fit lock strength as resistance to part separation. But the very first demand on a lock's strength is surviving the strain that occurs during assembly deflection. Locks that are damaged during assembly will not be able to perform their retention function.

Important!

Lock *decoupling* is the concept of independence between a lock's assembly and retention/separation behaviors. The degree to which decoupling is possible is a critical factor in a lock's performance capability. Decoupling was mentioned in the Chapter 6 introduction. Recall that lock efficiency (E_L), the ratio of retention strength to assembly force, was introduced in Chapter 6 as a way to quantify decoupling.

Lock features must satisfy conflicting requirements.

A fundamental snap-fit design challenge is that lock features must be weak and flexible for assembly but also stiff and strong to resist separation. This conflict can force design compromises that don't adequately satisfy either requirement. We must be certain that a lock feature will provide an assembly/retention solution set, or design window, that will allow the lock to do both. Some lock feature styles provide a larger design window than others, making them a preferred choice in many applications, see Fig. 7.1.

More variations of Fig. 7.1 are shown in Chapter 6, Fig. 6.35.

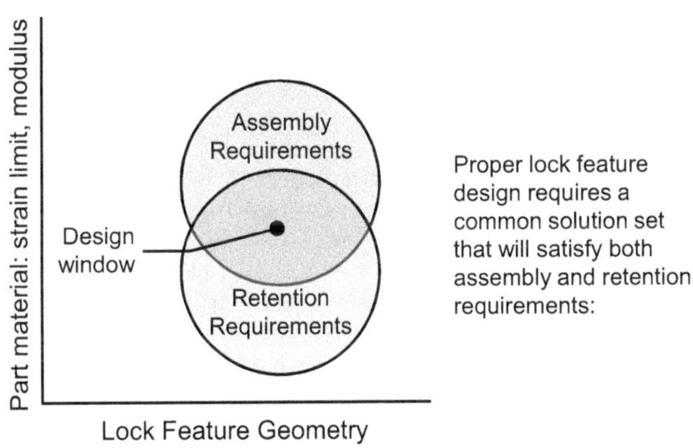

Figure 7.1 The lock feature assembly/retention solution set or design window

To maximize the design window, we would like the lock's assembly behavior to be totally independent of its retention performance. In other words, the two circles in Fig. 7.1 would completely overlap. This is possible with the highest level of decoupling. Most of the time, however, partial decoupling is sufficient.

We'll use a common extension ladder shown in Fig. 7.2 as a decoupling example. For the user's safety, the ladder's angle to the ground must be within a limited or safe working range. Therefore, a one-piece ladder would have very limited safe working height. Using the ladder outside of this safe range is possible but risky.

An extension ladder allows the user to adjust the base of the ladder's distance to the house independently of the desired working height so a safe angle is maintained. Likewise, we need some level of decoupling capability for proper use of lock features in snap-fits.

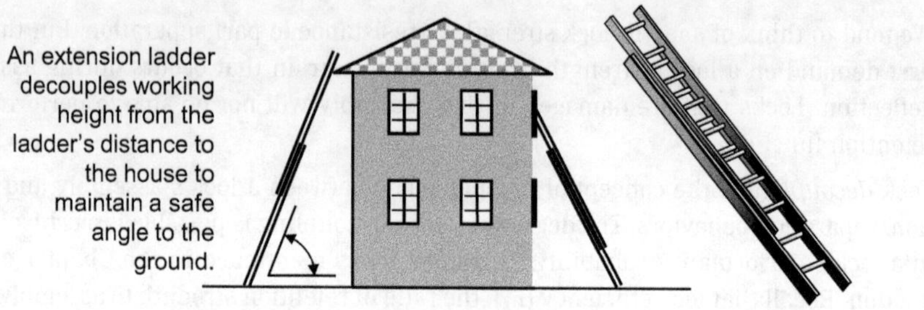

Figure 7.2 Extension ladder as a decoupling example

A lock's decoupling capability is explained by studying how the lock's assembly and separation behaviors are analyzed. Decoupling levels are ranked by their effectiveness in improving lock efficiency, see Eq. 7.1. The higher the lock efficiency, the stronger we can make the lock in retention relative to the assembly force.

$$E_L = \frac{F_R}{F_A} \qquad (7.1)$$

Where: E_L is lock efficiency.
F_R is retention strength.
F_A is maximum assembly force.

Table 7.1 Differences between the Levels of Decoupling

		The Elements of Assembly and Separation Performance Analysis			
E_L	Level	Values	Variables	Equations	Features
Lowest	0	same	same	same	same
	1	different	same	same	same
↓	2	N/A	different	same	same
	3	N/A	N/A	different	same
Highest	4	N/A	N/A	different	new

In Table 7.1, Level 0 represents no decoupling. Then as the decoupling level (and lock efficiency) increases, assembly and separation behavior progress through the elements of lock performance calculations: values, variables, and equations, and then finally, to the features themselves.

7.1 Level 0 No Decoupling

There is no decoupling in the hook shown in Fig. 7.3. The insertion and retention face angles are similar, and assembly and retention behaviors are virtually identical because changes made to the deflecting member will affect both assembly and retention.

The only *deflecting member* variable carried forward into the assembly and separation calculations is the beam deflection force, F_P.

For a closer look at beam deflection force, we can combine the basic strain equation, Eq. 13.2 from Section 13.2.4, and the basic deflecting force equation, Eq. 13.15 from Section 13.1.5, using the target strain variable, which appears in both, to establish equivalence. The result, Eq. 7.2, exposes all variables affecting the deflection force.

$$F_P = \frac{W_b T_b^{\,3} E\delta}{4L_b^{\,3}} \tag{7.2}$$

Where: F_P is beam deflection force. E is material modulus.
W_b is beam width. δ is deflection.
T_b is beam thickness. L_b is beam length.

A change to certain variables in the numerator will directly affect the deflection force. For example, making the beam thicker (T_b) will provide more deflection force and retention strength, but it will also increase the strain at the beam's base during deflection, as can be seen in Eq. 13.2. If, on the other hand, the beam's thickness is reduced for easy assembly and lower strain, it becomes weaker in retention.

Any change to beam length (in the numerator) will have an inverse effect on deflection force.

The only variables we can change on otherwise identical cantilever hooks are the retention and insertion face angles, and these variables do not appear in the above beam bending calculation. Thus, as was shown in Table 7.1, all the elements of assembly and separation performance analysis are the same.

In this example:

- We are ignoring the relatively small change in the deflection force moment arm as the contact point moves over the insertion and retention faces. We will also ignore the changes in the face angles as the beam deflects.
- Beam deflection force is the same for both assembly and separation.
- The insertion face angle (α) in the assembly force calculation and the retention face angle (β) in the retention strength calculation are identical.
- The retention and insertion face angles are not available for adjusting hook performance until we apply the equations for assembly force (F_A) and retention force (F_R). This will occur in Level 1 decoupling, described next.

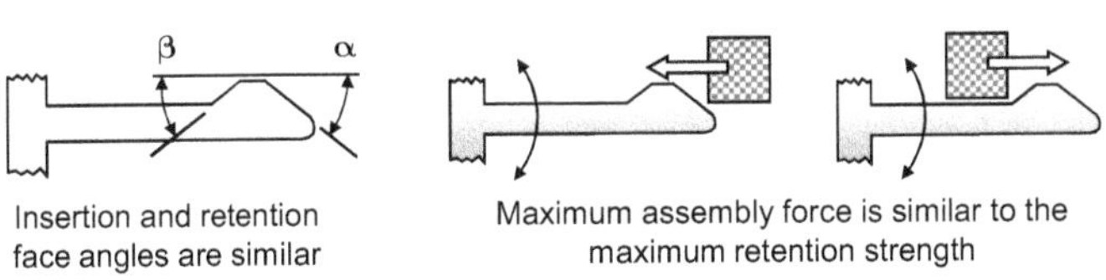

Figure 7.3 Level 0 decoupling

7.2 Level 1 Decoupling

For the hook shown in Fig. 7.4, we can only use the *values* of the insertion and retention face angle variables to adjust assembly and retention behavior. Hook performance can be partially decoupled by changing these angles.

In Chapter 13, these calculations are discussed in more detail.

The insertion face angle variable is found in the assembly force equation, Eq. 7.3. This equation includes the beam deflection force value (F_p) calculated in Eq. 7.2. By *reducing* the insertion face angle (α), we can get lower assembly forces without affecting retention strength.

$$F_{\text{assembly}} = F_P \frac{\mu_{\text{dynamic}} + \operatorname{Tan}\alpha_{\text{effective}}}{1-(\mu_{\text{dynamic}} \operatorname{Tan}\alpha_{\text{effective}})} \tag{7.3}$$

Where: F_p is beam deflection force.
F_{assembly} is assembly force.
μ is coefficient of friction.
α is insertion face angle.

In a separate calculation, we can also improve retention strength by *increasing* the retention face angle (β), which appears in the separation strength equation, Eq. 7.4. Like Eq. 7.3, this equation includes the beam deflection force value (F_p).

$$F_{\text{separation}} = F_P \frac{\mu_{\text{static}} + \operatorname{Tan}\beta_{\text{effective}}}{1-(\mu_{\text{static}} \operatorname{Tan}\beta_{\text{effective}})} \tag{7.4}$$

Where: F_p is beam deflection force.
$F_{\text{separation}}$ is separation force.
μ is coefficient of friction.
β is retention face angle.

With these independent changes to the insertion and retention face angle *values* (see Table 7.1), we have accomplished some assembly/separation decoupling, but the impact of these changes on lock performance is ultimately limited by the beam's deflection force (F_P), which appears in both the assembly and retention strength equations.

Sometimes ridges are added to the retention face in an effort to increase friction and retention strength, but they are largely ineffective and, in any case, performance is still limited by the beam's behavior.

Level 1 is the only level of decoupling available with the traditional cantilever hook.

Level 1 decoupling is the lowest and least effective level of decoupling, and it explains the relatively limited design space available with the cantilever hook style lock. There will be more discussion of the hook's limitations in Chapter 13.

In Level 1 decoupling, the retention face angle determines whether the lock is releasing or nonreleasing.

As an exception to the limits of Level 1 decoupling, we saw in Chapter 6 how a retention face angle greater than 90° on a buckle application can greatly increase retention strength by adding a tensile component to the lock's separation strength. But if sufficiently overloaded, even these hooks may release due to bending and/or gross distortion. Tensile/shear failure is also a possibility.

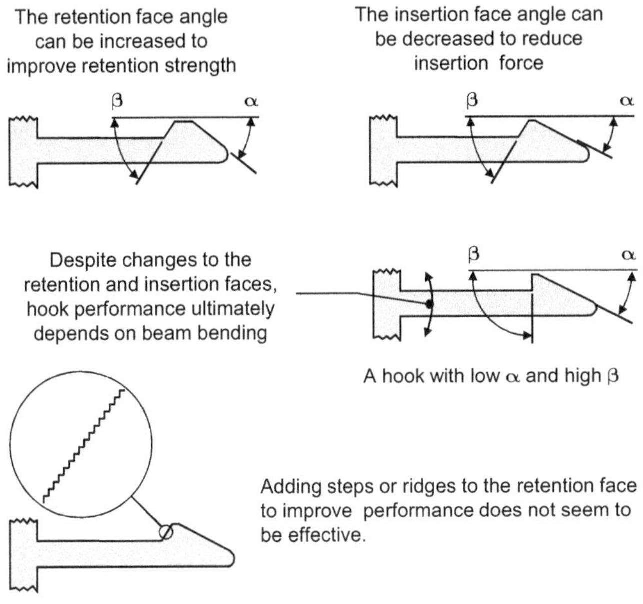

Figure 7.4 Level 1 decoupling

With Level 1 decoupling, the values of the insertion and retention face variables can be changed. For the next level of decoupling, the variables themselves will change.

7.3 Level 2 Decoupling

Redesigning the lock by rotating the beam 90° to the assembly and retention faces results in major changes to lock behavior. Level 2 decoupling makes use of different *variables* in the beam bending calculation resulting in significant improvements in lock efficiency. The locks shown here are called side-action locks.

In the beam bending force equation (Eq. 7.2), we see that F_P is directly proportional to the expression $W_b T_b^2$.

Referring to Fig. 7.5:

When we consider deflection for assembly, the beam has the values $T_b = 1$ and $W_b = 5$ for the thickness and width variables. Using Eq. 7.2 to calculate beam bending force, the value of the expression $W_b T_b^2$ is $(5 \times 1^2 = 5)$.

For determining separation strength, however, the variable that was width is now beam thickness ($T_b = 5$), while the variable that was thickness is now width ($W_b = 1$). With these new values for the variables, the value of the expression $W_b T_b^2$ is now $(1 \times 5^2 = 25)$.

In side-action locks, the ratio of (W_b/T_b) will always be the *maximum* potential improvement in separation strength.

With all other values in the calculations remaining constant, the ratio of the values of the expression $W_bT_b^2$ is 25/5 = 5. Carrying this ratio into the assembly and separation calculations, we find that this side-action hook could have as much as 5x the retention strength of the Level 1 hook.

With the side-action locks, however, beam distortion due to the side forces on the thinner section is possible and the actual strength effect in this case may not be as high as 5x, but the improvement will still be significant. Note that a beam length to thickness ratio near the (recommended) minimum beam L/T limit of 5 will be less likely to distort than a beam near the upper L/T limit of 10. Lower (W_b/T_b) ratios will also be less susceptible to distortion than higher ones.

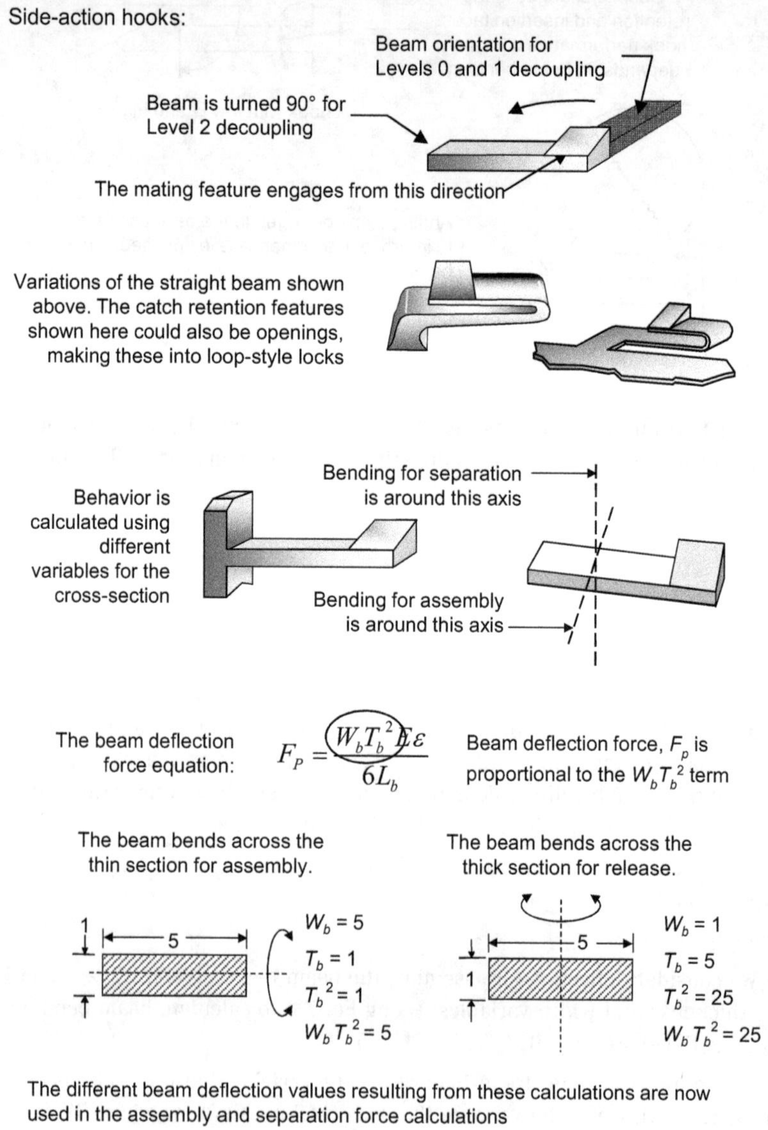

Figure 7.5 Level 2 decoupling

Loop style locks can also be used as side-action locks. Although nonreleasing loops are Level 3 decoupling, the side-action option can allow the use of both releasing and nonreleasing loops in short grip-length and low clearance applications.

Figure 7.6 shows two side-action hook applications. Both are short grip-length applications where a (Level 1 decoupling) hook style lock (often found in these kinds of applications) should *not* be used.

The speaker grille is from a panel/opening application. The panel has three T-style side action locks on three sides and a lug on the fourth side. The lug is engaged first to one edge of the opening and the grille is tipped into its final locked position. Because this is a speaker application, it is likely that locking features were needed on three sides of the panel to prevent possible vibration or buzzing. Otherwise, only one side-action lock would have been needed opposite the lug and simple pin/edge locators would have been sufficient on the other two sides.

In the author's opinion, this attachment and the second variation (one side-action lock) mentioned above should belong to a limited number of standard panel to opening attachment *concepts* for short grip-length applications. See Chapter 15, Section 15.7.5, for more discussion about the value of developing an in-house library of standard attachment concepts.

The chrome-plated plastic emblem is from a solid/surface application, where the surface is sheet metal. Chrome plating degrades the outer skin of a plastic part and is, itself, brittle so the minimum 5:1 length to thickness ratio rule-of-thumb does not apply to plated plastics. This L-shaped side-action hook provides sufficient deflection in this short grip-length application for assembly without damage to the hook. Note also the curved profile on the hook's insertion face for an improved assembly signature.

Speaker grille application:

T-shaped side action lock

Chrome emblem application:

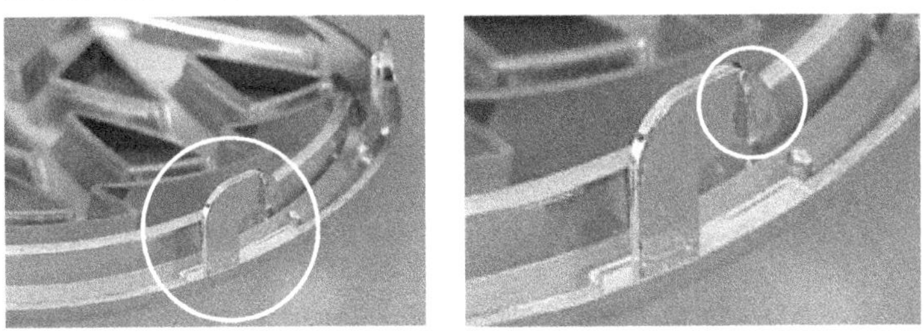

L-shaped side action lock

Figure 7.6 Side action locks in-product applications

In the side-action hook, turning the catch 90° exchanges the beam width and thickness *variables* in the calculation, see Table 7.1. However, the same (beam bending) equation is still applied to the same feature (the beam). With Level 3 decoupling, the *equations* themselves will change.

7.4 Level 3 Decoupling

Level 3 decoupling occurs when assembly and retention behaviors within the same feature require *different equations* for evaluation, see Fig. 7.7. This gives even greater independence between the assembly and retention behaviors and increases the developer's control over each of them.

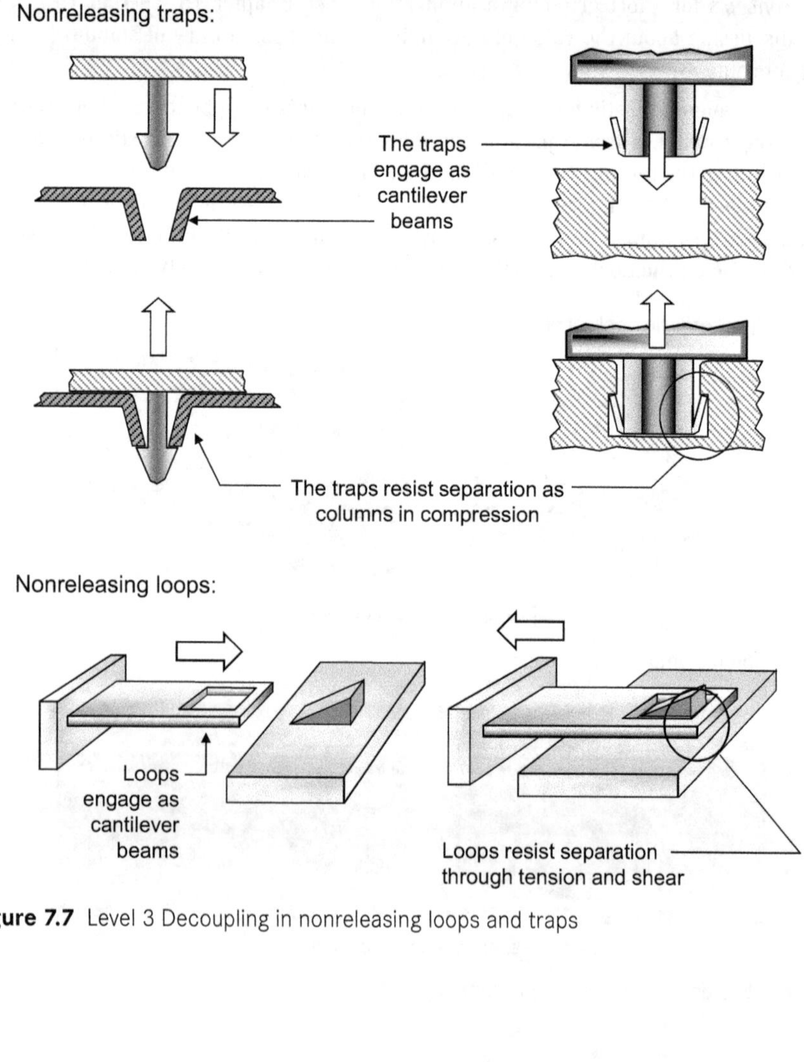

Figure 7.7 Level 3 Decoupling in nonreleasing loops and traps

Level 3 decoupling occurs naturally in nonreleasing trap and nonreleasing loop locks. For both styles, assembly engagement involves bending and is evaluated using the beam bending equation. However:

- Nonreleasing trap separation behavior is evaluated with equations for columnar beam behavior under *axial compression.*
- Nonreleasing loop separation is evaluated with equations for beam behavior under combined *shear and tension.*

Recall that trap locks are one of the more desirable lock features for ease of assembly and retention strength. A nonreleasing trap acting as a column in compression can be extremely strong.

Nonreleasing traps with a beam length to thickness ratio near the low *L/T* limit of 5 will be less likely to buckle under compression than a beam near the upper *L/T* limit of 10.

7.5 Level 4 Decoupling

Level 4 decoupling involves the use of a second feature as support for the primary lock feature to improve retention performance *without affecting assembly force.* Dramatic differences in assembly and retention performance are possible. Level 4 locks are non-releasing.

Note that the retainer enhancement used to improve hook retention as described in Chapter 6 does not represent Level 4 decoupling because it also increases assembly force and, most importantly, it does not change the hook's fundamental release behavior.

In the first example in Fig. 7.8, thin hooks on the mating part pass through a hole in a panel base part. Once the mating part is in place, a pin is added to prevent the hooks from deflecting and releasing. Often, the pin is preassembled to the part so it is in-place and ready for final assembly. The pin can also be molded in assembly position as part of the mating part. The pin is attached with thin runners that are broken when the pin is pushed into its final position.

Once the pin is in place, a simple locking mechanism or friction holds the pin in place. Retention strength of the mating part to base part attachment is a function of the hook beam's tensile strength.

In the second example, a feature on a third part fills the space behind the hook and prevents release. Again, retention strength is a function of the beam's tensile strength.

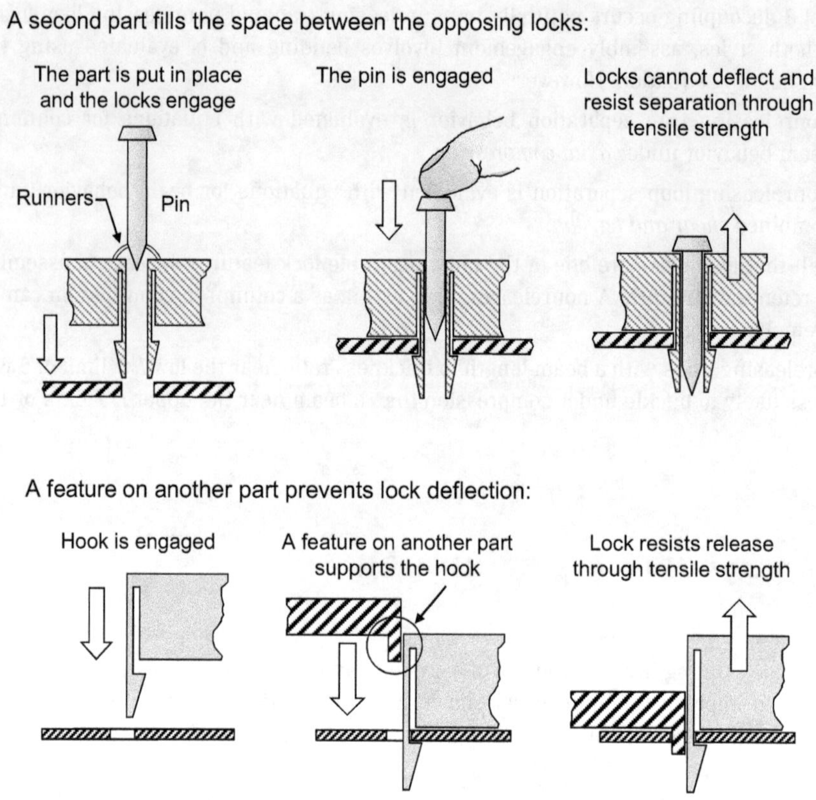

Figure 7.8 Level 4 decoupling with cantilever hooks in tension

Figures 7.9–7.11 illustrate a different strategy where lug locators on the mating part are engaged with a slide assembly motion. Other features (pins) carried on a bezel then fill the openings behind the lugs to prevent the mating part from sliding in the separation direction and releasing the lugs.

These applications require the mating part to slide and engage locators. It is easy to imagine applications where the lug/opening pairs would be replaced by pin/hook arrangements similar to those shown in Fig. 7.7, and sliding the mating part would not be necessary.

The cantilever hooks must still deflect for assembly while no deflection at all is required for the lug attachment.

1. The solid is placed on the surface with four lugs through the four holes.
2. The solid slides on the surface to engage each lug in the narrow area of a hole.
3. The bezel is attached to the surface with hooks and surrounds the solid.
4. Four pins carried on the bezel fill the holes behind the lugs and prevent the solid from sliding and releasing.

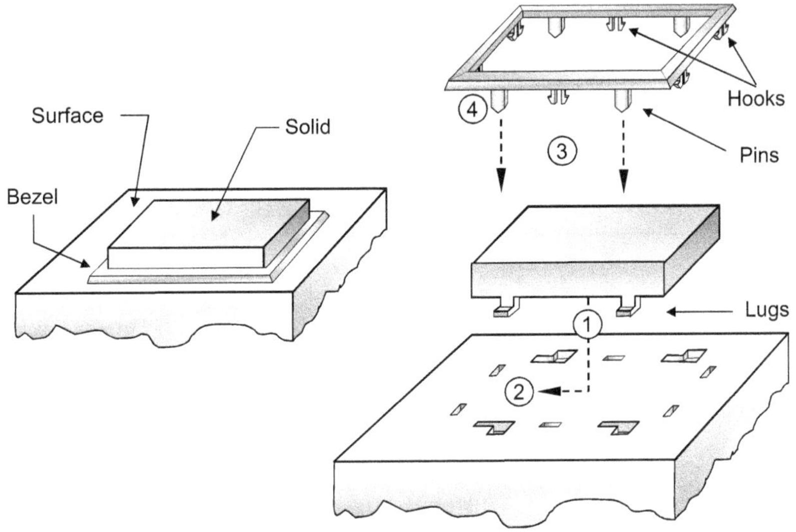

Figure 7.9 Level 4 decoupling application

Figure 7.10 is a variation of the application in Fig. 7.9. The solid's attachment interface has four lugs to engage four rectangular holes and two pins to engage two slots on the base part, a surface.

For assembly, the lugs are inserted into the rectangular openings in the surface and the solid is then slid on the surface so that the lugs engage the underside surface to resist 1 DOM in the $+z$ direction and assist in removing ±rotation around the x- and y-axes (4 DOM). The solid to surface interface removes 1 DOM in the $-z$ direction and interacts with the lugs to form couples and remove the x- and y-axis rotation.

Instead of lugs, prongs engaging keyhole shaped openings are also possible here.

The pin/slot locator pairs in the solid/surface interface constrain against rotation around the z-axis and lateral movement along the y-axis. Retention and positioning strength in the application can be very high because it is a function of the strength of the strong locator pairs: pin/edge and lug/surface.

A bezel carrying a strong pin and (a sufficient number of) locks is finally placed around the solid and against the surface. The lock features on the bezel hold it in place against the surface and carry no other loads. The bezel's pin engages the hole in the base part (a locator pair) and the bezel then prevents the solid from ±movement along the x axis and disengaging the lugs.

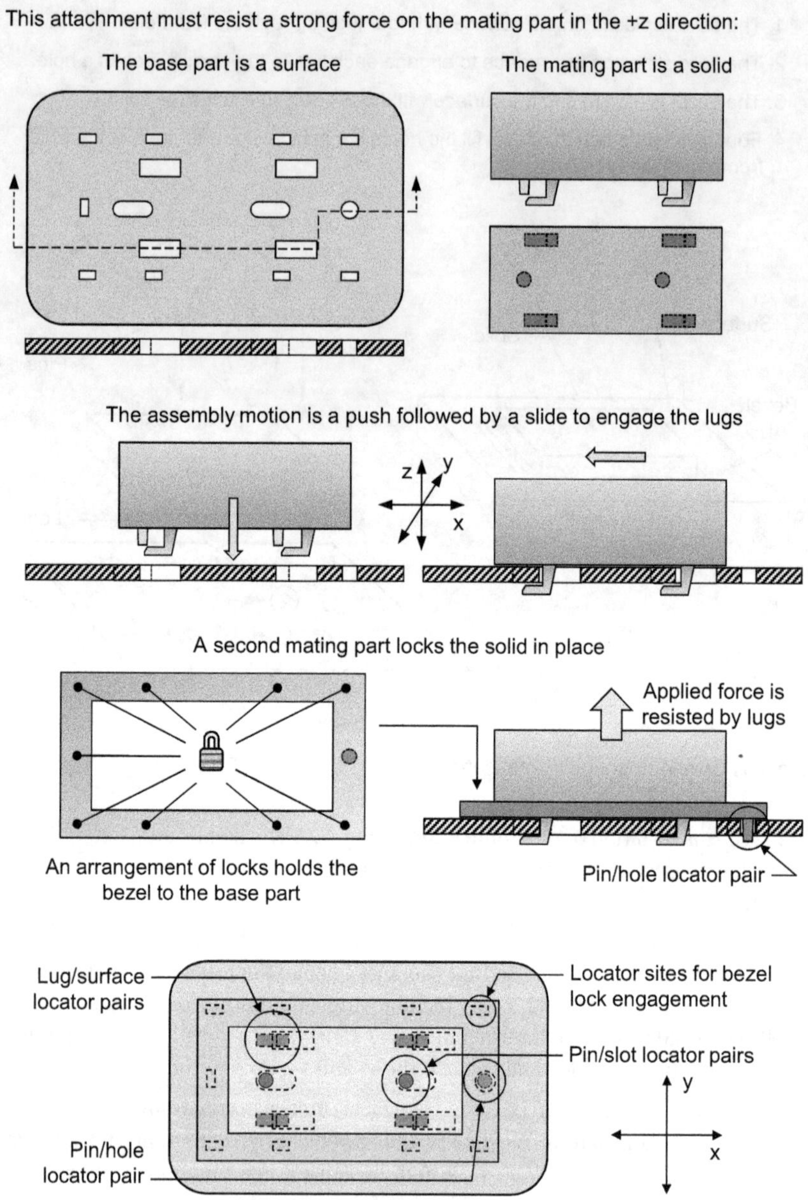

Figure 7.10 Level 4 decoupling application

Figure 7.11 shows a product application: an (automobile) interior door handle assembly that must resist multiple usage cycles to unlock and open a door. Strong pulls on the handle to close the door are possible and must be also be taken into consideration. On the demand-complexity matrix, this snap-fit application would be considered both complex and fairly high demand. Level 4 decoupling provides a solution.

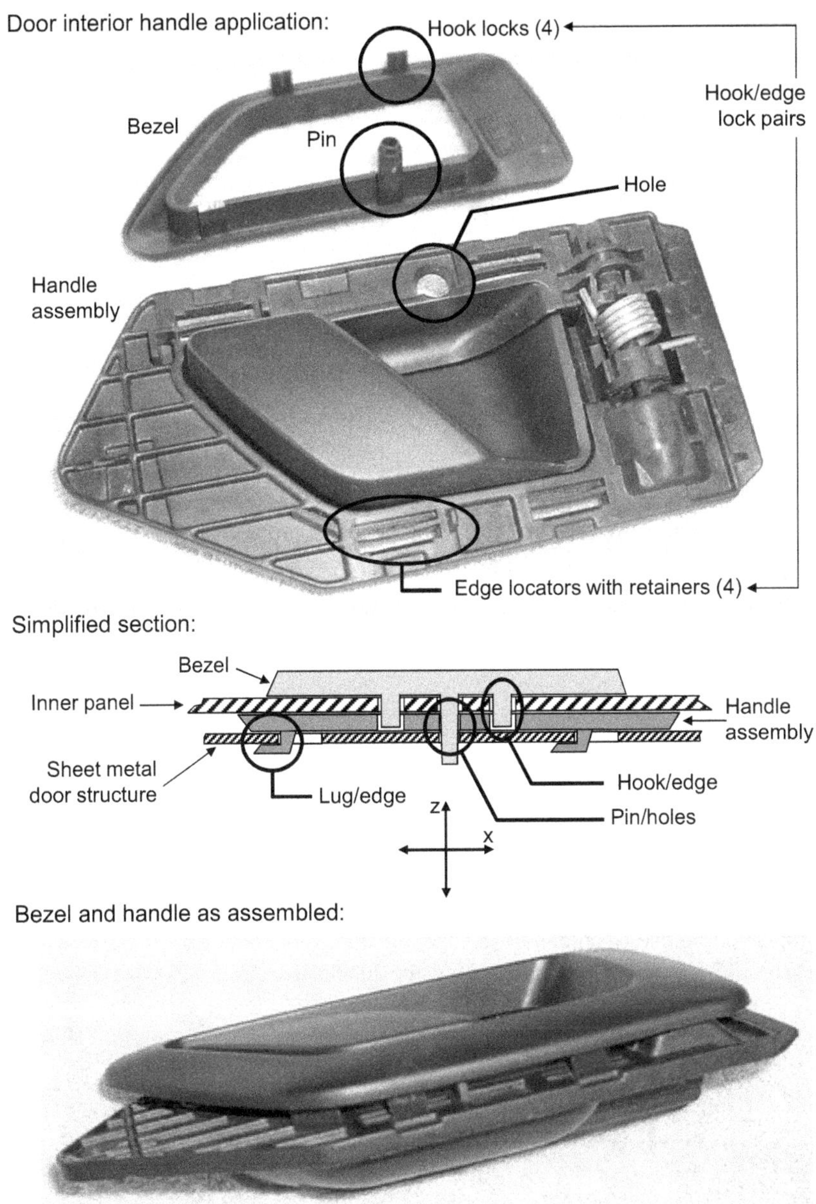

Figure 7.11 Level 4 decoupling

The bezel and the handle assembly are part of a sandwich consisting, in assembly order, of:

1. The structural sheet metal within the door, the *base* part.
2. The door handle assembly, the *mating* part: a strong frame, an internal spring and the handle itself.
3. The door inner (appearance) panel, held in place against the handle assembly by the bezel.

4. The handle trim bezel that, in this case, is more than just ornamental trim because it must hold the inner panel against the handle assembly. It also carries the locking pin for the handle to sheet-metal (Level 4 decoupling) attachment and cantilever hooks to attach the bezel to the handle assembly.

Lugs on the handle assembly engage slots in the sheet metal when the handle is set in place on the sheet metal ($-z$) and slid in the $-x$ direction. The hole in the handle assembly and a hole in the sheet metal are now aligned.

The inner panel (with a cutout around the handle assembly) is attached to the door structure.

A push assembly motion ($-z$) engages the pin on the bezel through the aligned holes in the handle assembly and sheet metal. Four hooks on the bezel engage edges of openings in the handle. Retainer enhancements support the hooks making bezel removal difficult, but not impossible.

The very strong pin/hole(s) locator pair resists motion on the x-axis and works with various lugs to form couples against rotation around the z-axis. The lug/edge locator pairs resist any outward pull ($+z$) on the handle and rotations around the x- and y-axes. (The y-axis is not shown, but is orthogonal to the page.)

Level 4 decoupling involves the use of different *features* for assembly and retention and is the highest form of decoupling. Dramatic differences in assembly and retention performance are possible.

The *push-pin* or *two-piece* plastic fastener (Chapter 10, Section 10.4.1.2), suggested as an alternative when integral locks will not work, uses Level 4 decoupling.

While a bezel is not a requirement for Level 4 decoupling, the author's observation has been that applications employing bezels along with the mating and base parts are good candidates for Level 4 decoupling. In some applications, addition of an otherwise unnecessary bezel to allow implementation of Level 4 decoupling is probably a cost-effective solution to an attachment issue.

Some may argue that Level 4 decoupling is not really snap-fit technology. However, it is a logical extension of the technology as a solution to very real issues that could otherwise prevent the use of a snap-fit in an application. As such, regardless of what one wants to call it, Level 4 decoupling belongs in this discussion.

7.6 Summary

This chapter used the concept of decoupling to explain why different lock styles have significant performance differences. An understanding of decoupling provides insight into lock selection decisions when balancing trade-offs between lock assembly and retention behavior.

Table 7.1 summarized the technical distinctions between the five levels of decoupling. With a better understanding of the subject now, reviewing Table 7.1 again is recommended.

Figure 7.12 ranks lock styles and decoupling by preference for separation strength. Note that *separation strength* is a relative term. Small parts with low applied forces may require Level 4 decoupling if dictated by the locking requirements of the design or by demand-complexity considerations.

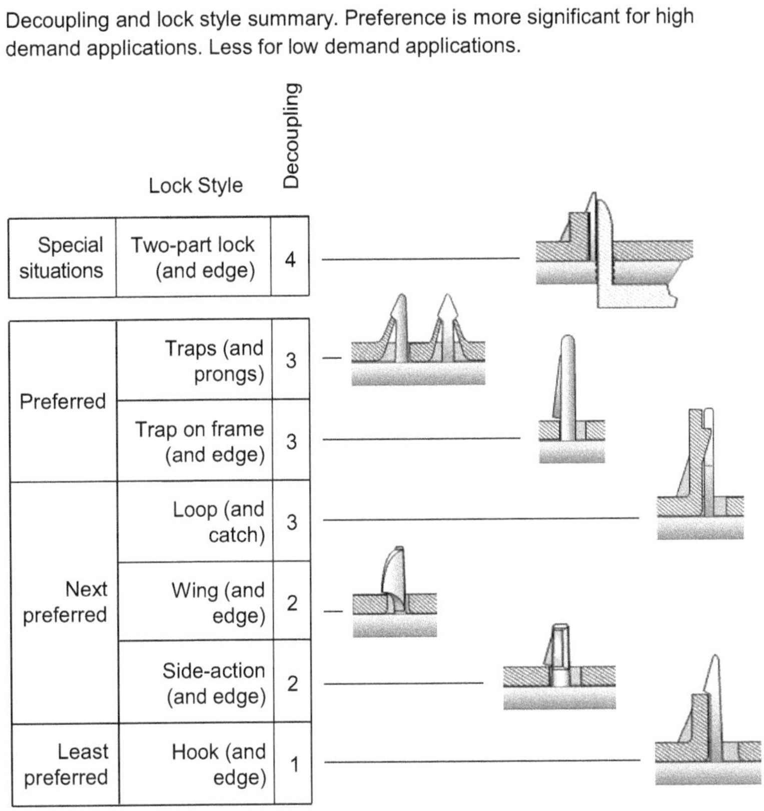

Figure 7.12 Decoupling and lock styles

Important points in Chapter 7:

- We are primarily concerned here with lock feature resistance to separation, but the applications shown in Figs 7.8 and 7.9 are also solutions to the short grip-length and low-clearance problems mentioned in Chapter 6, Section 6.7.2.
- The limitations of cantilever hook style locks are evident when the various lock styles are examined with an understanding of decoupling principles.
- Do not confuse decoupling with retainer enhancements. By our definition, retainers are not decoupling.
- The wing lock is shown as a Level 2 decoupling lock in Fig. 7.12 because its behavior is *similar* to that of a side-action lock, but plate-like behavior, as well as beam bending and distortion, are all present to some degree during assembly and separation. Accurate hand calculation of wing lock behavior is not possible but like traps, their robustness and high lock efficiency capability permit fairly reliable development, without analysis, by simply using hand-made prototype parts.

- Lock efficiency is the ratio of a lock's retention strength to its assembly force. Retainer enhancements can improve lock efficiency, but the higher levels of decoupling are by far the most useful and effective way to improve lock efficiency.
- Decoupling effectiveness is quantified by *lock efficiency*, the ratio of separation force to assembly force.

8 Constraint in Snap-Fit Applications

Constraint is strongly linked to the idea of a snap-fit as a system.

Constraint occurs across the attachment interface where locator and lock pairs provide positioning (locating) and locking to hold the mating and base parts together, see Fig. 8.1. Forces on the application are transmitted across the interface by the constraint features.

Constraint is a *key requirement* in the ALC and satisfying the other key requirements depends on a properly constrained snap-fit. The concept of constraint was introduced in Chapter 2, Section 2.1, and it is referenced throughout this book.

Readers may be familiar with the term *six degrees-of-freedom* where each *degree* includes both positive and negative movement in a Cartesian coordinate system. For constraint in snap-fit applications, it is more convenient to treat the positive and negative directions as separate movements. Rather than six degrees-of-freedom, we will use *twelve degrees-of motion* (DOM) to describe constraint.

In most snap-fits (*fixed* applications) the goal is to prevent movement in all DOM. In less common *moveable* applications, some motion between the joined parts is allowed and constraint may be less than 12 DOM. See Section 4.1.1, "Action," in Chapter 4.

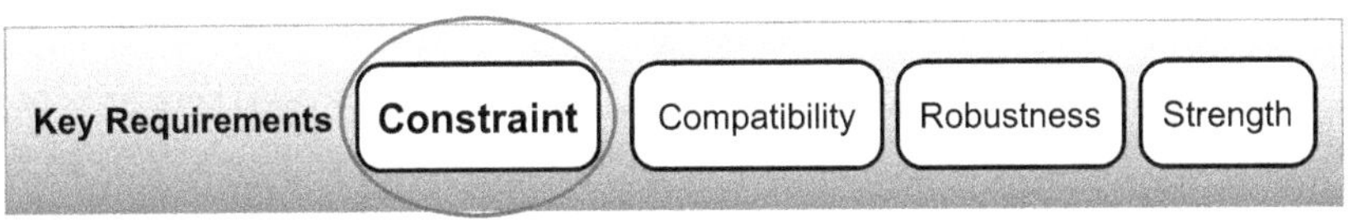

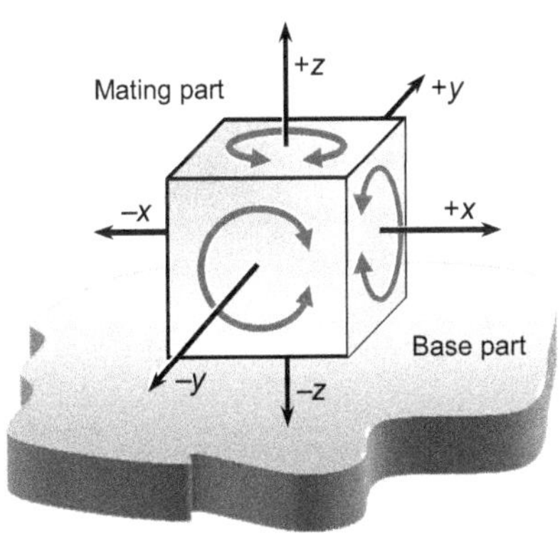

The mating part is an object in space and the base part is ground.

Motion of an object in space can be described by six ± translational movements along the axes and six ± rotational movements around the axes.

These six linear and six translational movements are called degrees-of-motion or DOM.

Figure 8.1 Constraint

The goal of snap-fit development is to design line-to-line fits, using constraint pairs, into the interface. Understanding the proper use of constraint makes it possible to balance the attachment's need for strength, ease of assembly and a line-to-line fit with the realities of part variation and tolerances.

We do not normally think of constraint during attachment development.

Attention to constraint in attachments is not common practice. Many developers are accustomed to designing attachments that use threaded fasteners. Threaded attachments achieve constraint in a simple manner: fasteners are tightened until the resulting clamp-load (and interface friction) prevents shear movement between the joined parts and the fastener's tensile strength resists part separation. Constraint between the parts happens automatically and conscious decisions about constraint are not necessary.

Unlike threaded fastener joints, it is not possible to develop significant clamp-load in plastic snap-fit attachments. Getting some low level of clamp-load through feature bending is possible and when these properly designed features do not break or yield, the residual clamp-load can ensure a line-to-line fit and help prevent part looseness. But, snap-fits cannot deliver clamp load levels comparable to threaded fasteners. Plastics tend to creep under stress and any significant clamp-load designed into a plastic snap-fit application will be greatly reduced or lost completely over time.

There are a few exceptions: the author has studied applications where a part with strong lug or pin features engages strong surfaces or edges on another part. One application is the speaker assembly shown in Fig. 1.2 in Chapter 1.

Low-deflection lugs used as lock features were discussed in Chapter 6, Section 6.3.4

In every case, these parts were assembled with a *twist assembly motion*, allowing locators on the mating part to engage locators on the base part. A lock feature then prevented rotation in the release direction. The clamp load in some of these applications compressed a gasket or O-ring. In other cases, very strong lugs designed to deflect without exceeding the maximum allowable material strain provided residual clamp load to hold the parts together. High total clamp load was achieved by specifying a sufficient number of locators.

Some design practices for attachments that use adhesives or other methods that do not rely on clamp-load are similar to snap-fit design but not identical. There are special issues with snap-fits that are not present in any other attachment. Developers must always be aware that many design principles associated with other attachment methods do not work for snap-fits.

Some people have an intuitive feeling for constraint. For others, an understanding of constraint is not intuitive and must be developed, (the author is in the second group). To assist with developing this understanding, a constraint worksheet was created, see Section 8.5 in Chapter 8.

8.1 Perfect Constraint

Improper constraint is a major cause of snap-fit problems.

Perfect constraint occurs when movement between parts is prevented by the minimum number of constraint pairs. After assembly, forces between all constraint pairs are statically determinate. In other words, we can calculate the interactions of those forces using principles of mechanics and statics.

We think of constraint features as locating and locking the mating part to the base part. Therefore, the convention in the figures is that the constraint vectors are shown acting on the mating part. All constraint vectors represent part positioning; some will also represent reactions to external forces acting on the application.

Figure 8.2 illustrates perfect constraint. The object's position is established by the plane, the line and the single point. Once the object is held in this position, its location is fully defined without redundancy between any of the locating points. *Perfect constraint* is the principle behind locating and holding parts in a known fixed position for operations like machining and precise dimensional measurement.

Understanding characteristics of perfect constraint provides a foundation for the more practical concept of *proper constraint* in snap-fit attachments.

Solid to surface application:

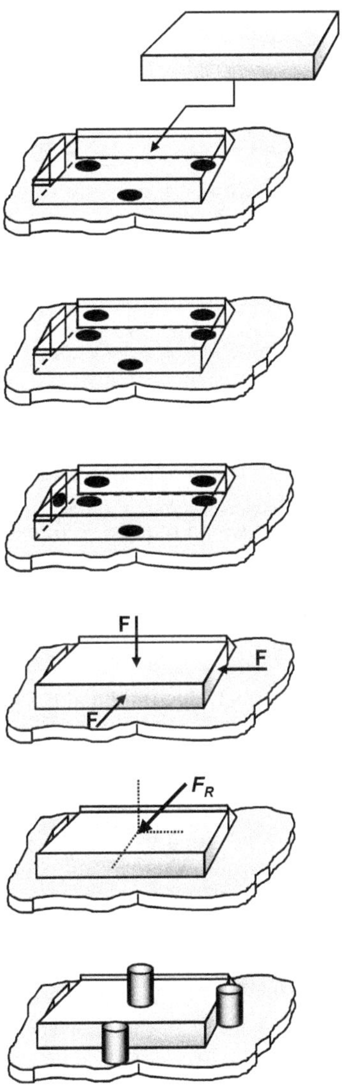

A rectangular solid must be positioned relative to another object, in this case, a surface.

First, three points define a primary plane (a surface).

Second, two points define a line to locate the solid in a secondary plane orthogonal to the surface.

Third, a single point in a tertiary plane orthogonal to the first two completes the locating.

Restraining forces orthogonal to each plane hold the solid in the located position.

The restraining forces can be composed into one resultant force to hold the solid in a perfectly constrained position.

In a snap-fit application, line-to-line contact at additional constraint sites replaces the restraining forces to hold the solid in position.

Figure 8.2 Perfect constraint

Good design will optimize the interface for maximum mechanical advantage (strength) and minimize dimensional sensitivity. To the extent possible, planar constraint should occur at the largest area of the mating part, linear constraint the next largest area of the mating part, and the single point at the third largest area.

Both part arrangements in Fig. 8.3 are perfectly constrained, but keeping the constraint sites in the primary and secondary planes as far apart as possible is preferred.

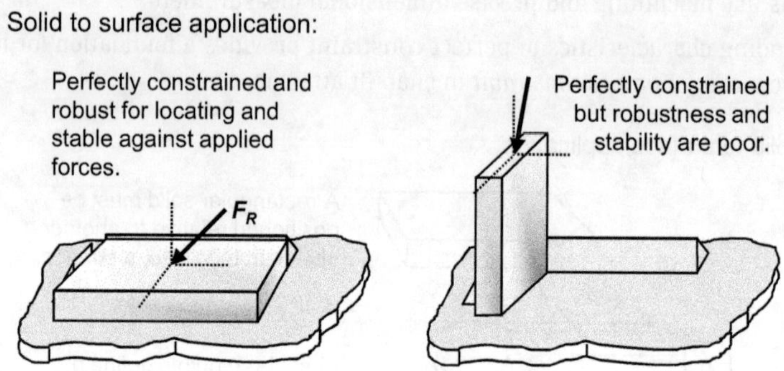

Figure 8.3 Stability, dimensional robustness, and strength

8.2 Proper Constraint

Constraint in an application can affect assembly, cost, reliability, squeaks and rattles, reliability, and even accurate attachment analysis.

For most applications, achieving perfect constraint would require no dimensional variation in either part. Zero tolerance is, of course, an expensive and impractical option. Close or fine tolerances are possible but costly. The realities of part geometry, compliance requirements and manufacturing variation will generally make perfect constraint in snap-fits difficult or impractical.

Proper constraint reflects snap-fit reality.

Snap-fit development is a compromise between perfect constraint and reality. When we have followed the constraint guidelines, we can say the snap-fit is *properly* constrained, meaning that the attachment is a reasonable approximation of perfect constraint.

Proper constraint exists when there are no *gross* violations of the rules defining improper constraint. It is total absence of *under-constraint*, eliminating *over-constraint* conditions where possible, and *managing* over-constraint where it cannot be eliminated.

When parts are properly constrained, they will have these desirable characteristics:

- Parts can be assembled without forcing them together.
- No significant residual forces exist between constraint pairs after assembly.
- Normal or loose tolerances between constraint features in the interface are possible; close or fine tolerances are not required.
- Static analysis of forces on the constraint features is possible.

- Applied forces will not cause relative part movement.
- Parts will maintain visual alignment.
- Thermal expansion or contraction, if it occurs, will not cause part distortion, damage, or buckling.

We will use fixed applications as examples of proper constraint so the mating part will be restrained in exactly 12 DOM. Recall that when the attachment's action is moveable (either controlled or free), proper constraint will exist with less than 12 DOM.

Level 4 decoupling also presents a special constraint situation. The primary mating part should be properly constrained to the base part in all DOM but the separation direction. The final DOM for the primary mating part will be removed by interference created by the secondary mating part. In turn, the secondary mating part must also be properly constrained to its own base part, which may be the original base part or the primary mating part.

Discussions of constraint in this book reflect our goal of proper not perfect constraint.

8.3 Under-Constraint

If in a fixed application, parts are constrained in less than 12 DOM, they are under-constrained, which can cause problems like:

Under-constraint conditions are easily identified and must be fixed.

- Lock damage because locks are improperly loaded.
- Part misalignment or movement because constraint features are weak.
- Looseness, squeaks, and rattles.
- Part separation when damaged constraint features release or break.

A common mistake leading to under-constraint is failure to use enough locators in the interface, thus requiring a lock(s) to carry forces in improper directions. Locks are weak in bending and can constrain *only* against separation. Locators must provide all other constraint. In Fig. 8.4, for example, with the exception of the surface, there are no locators on the mating-part, a panel.

The most important thing to know about under-constraint is that it can and must be fixed by modifying existing constraint pairs or adding more constraint sites to the interface.

There are more things wrong with this application than just under-constraint:

- The hooks violate the 5:1 length to thickness rule of thumb.
- Every hook on this part shows some evidence of damage.
- This is a short-grip-length application, so using this hook style is a mistake.
- The ribs added to strengthen the hooks increase assembly force and also increase strain in the beams; note the white stress marks on the hooks. The ribs add cost to the mold and part with no real benefits to the application.
- There are no guides for ease of assembly.

In Fig. 8.4, a *padlock* icon is used to indicate presence of a locking feature. This icon is sometimes used to represent a generic lock feature style or to simplify illustrations.

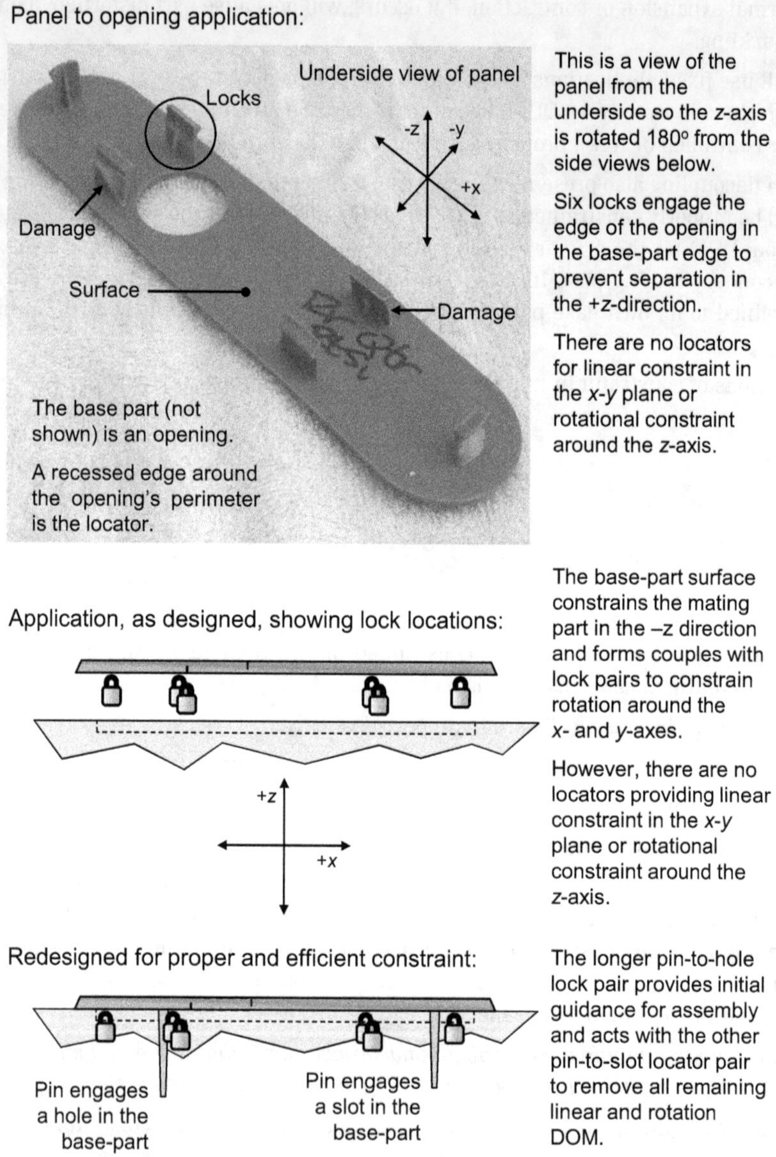

Figure 8.4 Under-constraint

A second common mistake is related to stability. Failure to arrange constraint pairs to maximize mechanical advantage for strength and stability is not technically under-constraint, but it can have similar effects. With complex part shapes, this can be a highly subjective area, and the difference between proper constraint and under-constraint can be a matter of degree, not an absolute. In Fig. 8.5, both examples of locator spacing are properly constrained, but the example with close locator spacing is closer to being under-constrained.

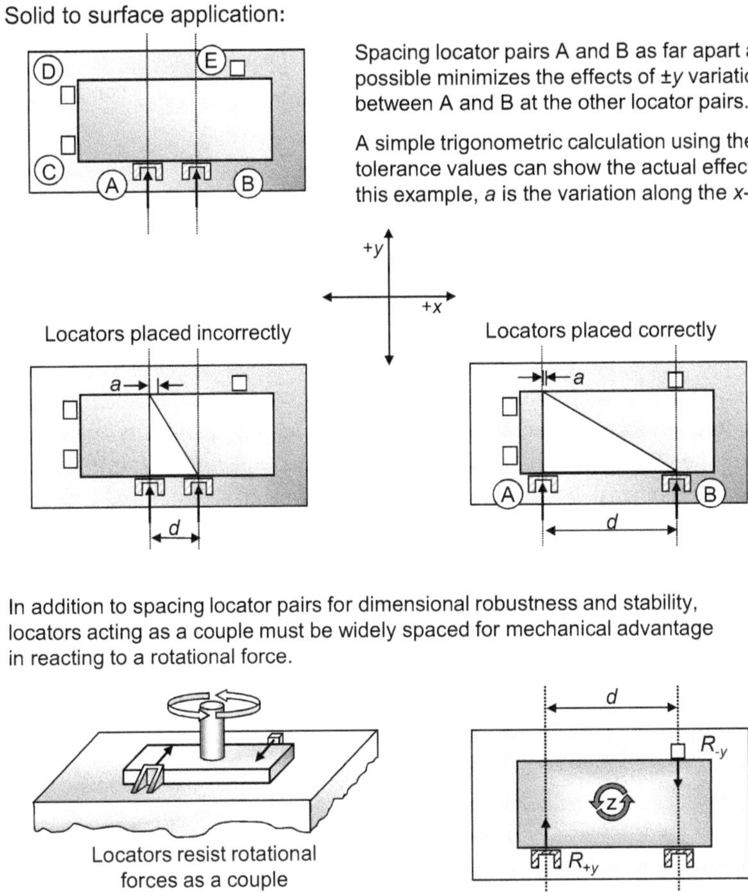

Figure 8.5 Dimensional robustness and mechanical advantage

8.4 Over and Improper Constraint

In snap-fits, the subject of over-constraint is a gray area and there are degrees of what is acceptable and what would be considered *improper* constraint. Over-constraint should be eliminated where possible and then any remaining over-constraint must be managed to minimize its effects. Symptoms of over-constraint include:

- Close tolerances are required.
- Difficult assembly when locator pairs must be forced together. High assembly forces result and immediate damage to constraint features is possible.

- Increased feature stress when assembly interference between constraint pairs creates residual internal forces. Short- or long-term feature damage and failure are possible.
- Part buckling and distortion as joined parts expand and contract due to temperature changes. This can be unsightly and may also cause feature damage and failure. Severity of this problem depends on the coefficient(s) of linear thermal expansion, (CLTE) of the joined parts and the distance across which expansion and contraction will occur.

8.4.1 Redundant Constraint Features

When multiple constraint pairs remove the same degree-of-motion with coincident lines-of-action, they are redundant, see Fig. 8.6. In other words, one of the constraint pairs can be removed without changing the system's overall constraint condition.

Sometimes redundancy can be relatively benign and it will not result in true over-constraint. But redundant attachments are inefficient, adding complexity and cost to the application, and they should be eliminated. Determine which constraint pair is least effective or more expensive to mold and eliminate it. Design the required strength into the remaining constraint pair.

Note that the six locks previously shown in Fig. 8.4 all constrain in the same direction but they are not redundant because their constraint vectors are not acting along the same line-of-action. An arrangement of multiple lock features around the perimeter of a flexible panel attached to an opening is a common and acceptable design situation because their lines-of-action are parallel, not coincident. Also see Fig. 1.1 in Chapter 1.

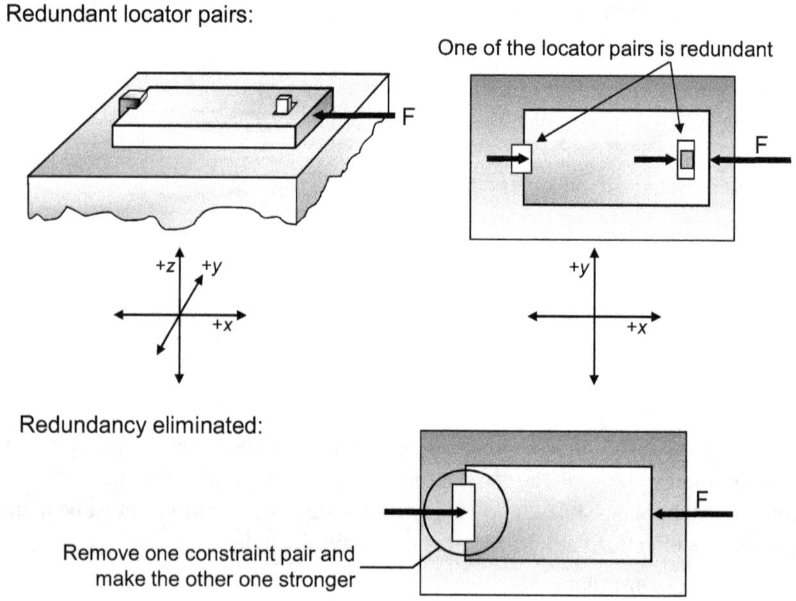

Figure 8.6 Redundant constraint example

8.4.2 Opposing Constraint Features

Opposing constraint features will have opposing constraint vectors on the same line-of-action, see Fig. 8.7. (These constraint features may also have interactions with other constraint pairs that are *not* causing constraint issues.)

Opposing constraint features may create a serious over-constraint condition. Also, like redundant constraint features, opposing constraint features can be inefficient, adding unnecessary complexity and cost to the application.

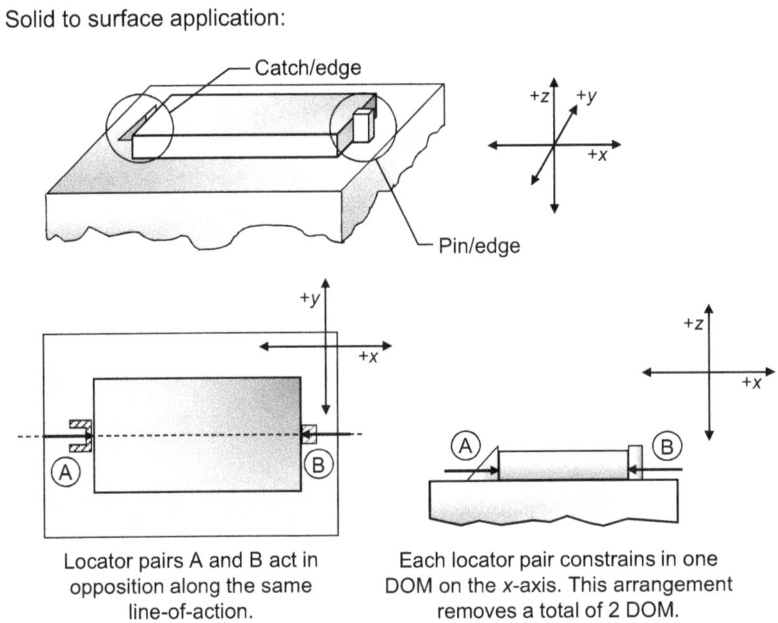

Figure 8.7 Opposing constraint features

Trying to fix over-constraint with close tolerances may be possible and worth considering if material CLTE values are low and the distance between the opposing features is small, but close tolerances can be problematic, see Fig. 8.8, and should not be a first choice.

Ideally, proper interface design will permit specifying normal or even loose tolerances. Close or fine tolerances add cost to the mold and may also mean more frequent mold maintenance and adjustment during the part's production life. This adds cost under the best of circumstances, but if the mating and base parts are from different or multiple suppliers, it can become even more costly.

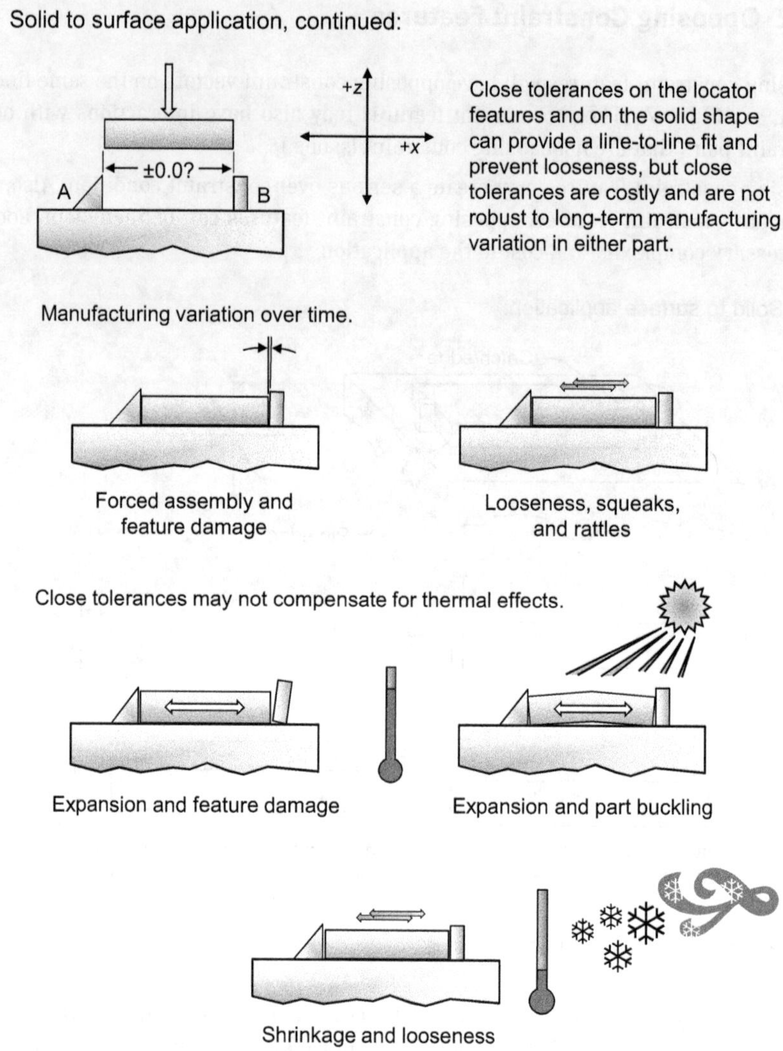

Figure 8.8 Fixing opposing constraint features using close tolerances

The preferred option is to *eliminate* the condition by selecting and arranging appropriate locator pairs, shown in Fig. 8.9.

Another option when force and alignment requirements permit, is to *manage* the over-constraint, see Fig. 8.10.

Avoid opposing constraint pairs whenever possible.

The two locator pairs introduced in Fig, 8.7 and carried through these examples only remove 2 DOM. In addition to the over-constraint issues created, this is an inefficient use of locators, which adds cost and complexity to the mold and to the parts.

Figure 8.9 shows locator pairs that solve the over-constraint issue we have been discussing and also make the attachment more efficient by removing more than just 2 DOM.

Also, review the discussion of constraint and locator pairs in Section 5.5, Chapter 5.

Tolerance requirements and variation between opposing constraint pairs are directly related to their separation distance. When opposing motions are removed within one constraint pair, the distance between the opposing constraint contact points is minimized so dimensional variation is also minimized.

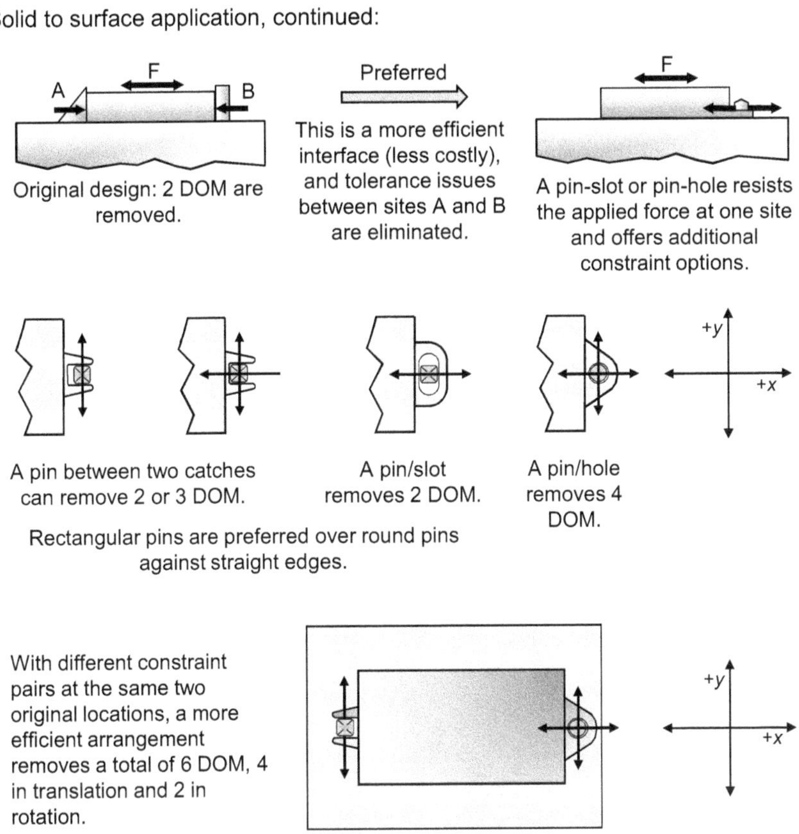

Figure 8.9 Fixing opposing constraint pairs by redesign

When opposing constraint pairs cannot be eliminated, adding compliance between the pairs can help maintain the desired line-to-line fit. Compliance can be added to the interface as local yield or elasticity, and it is discussed as an *enhancement* feature in Chapter 9.

Figure 8.10 shows how *darts* can provide compliance at noncritical constraint sites. Darts will increase assembly force and result in some residual force between the constraint pairs after assembly, but with proper design, local yield at the dart contact site will reduce the residual force to an acceptable level. The constraint features must, of course, be strong enough to resist the force and allow local yield without damage to the features. Some features may require strengthening. For example, in this case a rib may be needed behind the pin to strengthen it against the force of embedding the dart into the catch.

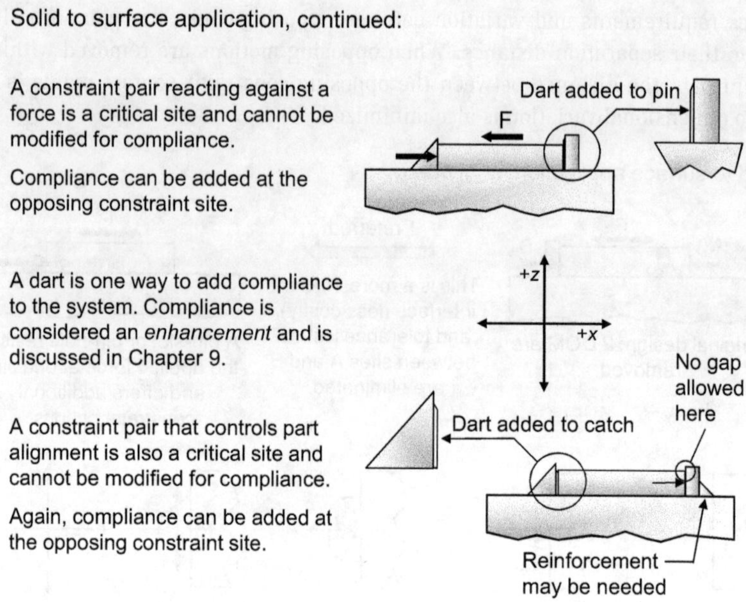

Figure 8.10 Using compliance to manage over-constraint

Figure 8.11 shows how separate constraint features with parallel lines-of-action of the same sense will behave as one larger feature. There may be times when it is desirable to either combine separate features into one larger feature or, conversely, divide a large feature into several smaller ones.

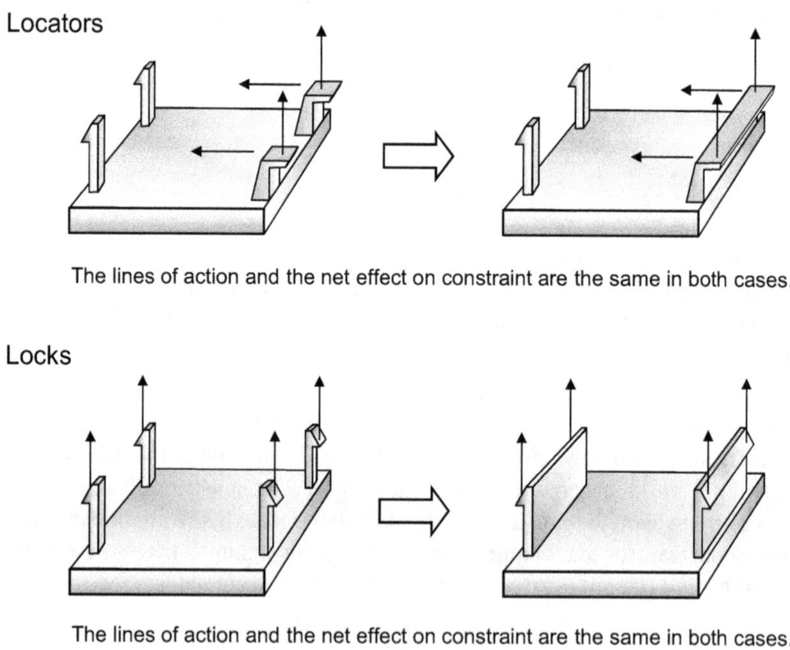

Figure 8.11 Combined and separate constraint features

8.5 The Constraint Worksheet

Constraint is a spatial concept and readers without a good understanding of the subject need a way of learning about it. Even if the reader already understands constraint in theory, read this chapter to understand the special issues associated with practical application of constraint principles in snap-fits.

The only way to understand constraint in snap-fits is to struggle with it.

Teaching oneself about constraint can be tedious, depending on how intuitive one finds the concept. Using the worksheet in Fig. 8.12 to evaluate constraint in a variety of sample applications can help [1]. If understanding constraint does not come easily, learning in a small group where constraint issues can be debated and discussed is generally more effective than trying to learn it alone. The important result is that the reader is able to recognize and avoid constraint violations in a snap-fit attachment.

Once constraint is understood, the worksheet may no longer be needed. However, it can be quite useful for explaining or justifying design decisions to others who do not understand the subject.

The author has found that trying to use the worksheet during early concept development and design is awkward, and the worksheet is best used first as a learning tool. The insights and lessons learned will then be applied naturally during the development process. Once a design is completed, the worksheet may be useful for review and verification.

The worksheet in Fig. 8.12 can be copied for use.

Some examples are provided here with completed worksheets. Once you understand the examples, fill out some worksheets for a few simple applications. Again, a team approach is suggested.

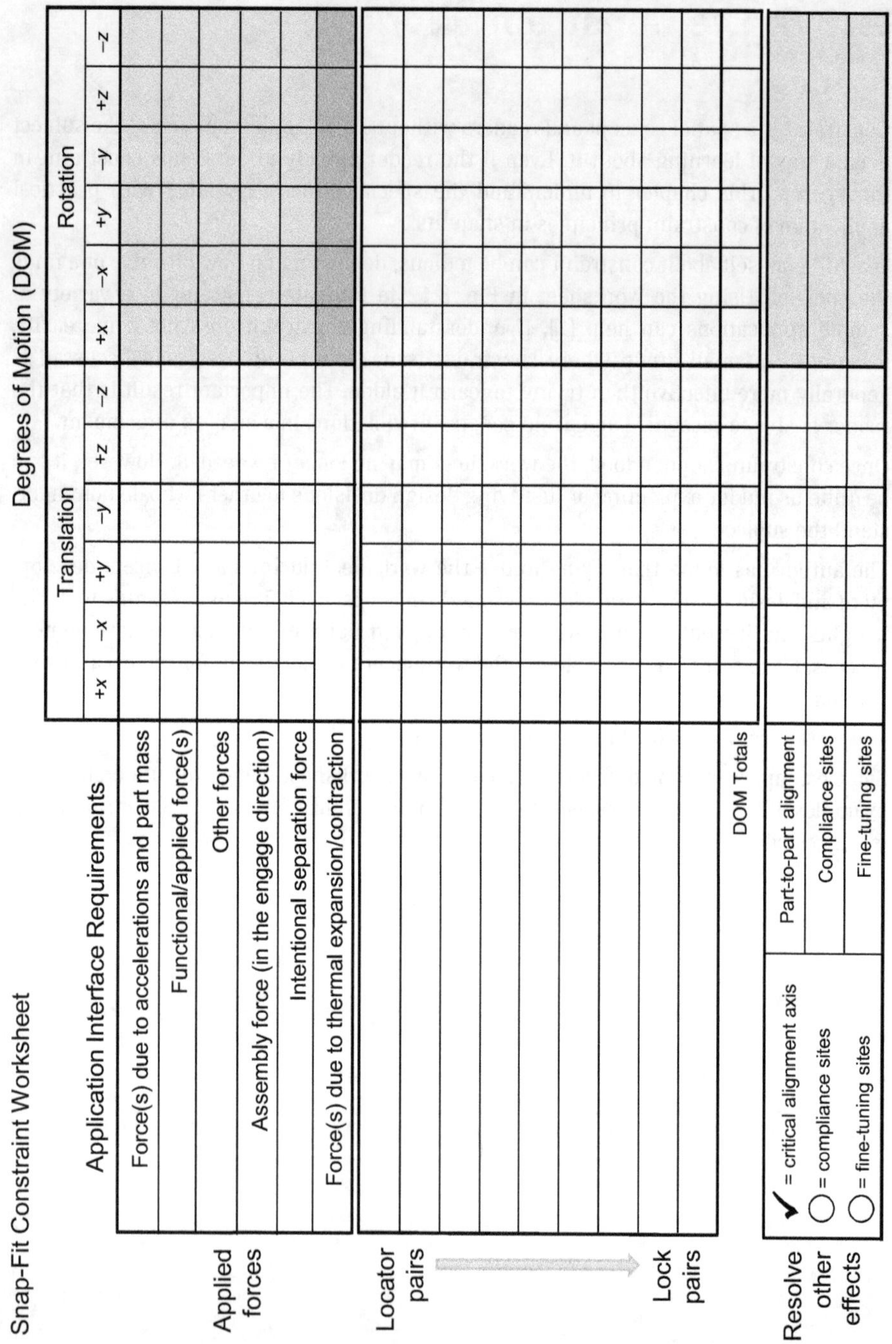

Figure 8.12 The constraint worksheet

Major areas of the worksheet are identified in Fig. 8.13. Actions for each area are described here. The reader should copy the worksheets in Fig. 8.12 or 8.13 for easier reference when reading the following material.

1. Identify forces in the application by marking the cell in the appropriate DOM column. Force vectors are identified as acting on the mating part. For example, if an applied force will move the mating part in the *-y*-direction, the *-y*-cell should be checked.

   Recall that constraint vectors, by convention, are defined as acting on the mating part to prevent motion in a given direction. Thus, in Step 3, the constraint feature resisting this force will also be marked in the *-y* column although the constraint vector itself is in the *+y* direction.

   Think about all the force effects that must be considered when establishing interface requirements. As a rule, these will only be translational effects so the rotation side of the worksheet is generally not used here.

   Thermal effects will always generate opposing forces, which will occur in both directions, (bi-directional) on an axis.

2. List all constraint pairs. They can be listed in any order, but one option is to order them from the most to least efficient locator pairs followed by the lock pairs.

   Another is to list them in this order:

   - All locator pairs that establish the primary interface plane: This is the three-point or planar interface from the perfect constraint example. Some may also be contributing to constraint in other planes, but list them here too.
   - Locator pairs (not already listed above) that establish the secondary interface plane: This is the linear (two-point) orientation.
   - Remaining locator pairs (again, not already listed above) that establish single point restraint for the tertiary interface plane.
   - List all lock pairs.
   - Label all natural locators with an *N* as a reminder that they may require special attention, if they are to be used as fine-tuning sites.
   - Label any locators that may require special attention as compliance sites with a *C*.

3. Removing DOM is about positioning the mating part. This step includes all translational DOM, not just those that resist force(s). For each constraint pair, identify all DOM *removed* by each constraint pair. To continue the example from (1), a constraint pair resisting mating part movement will have *1* in the *-y*-cell because it is resisting motion in the *-y*-direction, (however, its constraint *vector* is in the *+y*-direction). If it shares that job with another (nonredundant) constraint feature, each cell will contain the value *1/2*.

   When numbers appear for multiple constraint features in the same column, check for possible redundancy.

   Identify the contribution each constraint pair makes to removing translational DOM. Work across the top of the worksheet using the six columns of translation. The reader may wish to experiment with two ways to do this and choose the one that works best for them:

- Constraint pairs are considered one at a time and all DOM removed by that pair are identified or
- DOM is considered and the contribution of each constraint pair to removing that DOM is identified.

Use fractions to indicate the contribution of constraint pairs acting in parallel and of the same sense. If, for example, a panel is held in place by eight lock pairs acting in parallel, each pair would receive a value of 1/8 in the appropriate cell. It is convenient and generally accurate to assume equivalent strength and stiffness, thus equivalent contribution of each constraint pair; however, this may not always be the case.

As new entries are made, always check for translational under or over-constraint by studying the entries in the columns. Columns with a total less than 1 are under-constrained. Columns with a total of 1 are properly constrained. Columns with a total greater than 1 may be over-constrained; check the constraint pairs against the rules for proper constraint.

If an under-constraint condition exists, fix it and adjust the worksheet accordingly.

If over-constraint due to redundancy between constraint pairs exists, fix it by removing the least efficient pair (for mechanical advantage and dimensional robustness) and adjust the worksheet accordingly.

If over-constraint due to opposing constraint pairs exists, fix it if possible and adjust the worksheet accordingly, or record the condition for later review. Make note of the need for feature compliance along that axis.

One way to fix over-constraint due to opposition is by removing both directions of movement within one constraint pair (this is the preferred solution). Another is by providing feature compliance at one of the constraint sites.

Where compliance cannot be used or will not be effective, close tolerances between the opposing constraint pairs will be necessary, but this is the least preferred solution. Evaluate the effects of relative thermal expansion/contraction and the possibility of part warpage or feature damage.

If strength or alignment requirements are identified in both directions along the same axis, over-constraint in opposition should be avoided along that axis because it cannot be fixed using compliance. If over-constraint in opposition was noted in Step 5, it must be fixed.

4. This step is similar to Step 3, but applies to rotational DOM. Most applications will involve linear, not rotational forces, but rotational applied forces are possible. In any case, linear forces can cause rotation and rotational DOM must always be considered for part positioning.

   Rotational constraint requires constraint pairs to act as couples, so one constraint pair will have *1/2* in the appropriate DOM cell. The other constraint pair of the couple will also show *1/2* in a cell of the opposite sense on the same axis. Again, we will typically assume equivalent strength and stiffness of the constraint pairs in a couple, but that may not always be the case.

   Identify the contribution each constraint pair makes to removing rotational degrees-of-motion. Work across the top of the worksheet using the six columns of rotation.

Note that a single constraint pair of sufficient length can also act as a couple; a very long wedge in a slot is an example.

With rotational movement, effectiveness in both dimensional stability and in strength improves as the distance between the constraint pairs increases.

Check to verify there is no over or under-constraint in rotation. If there is, fix it and adjust the worksheet. Verify you have not changed any translational constraint conditions.

5. The values in the DOM columns are totaled in this row. A properly constrained fixed application will have a *1* in each column for a total of 12 DOM for a fixed application. (A moveable application will total less than 12 DOM with *1/2* or *0* in a cell.) A fixed application with less than *1* in the totals row is under-constrained and the condition must be corrected. A value greater than *1* indicates over-constraint which must also be addressed.

   Constraint pairs that resist forces will have check marks in the requirements section and must be designed to withstand those forces.

6. In addition to applied forces, other issues, particularly compliance and fine-tuning must also be addressed. Sometimes this will include DOM that also involve forces. This area reminds us to verify that these, sometimes conflicting, requirements are resolved.

   Identify all translational directions and corresponding constraint pairs where feature compliance is possible. Compliance sites should not resist external forces or provide critical alignment.

   There should be fine-tuning sites in each of the three translational directions, but not in opposing directions. For example, combinations like (+*x*; +*y*; −*z*) or (−*x*; +*y*; −*z*) are OK. A combination like (+*x*; −*x*; +*y*; +*z*) is not OK.

   Fine-tuning sites should control all critical alignment directions.

   Highly compliant parts (soft or flexible parts like panels) may require multiple constraint pairs (acting in parallel and in the same direction) to remove all possible flexure. Part compliance is often an issue in large and/or flexible parts with the *panel* basic shape.

   - Verify these constraint pairs are properly spaced to ensure against part flexure.
   - Adding stiffening features such as ribs to increase local part stiffness is often desirable.

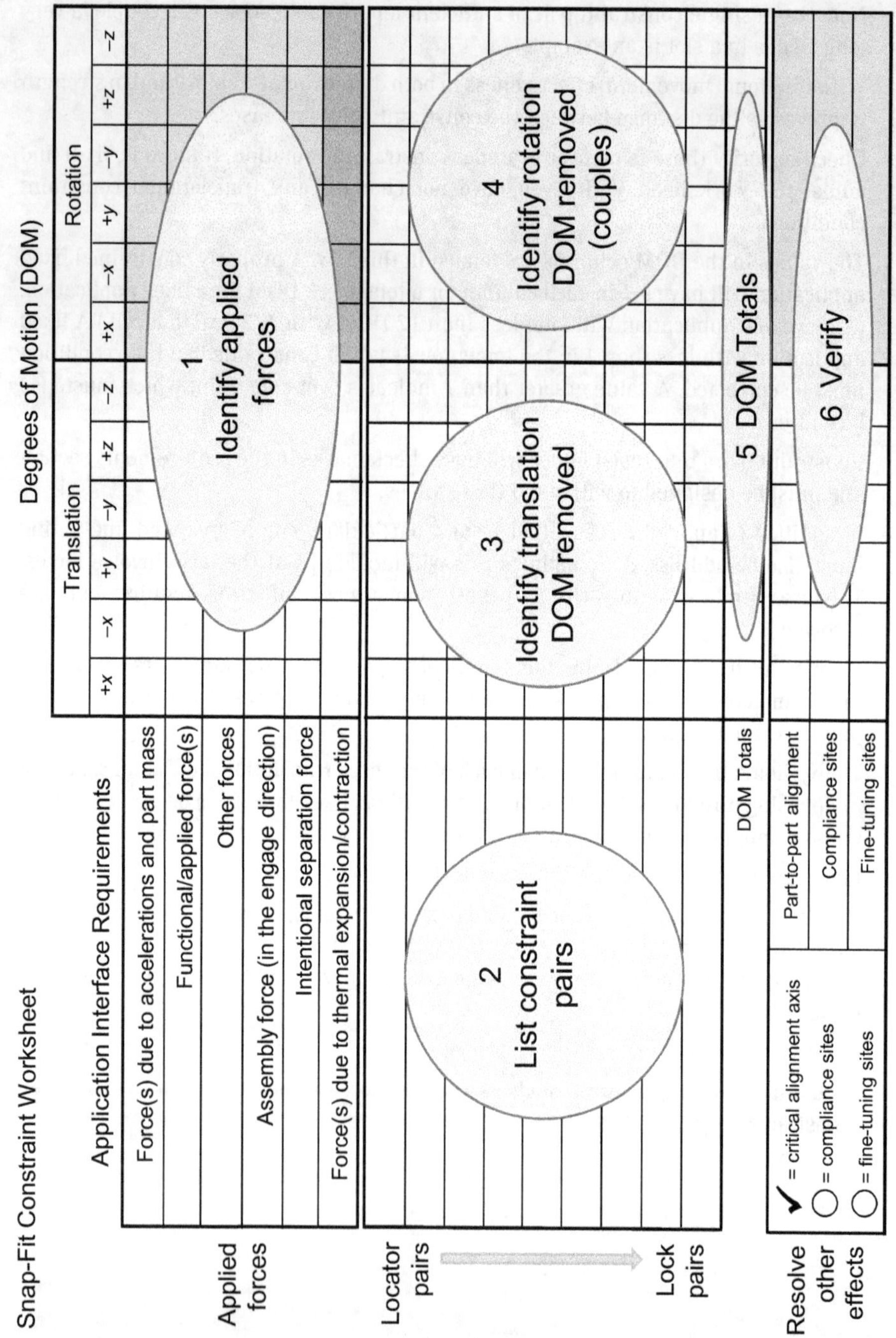

Figure 8.13 Major areas of the constraint worksheet

8.6 Using the Constraint Worksheet

The constraint worksheet helps us account for all the constraint action that is going on in the interface. Figures 8.14 and 8.15 illustrate how the worksheet would be completed for a simple solid-surface application. Study them and refer to the constraint rules in Section 8.7 as you do.

Note the padlock icons (introduced in Fig. 8.4) indicating the presence of locking features.

Figure 8.9 from Section 8.4.2 is used in these examples.

Worksheet – requirements area

Degrees of Motion (DOM)

Applied forces	Application Interface Requirements	Translation +x	Translation –x	Translation +y	Translation –y	Translation +z	Translation –z	Rotation +x	Rotation –x	Rotation +y	Rotation –y	Rotation +z	Rotation –z
	Force(s) due to accelerations and part mass												
	Functional/applied force(s)				✓								
	Other forces												
	Assembly force (in the engage direction)						✓						
	Intentional separation force					✓							
	Force(s) due to thermal expansion/contraction	✓		✓									

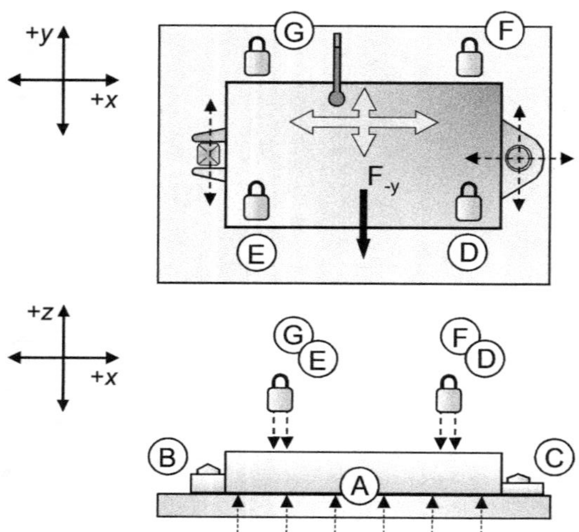

Assembly and separation forces are shown, but are not significant in this application.

An applied force (–*y*) acts on the mating part.

Thermal effects are expected in the x-y plane. (They will be more significant on the *x*-axis because of greater part length than on the *y*-axis.)

Constraint vectors are shown as dashed arrows acting on the mating part.

A is the surface-surface locator pair between the parts.

B is a pin-slot locator pair.

C is a pin-hole locator pair.

D, E, F, and G are lock-surface lock pairs.

Figure 8.14 Example: completing the requirements area

Constraint Worksheet – constraint pairs area

Degrees of Motion (DOM)

		Translation						Rotation					
		+x	−x	+y	−y	+z	−z	+x	−x	+y	−y	+z	−z
Locator pairs	(A) surface-surface (Mark it *N* as a natural locator)						1	1/2	1/2	1/2	1/2		
	(B) pin-slot (opposing catches acting as a slot)			1/2	(1/2)							1/2	1/2
	(C) pin-hole	1	1	1/2	(1/2)							1/2	1/2
	(D) lock-surface					1/2			1/4		1/4		
	(E) lock-surface					1/2			1/4	1/4			
	(F) lock-surface							1/4			1/4		
	(G) lock-surface							1/4		1/4			
Lock pairs													
	DOM Totals	1	1	1	1	1	1	1	1	1	1	1	1

The circled values in the −y column remind us this is the direction of the applied force.

Fractional values are constraint pairs with parallel lines-of-action. They share constraint without being redundant.

Resolve other effects		+x	−x	+y	−y	+z	−z	Rotation
✔ = critical alignment axis	Part-to-part alignment							
◯ = compliance sites	Compliance sites	(B)		(B)				
◯ = fine-tuning sites	Fine-tuning sites							

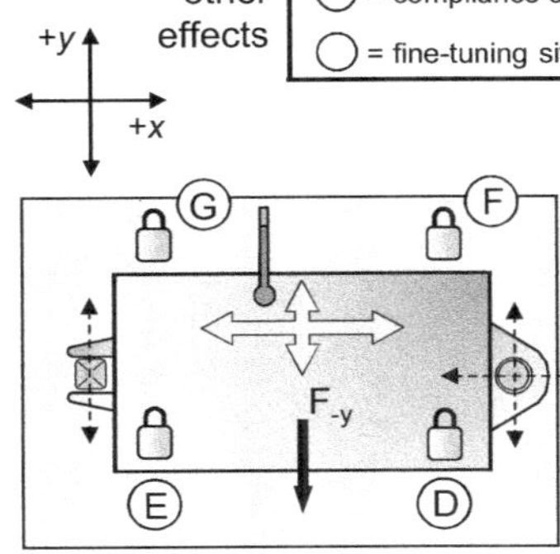

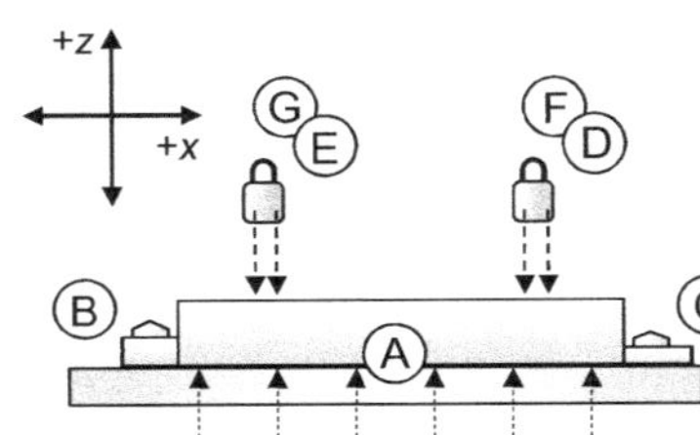

We will treat site *C* as the primary datum in the x-y plane and assume line-to-line fit at the pin-hole locator pair. This is not an unrealistic assumption.

There are no *critical* alignment requirements to consider, but compliance at site *B* can take up tolerance to prevent *looseness*. Darts are added to the pin at *B* as shown. The *y*-axis dart must be *opposite* the load bearing side of the pin.

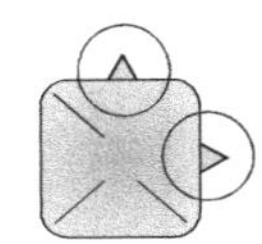

Figure 8.15 Example: completing the constraint and DOM areas

Figures 8.16 and 8.17 show a properly constrained toggle switch application and the completed constraint worksheet. The reader might want to complete a blank worksheet for this application before studying the completed worksheet in Fig. 8.17. This is also a good time to use some simple models to represent the application and to visualize how the constraint features are interacting.

This example is based on the product application described in Chapter 9, Section 9.1.4. Refer to that section for a detailed discussion and pictures of the application.

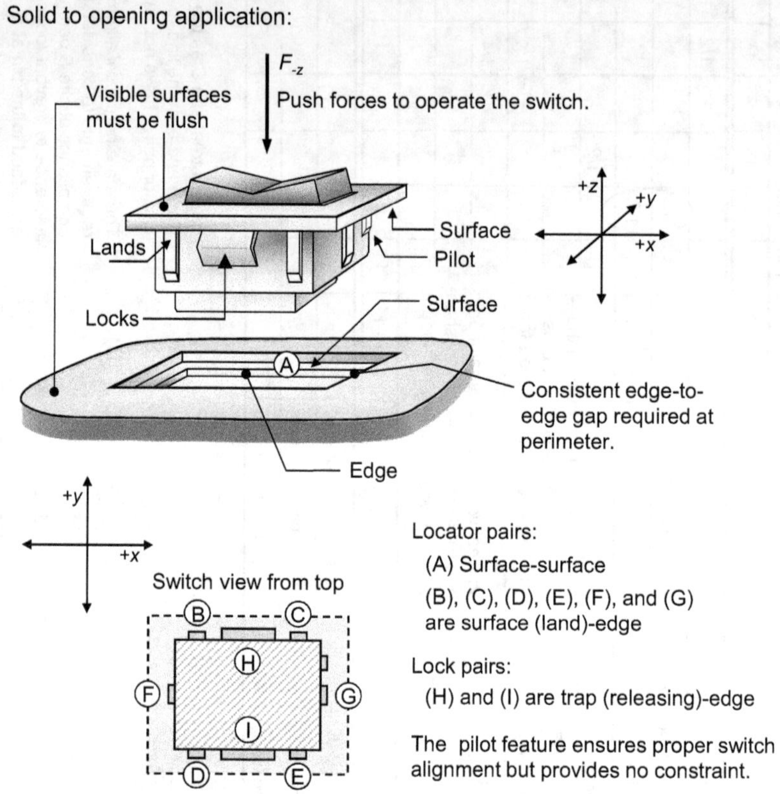

Figure 8.16 Solid-opening constraint example

In area 2 of the worksheet, the *z*-axis surface-to-surface locator pair, (A), is labeled with an *N* as a reminder it is a natural locator site that may require fine-tuning.

Locator pairs (B, C, D, E, F, and G) started out as natural locators in the original application and example shown in Figs. 9.5 and 9.6 in Chapter 9. In the improved application we are showing here, the natural locator surfaces of the switch body have already been redesigned as tunable lands. This is indicated by the *Fs* in the *fine-tuning sites* row in area 6 of the worksheet (see Fig. 8.17, bottom row).

Snap-Fit Worksheet – switch example

	Application Interface Requirements	Translation						Rotation					
		+x	−x	+y	−y	+z	−z	+x	−x	+y	−y	+z	−z
Applied forces	Force(s) due to accelerations and part mass												
	Functional/applied force(s)						✓						
	Other forces												
	Assembly force (in the engage direction)						✓						
	Intentional separation force				✓								
	Force(s) due to thermal expansion/contraction												
Locator pairs	N (A) surface-surface						1	1/2	1/2	1/2	1/2		
	(B) surface (land)-edge			1/2									1/2
	(C) surface (land)- edge			1/2								1/2	
	(D) surface (land)- edge				1/2							1/2	
	(E) surface (land)- edge				1/2								1/2
	(F) surface (land)- edge		1										
	(G) surface (land)- edge	1											
Lock pairs	(H) trap-edge					1/2		1/2		1/4	1/4		
	(I) trap-edge					1/2			1/2	1/4	1/4		
	DOM Totals	1	1	1	1	1	1	1	1	1	1	1	1
Resolve other effects	✓ = critical alignment axis — Part-to-part alignment	✓	✓	✓	✓	✓	✓						
	◯ = compliance sites — Compliance sites												
	◯ = fine-tuning sites — Fine-tuning sites	(F)	(F)	(F)	(F)	*	*						

* The z-axis is critical to the flushness requirement and tunable features *may* be required on one of the surfaces at *A*. We will identify this as a potential issue. The constraint pair is also marked with the letter N as a reminder.

Figure 8.17 Completed solid opening constraint worksheet

The reader will discover when trying to actually evaluate constraint and the feature interactions that it is an iterative procedure. Do not expect a linear thought process that will lead to a final answer in just one pass through the constraint evaluation process.

8.7 Constraint Rules

- The concept stage is the best time to recognize and avoid using opposing constraint pairs. Understanding how some constraint pairs can remove multiple DOM efficiently and using them in the design is the best course of action. When confronted with an existing situation, redesign of the constraint features may be necessary.
- Fixed snap-fits must be properly constrained in exactly 12 DOM.
- Moveable snap-fits will be properly constrained in less than 12 DOM.
- Locator features are strong so use them to remove as many DOM as possible. Minimize the DOM removed by the (relatively weak) lock features.
- The *tip*, *slide*, *twist*, and *pivot* assembly motions tend to maximize DOM removed by locators and are preferred for their high attachment strength potential.
- The *push* assembly motion generally maximizes DOM removed by locks and is not recommended when higher separation strength is required unless high strength locks are used.
- An under-constraint condition is unacceptable and must be fixed.
- Over-constraint due to redundant constraint pairs should be eliminated.
- Some locator pairs can remove up to 5 DOM. When developing a concept, use locator pairs that remove multiple DOM wherever possible for higher interface efficiency. Identify those locator pairs first before finishing with less efficient locator pairs.
- The distance between constraint pairs (with parallel strength vectors) affects both mechanical stability and dimensional sensitivity. Space these constraint pairs as far apart as possible when parts are *rigid* in the constrained direction.
- Proper constraint of *flexible* panels may require multiple constraint features on the perimeter. The goal is to create *local* stiffness to prevent looseness and vibration.
- As constraint pairs in a couple are moved farther apart, their mechanical advantage against rotation increases. Identify primary constraint pair(s) based on the alignment and/or strength requirements of the application. Plan to use these pair(s) as the datum point(s) for locating all other constraint features in the interface.
- Develop applications to avoid opposing constraint pairs because they can lead to an over-constraint condition. But, opposing constraint pairs are sometimes a necessity. To manage possible over-constraint use compliance enhancements. As a last resort, close or fine tolerances may be required, but thermal expansion may become an issue.

8.8 Summary

Constraint is the concept of preventing movement of an *object* in space. In snap-fits, that object is the mating part and the goal is to position it to the base part through a reliable attachment interface. Six translation and six rotation movements in a Cartesian coordinate describe the possible mating part movements.

A field that uses constraint principles is the use of fixtures to hold parts: where precise location of the subject part is essential. For example, locating and locking devices hold parts in a perfectly constrained position for machining operations or for dimensional checking

Most of the constraint principles expressed in this section and throughout the book lend themselves to expression in mathematical terms. Tools and methodologies for optimizing a snap-fit interface in terms of constraint, strength, compliance, and tolerances can and should be developed [2].

Important points in Chapter 8:

- Conscious or explicit consideration of constraint in attachments is not common practice because many product developers are accustomed to specifying threaded fasteners.
- Do not rely on clamp-load in a snap-fit and do not try to design clamp-load into a snap-fit attachment. Design instead for a line-to-line fit.
- Clamp-load is possible in certain applications but requires a high degree of care in design, (generally) a twist assembly motion, strong locator features as locks, and validation in a thorough test program.
- Perfect constraint is a theoretical ideal. By avoiding constraint mistakes and minimizing certain non-preferred conditions, the developer can ensure a snap-fit with proper constraint. *Proper* constraint is essentially the absence of *improper* constraint.

A practical approach is to design for proper constraint by:

- *Eliminating* all under-constraint conditions.
- *Eliminating* redundant constraint pairs.
- *Replacing* inefficient constraint pairs with more efficient ones.
- *Managing* over-constraint where it cannot be eliminated.

A worksheet was introduced as a learning tool and for practical use in verifying proper constraint in product applications. If you do not have a high comfort level with your understanding of constraint, use the constraint worksheet until you do.

In many constraint situations, there is a continuum between improper and proper constraint. Move the design as far as possible in the direction of proper constraint.

References

[1] Luscher, A.F., Bonenberger, P.R., *Part Nesting as a Plastic Snap-fit Attachment Strategy*, DETC97/DTM-3893, Proc. of DETC '97, ASME Des. Eng. Tech. Conf., Sept. (1997)

[2] Bonenberger, P.R., *A New Design Methodology for Integral Attachments*, ANTEC '95 Conf. Soc. Plast. Eng., Boston, MA, May (1995)

9 Physical Elements: Enhancements

Enhancements are the third physical element in the ALC, see Fig. 9.1. They may be distinct interface features or they can be attributes of other interface features. Most enhancements do not directly affect reliability and strength but they can have important indirect effects on strength, quality, reliability, and cost by improving the application's robustness to manufacturing, assembly and usage variables.

Without enhancements, a snap-fit application may be functional, but it will *not* be complete.

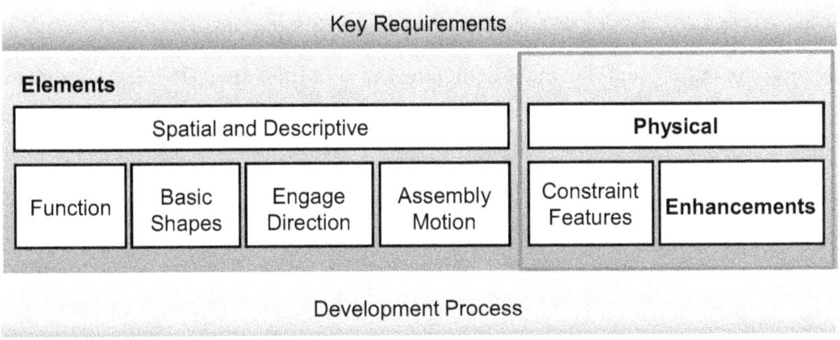

Figure 9.1 Enhancements in the ALC

Enhancements are a relatively undocumented aspect of snap-fit design. Some are tricks-of-the-trade learned by the developer through trial-and-error. They are a big part of the *attention to detail* aspect of good snap-fit design.

If you have examined some snap-fits and found features you could not identify or wondered *why did they do that?* you may have been looking at an enhancement. If you have assembled and disassembled similar snap-fit applications with interface designs from different sources and marveled at how the applications could behave and feel so different, the difference may have been due to enhancements.

During product benchmarking studies, some of the best ideas and creative hints will not be dramatic or highly interesting features. They will be subtle and rather mundane details - much like those described in this chapter. By studying these details (enhancements) on real products, the reader will find important clues to design problems which had to be overcome. You can then avoid those problems in your own design. Benchmarking is discussed in detail in Chapter 10, "Applying the Snap-Fit Development Process."

This has been said before, but it is worth repeating: *Some of the snap-fit problems discussed here may seem trivial or obvious and readers may say to themselves, I would never do anything like that! The author can assure the reader that even the trivial problems or mistakes discussed in this book were found in multiple real-life applications.* In some of these applications, if an enhancement(s) had been included in the design, the problem would have been avoided.

Certain enhancements belong in *every* snap-fit application; we'll identify them in side notes as we proceed through the chapter. Other enhancements are required only to meet specific application needs.

- *Assembly Enhancements* are features or attributes that support the assembly process. Some of them are already well documented in design-for-assembly/manufacturing (DFM) practices.
- *Activation Enhancements* enable release or operation of the attachment.
- *Performance Enhancements* provide extra support to the locks or affect interface conditions to ensure that the attachment performs as expected.
- *Manufacturing Enhancements* are design practices and part features that support manufacturing and part consistency. Many are documented in standard design and manufacturing practices for injection-molded parts and are already recognized as important guidelines for plastic part design. They fit neatly into the ALC as enhancements and, because of their importance to snap-fits, are included here.

Some developers may feel they have somehow failed in their snap-fit design if they must add various enhancement features. That impression is the result of applying traditional threaded fastener thinking to snap-fits. Remember that a threaded fastener attachment represents a brute strength approach to fastening. The interface details required for a good snap-fit design go beyond those necessary for a threaded fastener attachment. In reality, enhancements belong in every snap-fit and an application without them will not be the best possible design.

As with the other physical features of snap-fits, locks and locators, the author does not pretend to have invented enhancements. All the enhancements discussed here were found on products, often on many different products in many different variations. Certain enhancements, particularly those related to manufacturing and design-for-assembly, are well documented elsewhere; others are not. In all cases, the Attachment Level Construct effectively captures and organizes them, providing a means of describing and classifying them for use.

9.1 Assembly Enhancements

Assembly enhancements help ensure the assembly process will consistently and efficiently produce a good attachment.

Initial mating-part to base-part alignment followed by a simple assembly motion (push, slide, tip, twist, or pivot) should be all that is necessary to accomplish assembly. This is the purpose of assembly enhancements. (The author likes to describe the ideal assembly situation as one where the assembler can simply drop or toss the mating-part in the direction of the base-part and have it settle into position, ready for lock engagement.)

Imagine a worst-case scenario for assembling snap-fit parts:

1. Initial alignment - Large motor movements orient the mating-part to the base-part.
2. First adjustment - Small motor movements engage the first locators.
3. Second adjustment - Small motor movements engage additional locators and overcome minor part feature interference as the mating-part is moved to final locking position.
4. Third adjustment - Small motor movements align locks.
5. Locking - Force is applied to engage the locks and any remaining locators. The assembly operation is complete.
6. Verification - The assembler may or may not be certain a good attachment has been made.

Most applications are not quite this bad, but many are not very good. Each part positioning step takes time and effort. Unnecessary movements cost time and money and can also contribute to cumulative trauma injury.

Steps 1-4 above are addressed by *guides*, *clearance*, and *pilot* enhancements (collectively called *guidance*). Step 5 is the actual locking step. Step 6 involves *feedback*.

The process simplifies to:

1. Initial alignment - Large motor movements orient the mating-part to the base-part.
2. First adjustment - Small motor movements engage the first locators or guides. Guidance enhancements can replace all three adjustment steps in the first scenario with a single engagement operation.
3. Locking - Force is applied to engage the locks and any remaining locators. The assembly operation is complete.
4. Verification - Positive feedback signals the assembler that the attachment operation was successful. Unsuccessful attachments are easily identified and fixed.

9.1.1 Guides

Required Enhancements: Guides

Guides are physical features, usually protrusions that stabilize the mating-part to the base-part so the assembler can easily bring the parts together for locking without feature damage, wasted time, or extra movements.

Some common guide features are shown in Fig. 9.2. The guide function can be carried out by special features dedicated to that purpose, but it is always more efficient to use features that already exist in the interface. Note that some of the guides in this figure are protrusion locators. When possible, the guide function should be incorporated into pre-existing part features (corners and edges) and locators. When locators are to be used as guides, add guide functions to the first and, if necessary, the second locator pair(s) to be engaged.

In situations where precise lock pair alignment is required for ease of assembly or to prevent lock damage, guides should be placed very close to or built into the lock pairs. This may be necessary if a lock feature or the wall, for example, on which it is mounted is subject to warpage and the lock's final position is somewhat variable.

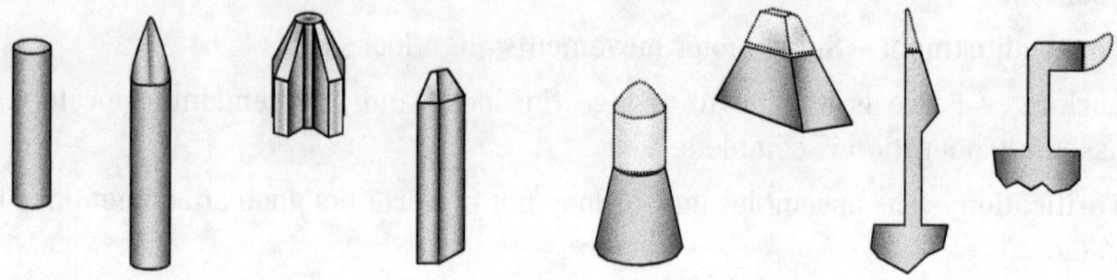

Figure 9.2 Guide features

General rules for guide usage:

- Guides or locators used as guides must be the first features to make contact with the other part, see Fig. 9.3. Lock features must never make first contact.

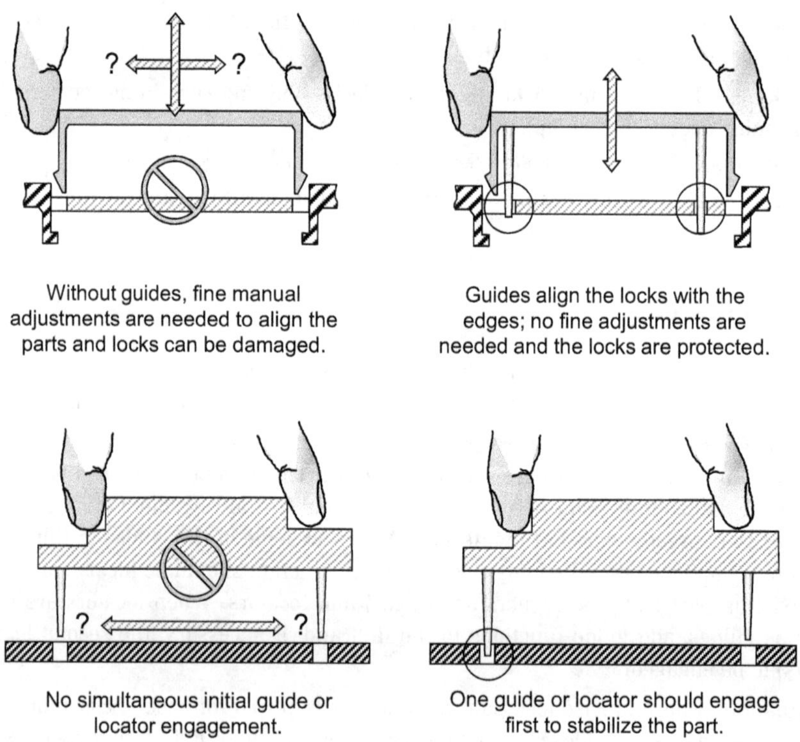

Figure 9.3 Guides in an application with a push assembly motion

- Guides must engage before other parts interfere with the assembler's hand or mating-part movements.
- A *tip* assembly motion is preferred over a push motion. A tip motion can eliminate simultaneous engagement because it requires initial engagement of a locator pair(s). This will stabilize one end of the mating-part in multiple DOM. The mating-part is then tipped to engage the remaining features.
- Avoid simultaneous engagement of multiple features. One or two guides (or locators) should engage first to stabilize the mating-part to the base-part, particularly when the features are protrusions, like pins, engaging with holes or slots.
- Build guide and pilot functions into existing constraint features whenever possible.

9.1.2 Clearance

Required Enhancements:
Guides
Clearance

Clearance is not a physical feature, but a characteristic or attribute of other part features.

Clearance ensures all adjacent part features, including snap-fit features in the interface can pass by each other during the assembly process without interference, see Fig. 9.4. As with guides, wasted motions are eliminated, this time because minor part positioning adjustments are unnecessary.

Clearance is not difficult. It is simply thinking about all possibilities for part-to-part interference and eliminating them. In general, clearance is achieved by designing generous radii or bevels on part edges and by tapering or rounding-off edges on locators and guide features. This is a very simple concept but it is often overlooked in practice. Some clearance rules are:

- Always specify a taper, bevel, or radius on corners and edges of the mating and base-parts that pass by or come close to each other as well as on all the constraint features.
- Clear and *explicit* taper, bevel, and radius callouts for each critical site will have more meaning and are preferred over a general note on the part drawing.
- For initial engagement, always provide generous clearance between the mating and base-parts as well as the features.

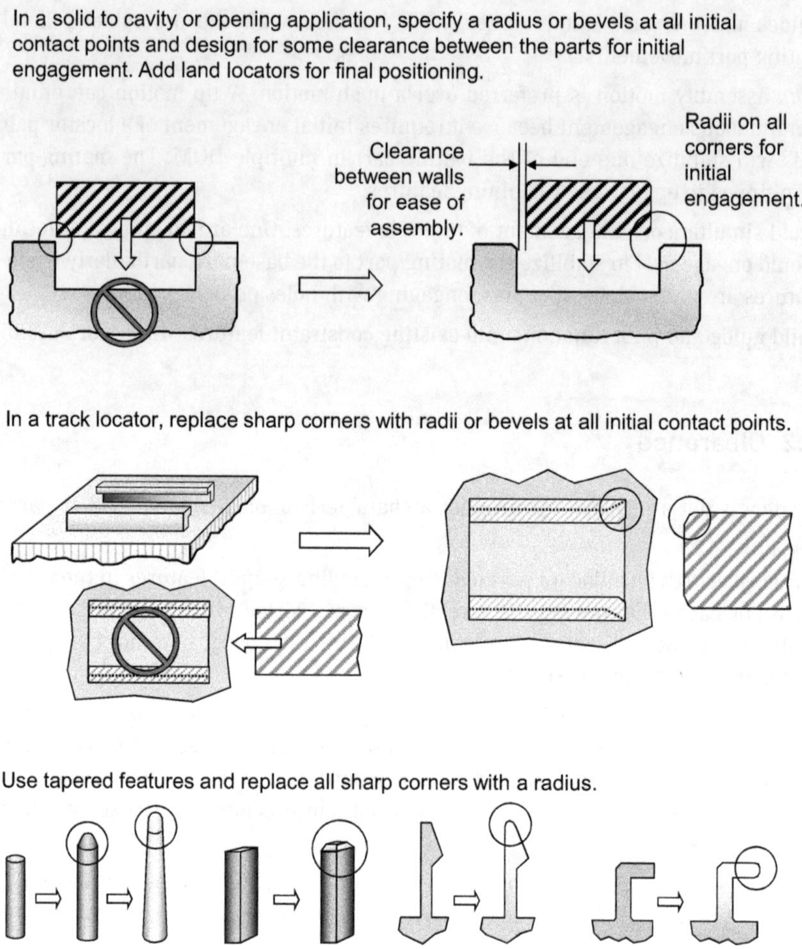

Figure 9.4 Clearance

9.1.3 Pilots

Pilots ensure proper orientation of a mating-part that could be assembled incorrectly. This can occur when symmetric parts with symmetric interface features can be assembled more than one way.

Pilots may be distinct and separate features but guides or locators can usually be used to perform the pilot function, avoiding the cost of adding a special pilot feature. A pilot is included in the switch assembly application and example that follows. Note that in the application itself, the lock features provide the pilot function.

9.1.4 Example: Switch Application

The application shown in Fig. 9.5 inspired the example that follows. The switch, a solid, is the mating-part and the base-part is a cavity: an opening within a recess in the panel that carries the switches. Very similar switch applications from other manufacturers, when compared to this application, did not have the assembly issues found in this particular design. (One of those in particular was extremely easy to assemble.)

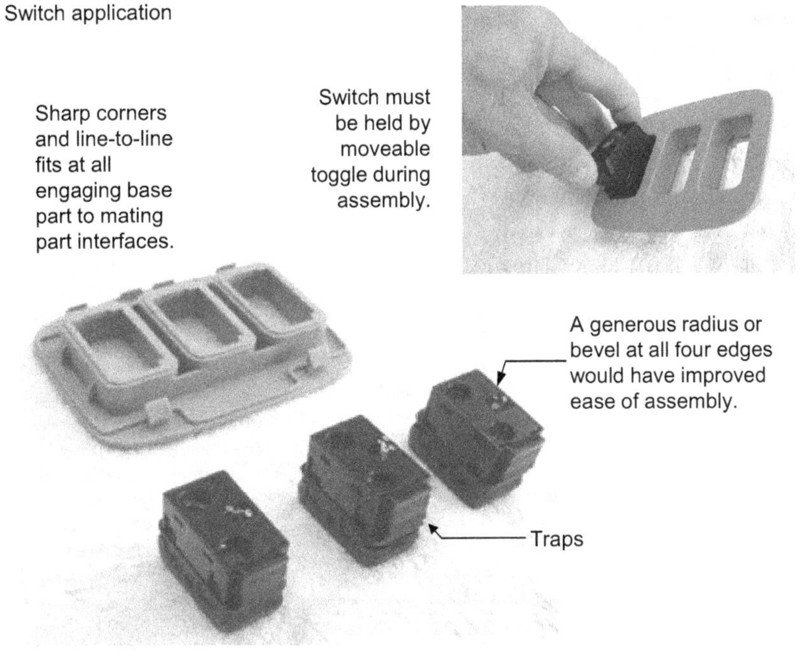

Figure 9.5 Toggle switch application

This application requires multiple small motor adjustments to align and engage the switch to the opening before final locking is achieved. Hand assembly of these three switches is unnecessarily time-consuming.

The discussion assumes manual assembly, however, even if these parts were to be assembled in an automatic (robotic) process, the close fits and sharp corners would require very precise (expensive) handling for initial locating/alignment prior to locking.

Application negatives:

- Sharp corners on both the mating and base-parts in addition to the corner/edge line-to-line fit around the opening's perimeter make these parts hard to assemble. This is the primary design flaw in the application.
- The need to hold the switch by the (moveable) toggle makes it unstable in the assembler's fingers and magnifies the alignment and assembly difficulties caused by the sharp corners and line-to-line fits.

- The engaging areas are buried in the cavity, which blocks visual alignment. This makes the need for guidance enhancements even more important.
- No guide or clearance enhancements are included in this design.

Simply providing a generous bevel or radius on the engaging edges of the switches would have greatly improved ease of assembly. Including such a radius or bevel in the original design would cost nothing or very little.

Application positives:

- The pilot function is accomplished through differences in lock feature width corresponding to differences in the width of the engaging areas on the base-part.
- The lock features are traps at the narrow ends of each switch. Traps are a good lock feature choice because this is a short-grip-length application.
- The switches are difficult to install but, once in place, they are locked firmly in place with no looseness or movement when the toggle is operated.

Figure 9.6 shows an example that is similar to the above toggle switch application. The example features the sharp corners and line-to-line fit of the switch application. For clarity, lock features are not shown so, unlike the application above, the pilot function is provided by a separate feature.

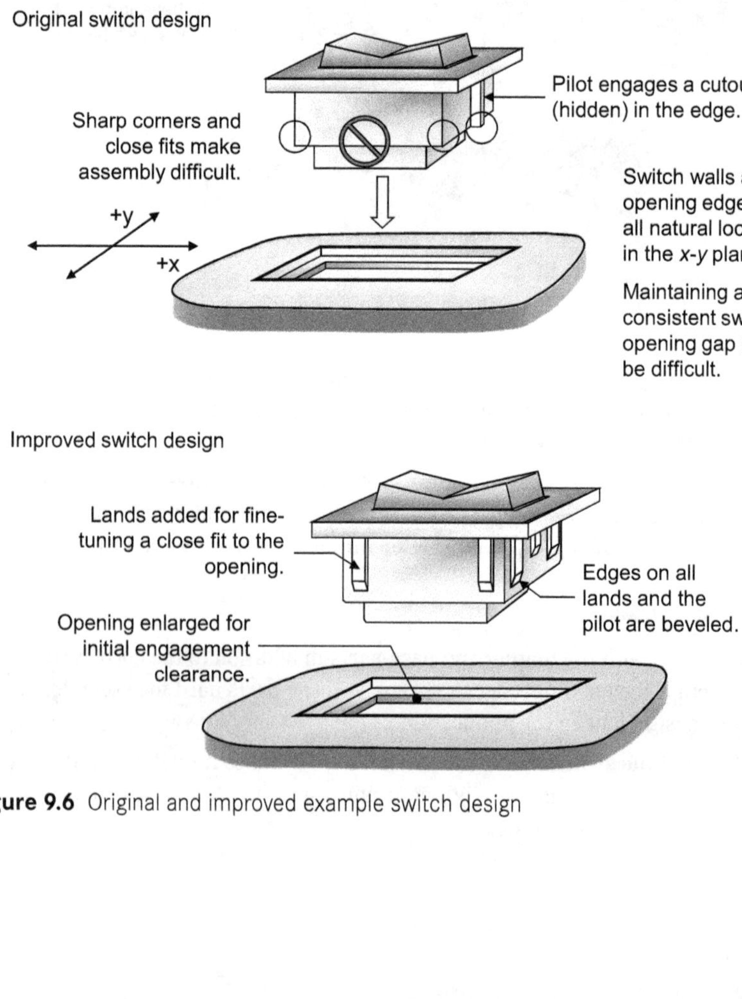

Figure 9.6 Original and improved example switch design

In the improved design:

- Relief is provided for easy initial engagement by making the opening larger relative to the switch body.
- Once initial engagement of the mating-part occurs, the required line-to-line fit is provided by lands acting as locators on each switch body surface.
- The lands also provide for easy fine-tuning of the switch position in the opening.
- Beveled faces on the lands and around the opening and leading corners of the walls provide additional clearance so that no additional small motor movements are required.
- A pilot feature, also beveled, ensures correct switch orientation in the opening. The pilot feature could possibly be integrated into one of the lands or into the lock features, which are not shown.

The difference in average assembly times between these two designs is *only seconds.* Nevertheless, the cost in assembly time can become significant. Table 9.1 shows the cost of four seconds of wasted time for different labor rates and part volumes. Note too that the product application above has three switches that must be installed, representing three units in the table below.

Additional quantifiable costs, like burden, could be added to these numbers, but there are other costs associated with difficult assembly that may be difficult or impossible to measure. *These costs have the potential to be much higher than the assembly time cost.* They include:

- Assembler frustration from struggling to assemble the parts. This frustration might, in turn, cause quality problems. Regardless of the quality aspects, assembler frustration itself is undesirable.
- In the long-term, the extra finger and wrist movements required during installation might result in the added cost of workers' compensation for cumulative trauma injuries.
- If the product is intended for automatic or robotic assembly, more costly equipment would likely be needed to get the precise control required for assembly. In this case, the developer would likely look for ways to reduce the precision required for assembly and employ the guide and clearance techniques discussed here. If one would try to design this product to be easy for a robot to assemble, why not design it to be just as easy for a human being?

If you want to learn how to design products for people to assemble, hang around with robots [1].

The point is that for just a little more effort in design and little or no increase in piece cost, an attachment that is easier and cheaper to assemble can be created.

Table 9.1 Cost of Four Seconds of Assembly Time per Unit

	Labor Rate $/hr				
Units per year	8	10	15	20	25
20,000	178	220	330	440	550
50,000	444	556	834	1112	1390
100,000	889	1111	1666	2222	2778
200,000	1760	2222	3333	4444	5555

There is another hidden cost associated with every poorly designed snap-fit application. This cost is largely unknowable but, in the author's opinion, it can be very significant, affecting a company's products for many years. This is the cost of having poor design practices in the organization's technical memory. In the author's opinion, based on observation and experience:

The most powerful and durable form of corporate memory is the company's own products.

Do you really want poor designs in the corporate memory?

Any poorly designed application may contain multiple violations of snap-fit design rules. If and when a poorly designed application goes into production, it may become a liability or it may not. In any case, it also becomes an example that can be studied by future developers who may then copy those poor design practices into their own designs. In this manner, bad snap-fit designs are propagated into future products. Likewise, good designs can also be copied into future products. This, in itself, is a compelling reason to ensure that good design practices are always followed.

9.1.5 Example: Reflector Application

This application first appeared in the discussion of strength and robustness in Chapter 2, Section 2.4. It is an excellent example of how failure to use enhancements can affect snap-fit performance. The simplicity of the parts and the application itself may have given the impression that the application was *easy* and not worthy of too much attention.

The reflector application is used here to support discussion about guide features and operator feedback as necessary enhancements. However, the reader will see there was much more wrong with this application than just missing enhancements.

This is a panel-opening application in which a small plastic reflector attaches to an opening, see Fig. 9.7. The mating-part, as originally designed, used four hooks as locking features. Panel-opening applications are a common snap-fit design situation, and cantilever hooks are often used as the locking feature in this kind of application.

The problem with this application was that the small reflector (about 30 by 80 mm) would sometimes fall out of the opening.

At first glance, the cause of the problem would appear to be the lock features because failed reflectors always had one or more broken or damaged hooks. A logical conclusion, based on feature-level thinking, would be that the hooks are weak. The solution, therefore, would be to design stronger hooks.

Diagnosing snap-fit problems is discussed in Chapter 14.

However, an *attachment level* diagnostic approach is to look at the application interface as a system before reaching any conclusions.

By thoroughly examining the application for systemic problems before simply redesigning the hooks, we find that the hook locks can be damaged during assembly and that several enhancement-related aspects of snap-fit design must be fixed before addressing the hooks themselves.

To properly diagnose any snap-fit problem, one must get parts and, ideally, observe the assembly operation itself. Without actual parts to work with, one cannot fully understand the problem. This application was no exception.

The original reflector design includes:

- Hook locks with short L/T ratios, steep insertion face angles, and excessive deflection for engagement.
- No locators and no guide enhancements.

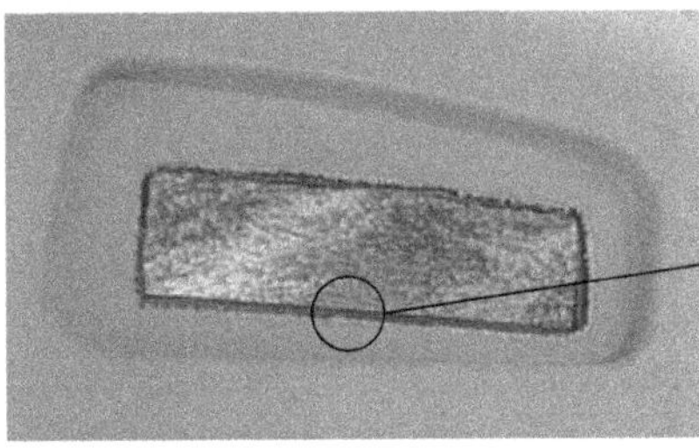

The base part is an opening with a recessed surface and edge in a larger part.

The opening can have a poorly defined edge due to imprecise trimming of cover material.

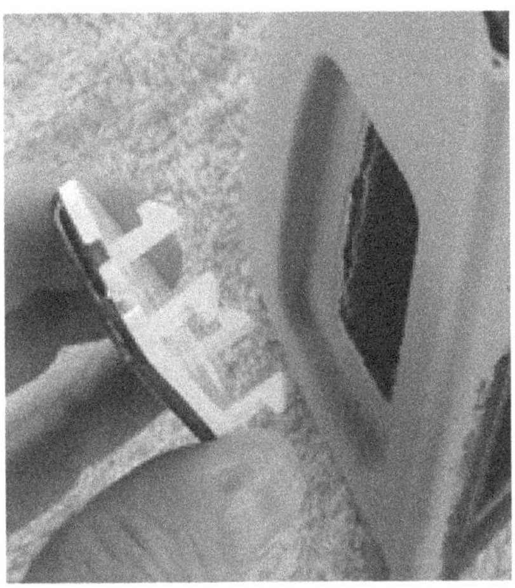

Due to the recess, the part surface interferes with the assembler's fingers before the hooks can begin to engage.

The assembler's hand interferes with their view of the area, making it a blind assembly.

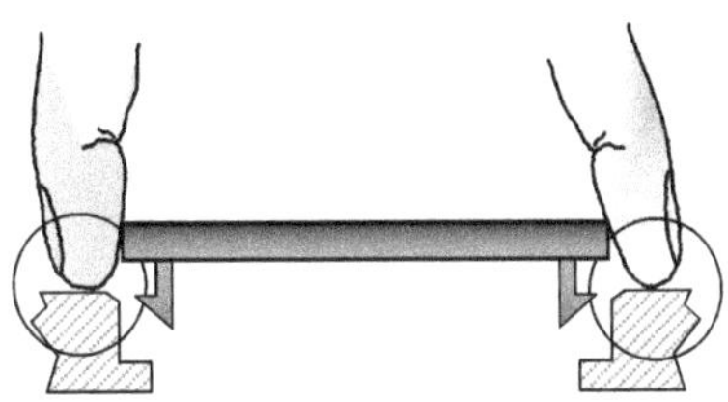

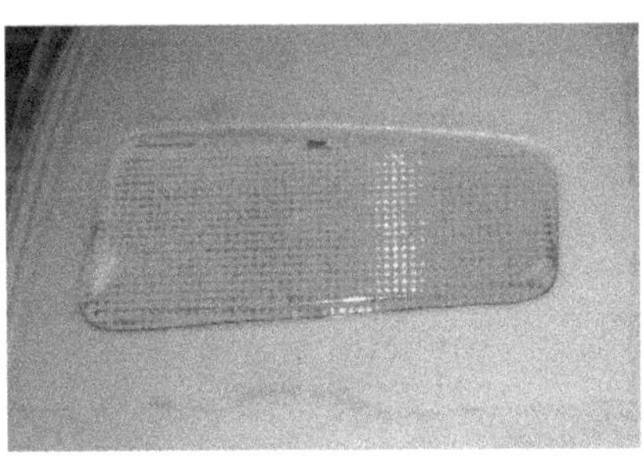

Once assembled, the reflector appears to be firmly locked in place.

Damage to one or more locks, however, can allow the reflector to eventually fall out.

The observed lock damage could lead to a conclusion that the locks require strengthening.

Figure 9.7 Reflector application

Figure 9.8 illustrates some options for the application. In one, locators (missing from the original design) can also be used as guide enhancements to align the reflector to the opening during assembly.

As designed, the application uses a push assembly motion. This requires more locks and more locators/guides in the interface than would be required with a tip assembly motion.

As a rule, when access and part shapes permit, as with this example, a tip assembly motion is always preferable to the push motion. Another possible design, using a tip motion and a lug, is also shown.

Yet another option (not shown) could be a tab extending in-plane from one end of the reflector and engaging a slot or cutout at the bottom of the recess wall. The tab option might be a simpler solution than the lug because a tab may be easier to mold and the mating locator in the recess wall would not incur the cost of die action.

In Figs. 9.8 and 9.9, the reflector is represented as symmetric across the *x*-axis so a pilot might be expected in the interface. In the actual application, the reflector is not symmetric and incorrect orientation during installation is not possible, therefore a pilot is not needed.

Adding the locator/guides improved the assembly process, but failures were still likely due to poor lock design. This does not mean the guides were a bad idea. Guides are *required* in this, and every, application. It just means there was more than one problem, including:

- Issues with lock performance in a short grip-length application. The cantilever hooks were a very poor choice for this application.
- Poor assembly feedback.

Adding guide features

The original design has no locators or guides except the natural locator pair formed by the reflector and panel surfaces.

Remember–locators are a required interface feature in any snap-fit. Locators used as guides can align the reflector with the opening and prevent hook damage during assembly. At a minimum, locators/guides are required at **A** and **B**. Depending on part size, locator/guides represented by the four **C** features may be necessary in addition to or in place of the **B** features.

Figure 9.3 shows an alternate solution using a pin/hole and a pin/slot locator pair in a similar application. If feasible, this would be a more efficient design for this application.

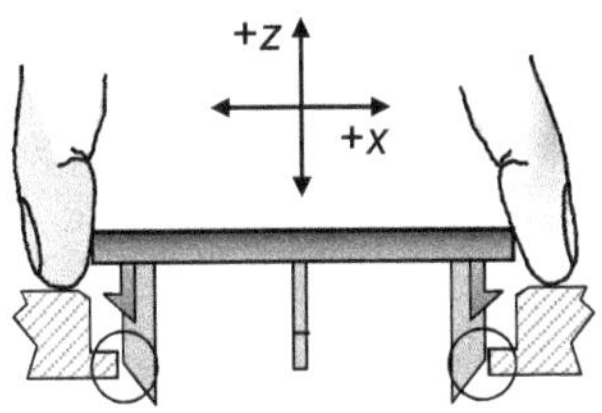

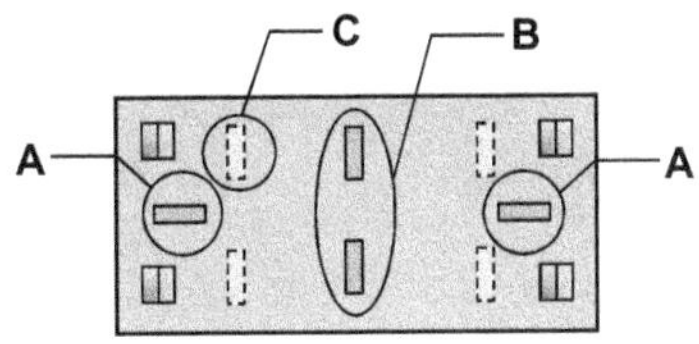

View of the reflector's interface surface looking in the +z direction

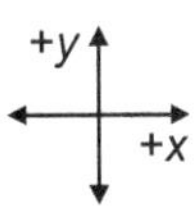

View of the opening looking in the −z direction

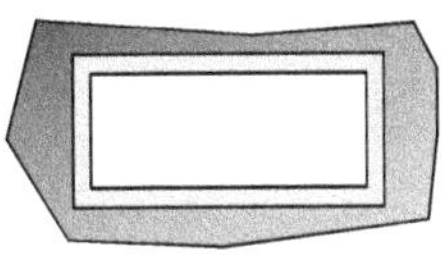

With a tip assembly motion:

- The original design uses a push assembly motion.
- Using a lug(s) and a tip motion eliminates the need for some guide features. This is a preferred design, however, lock feature redesign is still necessary.

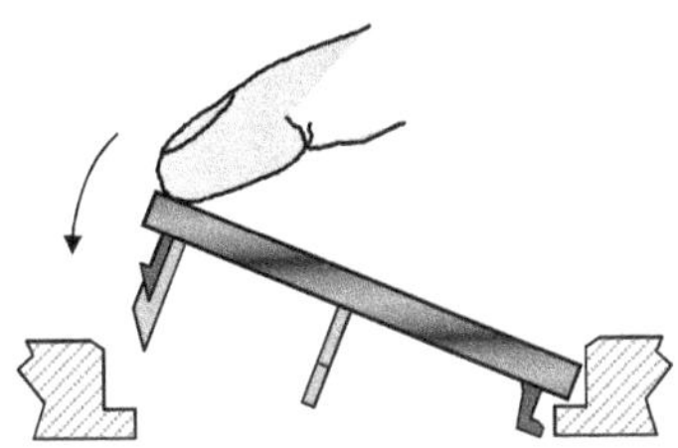

Figure 9.8 Improved panel-opening application

Figure 9.9 shows the lock design that solved both the lock and feedback issues [2].

The solution shown in Fig. 9.9 was a fix for an existing part problem and is not what would have been recommended for an original design, which would have included:

- A tip assembly motion would have been the starting point for the application concept.
- The tip motion would have allowed a more efficient arrangement of locator/guide features on the reflector.
- Those locator/guide features would have required some very simple changes to the edge locator around the opening.
- A lock feature(s) more appropriate for short grip-length application would have been used.

In the real-life situation, the solution in Fig. 9.9 was preferred because:

- Time was of the essence and the recommended trap lock style, being very robust, guaranteed success with very little testing required.
- Open space behind the reflector allowed room for the deep frames that carry the traps.
- It required no changes to the opening's edge so that no mold redesign for the base-part was needed.
- The trap locks could accommodate the short grip-length application and were robust enough to allow for the highly variable edge conditions at the opening.
- The traps provided excellent feedback of a good attachment.

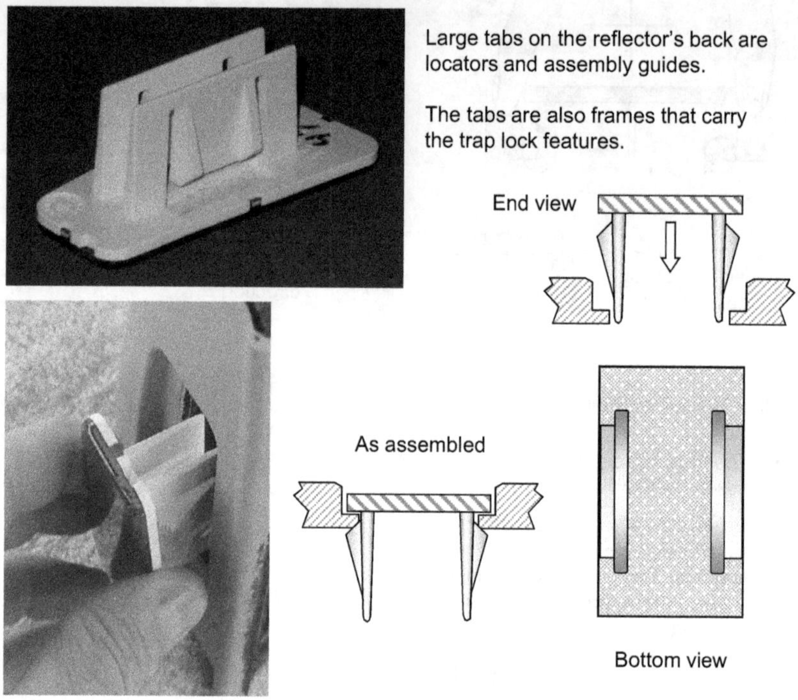

Figure 9.9 Final reflector redesign

9.1.6 Feedback

Required Enhancements:
Guides
Clearance
Feedback

When an assembler works with snap-fit applications, they have no calibrated tool or electronics providing feedback when a good assembly has been made. The snap-fit assembler has sensitive fingers, eyes and hearing all connected to a powerful processor–the human brain. The assembler relies on direct feedback to indicate assembly success. Designing the snap-fit to ensure consistent and positive feedback to the assembler helps ensure that properly assembled attachments occur every time.

In the reflector application discussed above

- The soft covering and short (low deflection) hooks prevented positive feeling of mating-part seating, and the assembler received no feedback of proper locking. The assembly signature looks something like those at the bottom of Fig. 9.10.
- High assembly force, due to poor lock style selection, interfered with assembler sensitivity to tactile feedback.
- A reflector with broken or damaged hooks could remain in place, appearing to be properly assembled, at least for a while.

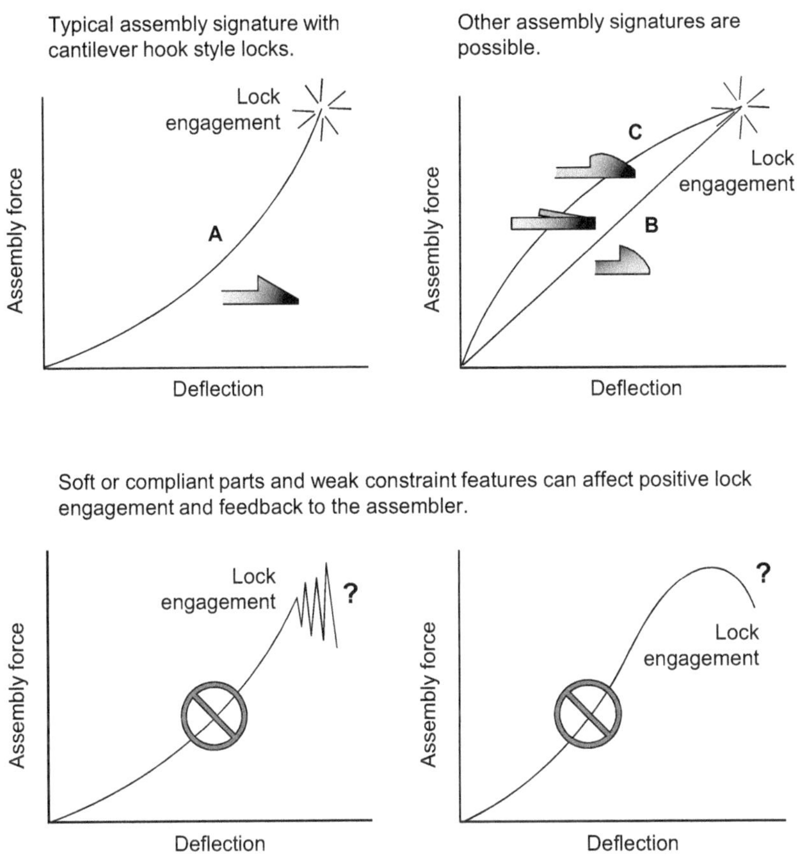

Figure 9.10 Force-deflection signatures and assembly feedback

The goal during snap-fit development is to ensure direct feedback to the assembler while eliminating or minimizing other factors that can interfere with the direct feedback. We can think of these interfering factors as noise in the system. Direct assembly feedback has three forms: *tactile, audible,* and *visual.*

Tactile feedback results from the sudden release of energy, usually the lock(s) snapping into place. It is enhanced by the shape of the assembly force-deflection signature and by the solid feeling created when locator pairs come together. Tactile feedback is generally preferred over the other forms of feedback because it is not subject to audio or visual interference.

Audible feedback is also the result of a sudden release of energy. Ambient noises and possible hearing limitations may reduce its effectiveness.

Visual feedback involves alignment of visible mating and base-part features. Ambient lighting conditions and line-of-sight interference may reduce its effectiveness. It may also require a subjective judgment on the part of an assembler or inspector.

Ideally, more than one source of feedback should be available to the assembler. The sudden release of energy that gives a tactile signal may also cause an audible signal. Position indicators may provide a visual indication to supplement an audible or tactile signal.

Tactile feedback can be understood if we think in terms of the assembly force-deflection signature introduced during the lock discussion in Chapter 6. The signature represents what the assembler feels as the mating-part is installed. Some common snap-fit assembly signatures are shown in Fig. 9.10 along with some of the lock insertion face contours that produce them.

The concave curve **(A)** in Fig. 9.10 is typical of many attachments. It has a geometrically increasing force as the insertion face to mating surface contact angle increases with (cantilever) hook deflection. The parts then make solid locator contact as the lock(s) engage. In many cases, this is acceptable and provides adequate feedback but in applications with high feedback interference, it may not be sufficient.

Improved feedback and assembly feel occur when the insertion face profile results in either a flat **(B)** or a signature with decreasing slope **(C)**. The maximum assembly force is generally lower for the same deflection, which means that lock deflection may be increased for a stronger feedback signal. (Remember that strain limits in the lock material must also be considered before increasing beam deflection.) A flat signature is produced when the instantaneous insertion face angle remains constant with respect to assembly deflection. A signature of decreasing slope is produced when the instantaneous insertion face angle decreases with respect to the assembly deflection and is inherent in the trap lock feature. A discussion of the insertion face profile can be found in Chapter 13, Section 13.6.1. It was also discussed in Chapter 6, Section 6.3.1.1.

The signatures shown at the bottom of Fig. 9.10 represent applications where soft materials in the interface, structurally weak components, or compliant locators may require the assembler to hunt for engagement because the locator contact and lock engagement points are not well defined.

Similar feedback issues exist in moveable applications where the customer operates the snap-fit application. These are discussed in the next section.

Most design characteristics that support good assembler feedback are related to tactile feedback. Using assembly enhancements to make the assembly process as easy as possible will improve the quality of that feedback

Ergonomic factors are one form of background noise that can interfere with assembly feedback. Numerous sources of ergonomic information and data are available to the developer.

Some general rules for good ergonomic design include:

- Assembly forces must be within an acceptable range. Avoid high forces on fingers, thumbs, or hands to install a part. High cumulative assembly forces (as multiple locks are engaged) can interfere with feedback as the assembler struggles to overcome them. Lock designs to reduce assembly forces are discussed in Chapters 12 and 13.
- Avoid forces of long duration.
- Avoid awkward body positions and extreme reaching or twisting motions.
- Design for top down, forward, and natural motions carried out from a comfortable body position. Avoid reaches over the head.

Anything that creates a difficult assembly operation will interfere with the assembler's ability to make a good attachment in the first place and then to recognize a poor attachment if it occurs.

Other factors that support improved assembly feedback include:

- Make the assembly area visible.
- Provide solid pressure points and a rigid path from those points to the lock features. Weak or flexible parts may require local strengthening.
- Positive and solid contact between strong locator features will send a clear, unmistakable signal that parts are positioned properly against each other.
- A rapid lock return can give a good audible and tactile signal that the lock is engaged. A lock with high deflection is generally more effective than one with low deflection. High deflection does not necessarily mean high assembly force, however. The audible feedback signal is generated by lock speed as it snaps into place, not by lock force.
- A strong over-center action as a lock engages will give a feeling that the part is being pulled into position.
- Consistency in part assembly performance. Consistency allows the assembler to acquire a feeling for a good attachment. Once this feeling exists, anything out of the ordinary will signal the assembler to check for problems. Consistency of performance is a function of the design's robustness to manufacturing and material variables.
- Provide highly visible features that are clearly aligned when the assembly is successful.
- Sometimes a failed assembly can look good and parts can stay together for a while, escaping detection. Design for go/no-go latching so a mating-part that is not properly attached will easily fall out of position to create an obvious failure that can be fixed immediately.

Poor feedback is caused by poor execution of the characteristics that provide good feedback and by failure to eliminate background noise that interferes with feedback.

9.2 Activation Enhancements

Activation enhancements make it easier to use a snap-fit.

Activation enhancements are mechanical and informational features that enable attachment disassembly or usage.

Most of the time, activating a snap-fit means releasing it, either to separate parts or to operate a movable snap-fit. In the case of a movable snap-fit, activation can also mean re-locking the attachment after use. Enhancements for activation are *visuals, assists,* and *user-feel.*

Visuals provide information about attachment operation or disassembly. Assists provide a means for manual deflection of nonreleasing locks. User-feel refers to attributes and features that ensure a high quality feel in a moveable snap-fit.

9.2.1 Visuals

Most of the time, snap-fit operation is obvious. When operation is not obvious, visuals provide a message to the user about how to release the lock and avoid damage due to misuse.

Recall that part separation is accomplished by reversing one of five simple assembly motions, (push, slide, tip, twist, or pivot). Thus, a simple visual indication of the mating-part's separation direction and motion may be sufficient when the application uses a releasing lock. When a lock feature is nonreleasing, both an indication of the manual deflection to release the lock and an indication of mating-part separation motion may be necessary.

Examples of common visual enhancements are the arrows on battery covers of most television and other electronic device remote controls. Many children's toys have visuals indicating how to open, move, or remove parts. Visuals may also be instructional text located close to the attachment's activation point.

Visuals should be large so they are easy to find and understand when they are in an area of the part where appearance is less important. A visual on an important appearance surface, however, cannot be obtrusive or unattractive, but customers and service personnel must be able to find it and understand its meaning; see, for example, the courtesy lamp application in Fig. 9.11. As snap-fits become increasingly common in products, both consumers and product service personnel must learn to expect and look for visuals.

Visuals can also point to detailed text instructions on nearby labels or in product owner and service manuals.

The primary purpose of visuals is to avoid part and feature breakage during the useful life of the product. However, material recycling and reuse are also becoming an important product concern. Parts not intended for disassembly during the life of the product must still be efficiently disassembled for recycling or reuse. Visuals can indicate a breaking point or a critical pressure point for efficient part separation. The common recycle symbols that use a number indicating a polymer family for separation and reprocessing are visuals that support recycling.

When designing snap-fit visuals, keep the customer in mind. Locking and releasing methods should be as obvious as possible and any supporting visuals intuitive and very visible. Remember that the typical customer will be unfamiliar with the parts and the attachment method and even experienced service technicians will need to become familiar with new designs.

While standards exist for many symbols, to the author's knowledge, no set of standard international snap-fit symbols has been developed. In addition to very recognizable visuals like arrows, other visuals are also needed for general use as well as in areas of limited space or in appearance areas where large obtrusive marks would be unacceptable. Industry leaders in plastic products should take steps to establish an international set of standard symbols.

Some possible symbols are shown in Fig. 9.11. These shapes are proposed to describe snap-fit activation (release or operation) when space or appearance considerations prevent more detailed information [3]. Standards for symbol geometry exist and should be applied to determine actual symbol dimensions. The SAE Recommended Practice J1344 describes a system for marking plastic parts with material identification symbols [4]. The SAE practice is based on the standard symbols for plastics [ISO 1043] published by the International Organization for Standardization. The SAE practice includes standards for text but all symbols, not just text, require dimensional standards.

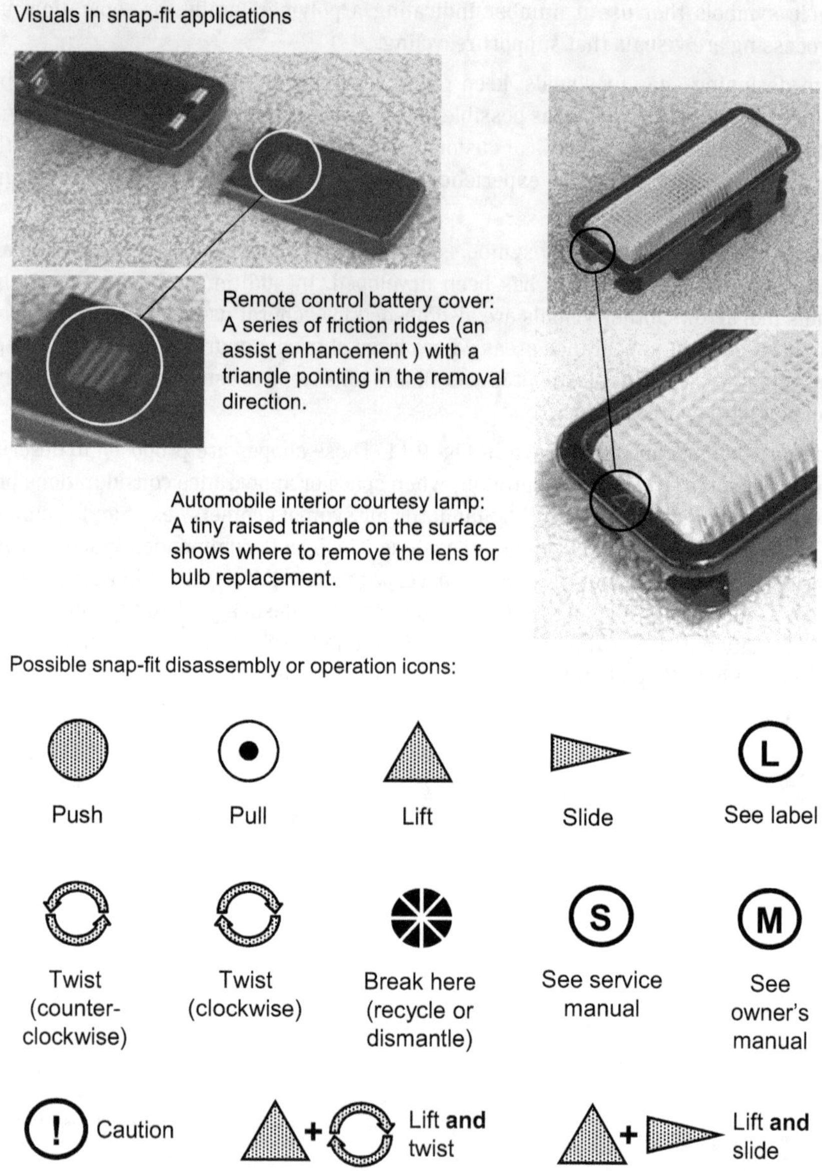

Figure 9.11 Application examples of visual enhancements and suggestions for standard snap-fit symbols

9.2.2 Assists

Assists help the user engage or release a locking feature or operate a movable snap-fit. When lock operation is hidden or not obvious, to prevent damage, an assist should be accompanied by a visual to indicate how the assist is to be used.

Examples are shown in Fig. 9.12. Finger activation of the assist is preferred, but sometimes tool activation is necessary. Any form of manual lock deflection requires caution because over-deflection and damage to the lock feature can occur. This is particularly true if a tool is used and/or the lock feature is in an area that is difficult to see or reach. Enhancements called *guards* can help prevent lock damage due to over deflection and are sometimes used with assists. Guards are described in an upcoming section.

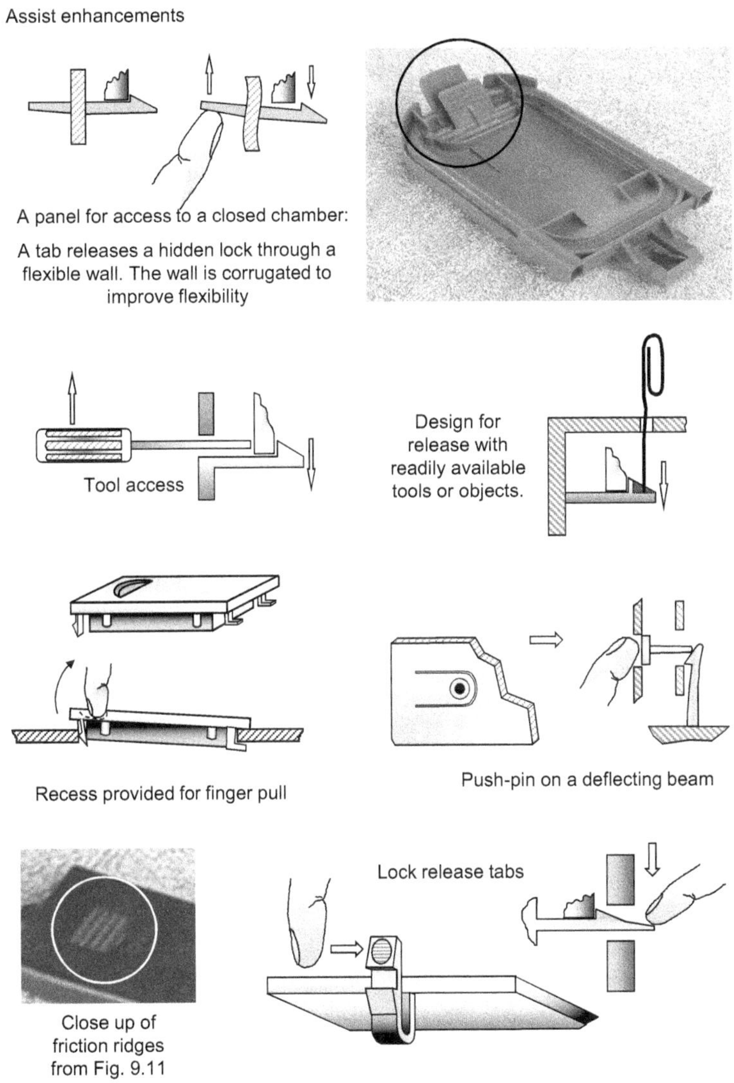

Figure 9.12 Assists

A battery cover for a remote control was shown in Fig. 9.11. This cover incorporates both a triangle shape as a *visual* pointer and friction-ridges as an *assist*. Remote control devices like this are readily available and should be studied as a snap-fit learning exercise, not just for enhancements but for execution of the entire interface.

Assists can also be used for part assembly, for activating a movable snap-fit, or for assisting part movement to unlatch a releasing lock. An access opening can even be skinned over with an indication, a visual, on a nearby surface to drill or punch through at that point to reach the lock. A surface or exterior operated assist feature can be used to activate a lock buried in the part's interior. Obviously, these features can add expense and complexity to the part and mold. The author is not advocating making snap-fit application more complex than necessary, but options like these are available if needed.

Rules for assists:

- If necessary, use visuals to show how to operate the assist.
- If tools are required, design the assist so readily available tools can be used. Screwdrivers, thin blades (as on a knife or paint scraper), and steel rods (nails, paper clips, etc.) are common tools and will generally meet disassembly needs. A visual or the access hole shape itself and size can sometimes indicate the kind of tool required.
- Use guards to protect the lock feature against over-deflection during disassembly, particularly when the feature is hidden. Any kind of manual deflection, with tools or fingers, can easily result in over-deflection and feature damage, so a guard should always be considered.

9.2.3 User-Feel

User-feel is related to customer impressions or perceptions of product quality and includes concepts similar to assembly feedback. The tactile and audible signals that make assembly easier for the assembler can also improve the customer's perception of quality in a moveable application. (Recall that *moveable* is a function element.) Movement can be either free or controlled. User-feel is more important in applications used frequently and involving higher forces. For example, user-feel in a battery access panel on a remote control is much less important than it is for a frequently used appliance door.

The concepts of assembly and separation force-deflection signatures, see Fig. 9.10, also apply to user-feel. A solid and firm feeling of engagement accompanied by a smooth, over-center feel for both assembly and disassembly will give an impression of quality.

A good application example is a center armrest cover in an automobile that opens to a storage compartment. Sometimes the latching mechanism is a manual release snap-fit. A console door gets a lot of use; every time it is opened and closed, it can be a reminder, for better or worse, of product *quality*. It is a simple matter to design the lock feature to provide good user-feel and an impression of high quality to the customer.

If the application uses a releasing lock, the developer must also pay attention to user-feel during separation. Design both the insertion and retention faces to give high-quality tactile feedback to the user. Design moveable snap-fits to close with a solid and reassuring lower frequency sound like a *thud* or *thunk* rather than a high-frequency and cheap sounding *click*. Obviously this is somewhat subjective and hard to quantify but, for the consumer, it is easy to recognize.

As a side benefit to improved assembly feel, when a contoured insertion face profile is used, the assembly force is reduced. This means stresses on the lock pair are reduced. In a frequently used application, this may help prevent long-term lock feature failure.

An application most readers are familiar with is the common plastic buckle, see Fig. 9.13. Most have excellent assembly and disassembly behavior and they are extremely strong and durable, as they must be. Some, like the one shown here, use cantilever hooks with a retention face greater than 90°, others use trap locks. It is worthwhile to study these buckles because most of them represent excellent snap-fit design, and they are very user-friendly.

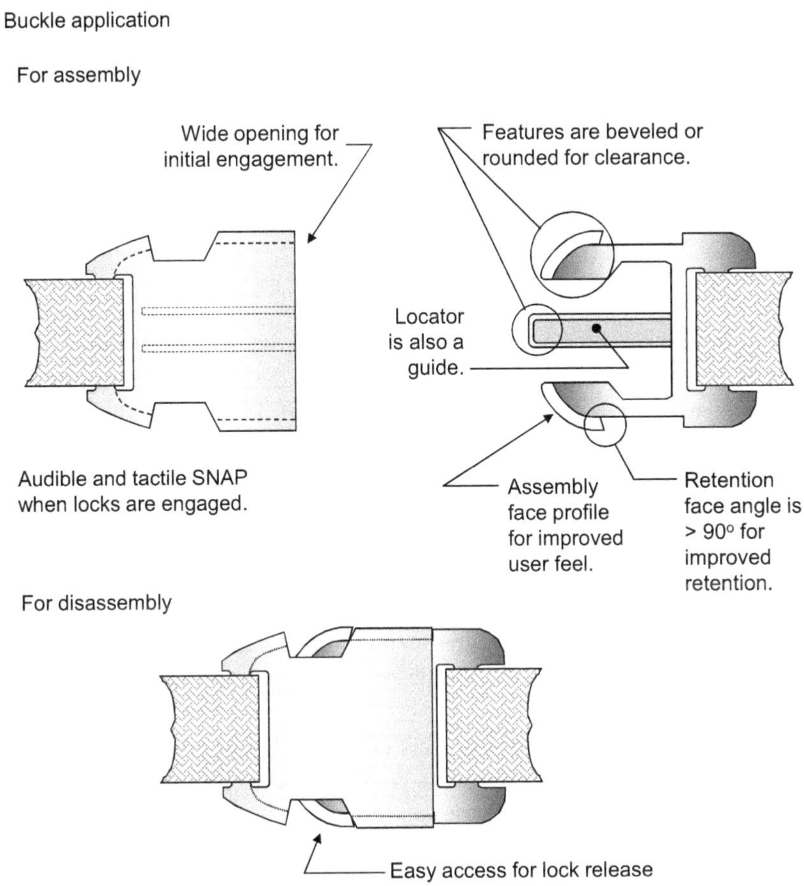

Figure 9.13 The common plastic buckle

9.3 Performance Enhancements

Performance enhancements ensure the snap-fit attachment works as expected. While we can perform feature analysis and other evaluations to ensure strength and reliability, sometimes product design parameters, such as material requirements or wall thickness, can severely limit a locking feature's retention strength. No matter what we do to the lock itself, we simply cannot make it strong enough. Other times we would like to prevent damage to a lock feature or provide insurance that a costly part is not ruined if a lock feature breaks.

Performance enhancements include:

- **Guards** to protect sensitive lock features from damage.
- **Retainers** to provide local strength and improve lock performance.
- **Compliance** provided by attributes and features that take up tolerance and help maintain a close fit between mating-parts without violating constraint requirements.
- **Back-up locks** that provide a second means of attachment if the lock feature should fail to work or suffer damage.

9.3.1 Guards

For efficient design, use guides or locators as guards.

Guards protect other features and were mentioned in the above discussion of assist enhancements, see Fig. 9.14.

Because locking features are flexible and usually weak in bending, guards are used when it is necessary to protect the lock, Fig. 9.14. Conditions that create the need for guards should generally be avoided, but design constraints may force those conditions. The principles behind the use of guards, apply to any locking feature or other sensitive part feature.

A number of situations may call for guard features. Snap-fit features (or other part features) may be in exposed locations and susceptible to damage when parts are stacked for shipping and handled before or after shipping. They may be subjected to short or long term deflections. Guards can provide protection and prevent permanent set or breakage.

When nonreleasing locks must be manually deflected for part separation, over-deflection and damage is possible. When the lock is hidden and a tool must be used instead of a finger, the chance of damage increases. Because plastic performance is time dependent, a lock that survives rapid short-term deflection during assembly may not survive a similar deflection of longer duration as might occur during slower manual disassembly. Guards can prevent permanent damage by limiting lock deflection to just what is needed for release.

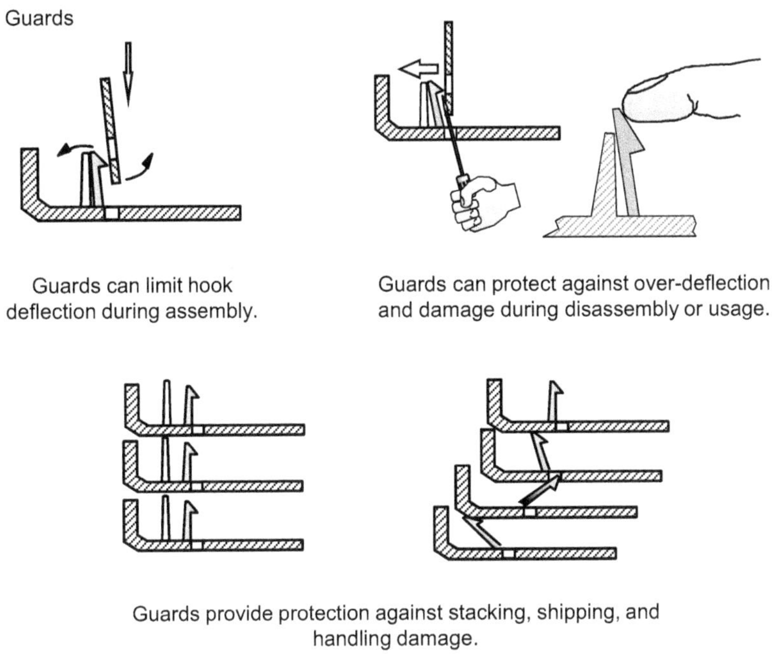

Figure 9.14 Guards protect relatively weak features from damage

Sometimes during assembly, a hook can be deflected beyond a safe strain level. In a manner similar to that for preventing over-deflection during disassembly, a guard can limit hook assembly deflection before it becomes excessive. The increased hook stiffness transfers deflection to the other part. This will, however, come at the cost of higher assembly force because a second part/feature is now being deflected.

9.3.2 Retainers

Preferred practice is to design the attachment's separation performance directly into the lock feature(s). However, design constraints, material requirements, or compliance in the parts themselves may result in an inherently weak lock and retention capability cannot be guaranteed through lock selection or design.

Retainers can improve a lock's retention strength by increasing its bending spring rate, or by providing positive interference against deflection, see Fig. 9.15. It is appropriate to use retainers for improving both releasing and nonreleasing locks. Even nonreleasing locks can release under very high load conditions due to gross distortion of the part or the lock itself. A retainer can be positioned to prevent that gross distortion. Retainers behave somewhat like guard enhancements, but they improve retention strength.

Lock features mounted on weak, flexible walls will have limited strength. Retainer enhancements can add local strength or stiffness at the lock pair.

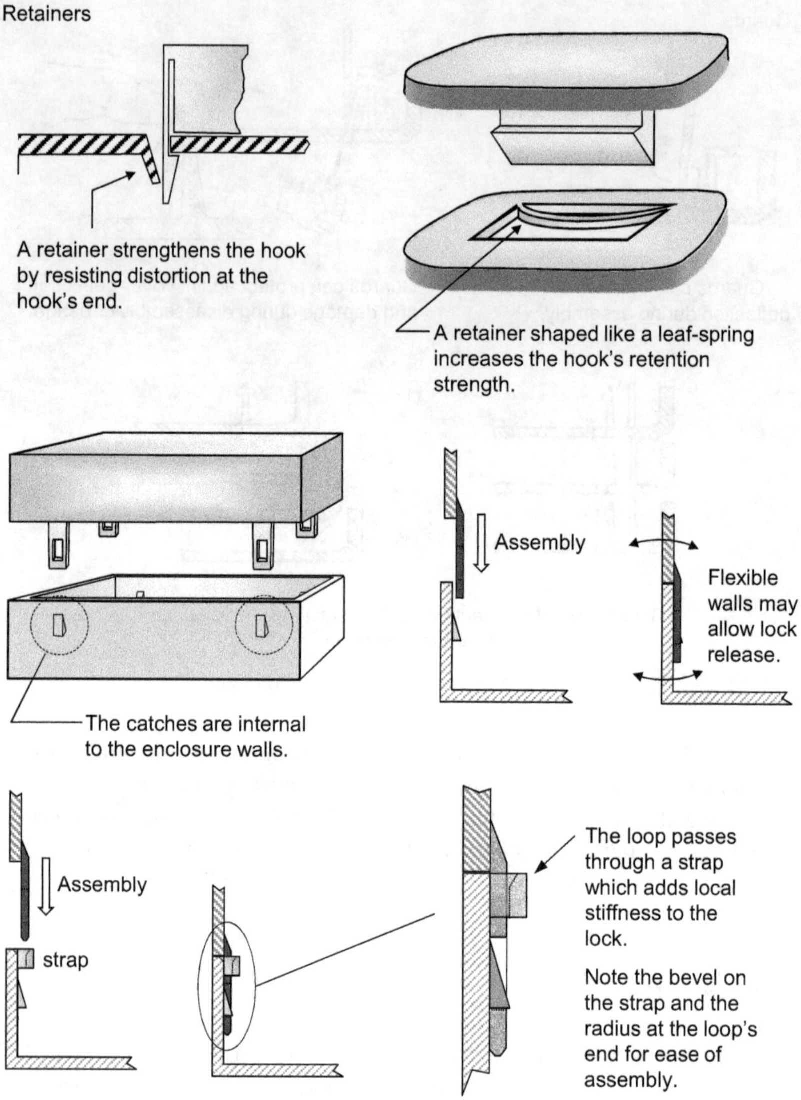

Figure 9.15 Retainers provide local strengthening or stiffness at the lock pair

9.3.3 Compliance

Compliance is the attachment's ability to accommodate dimensional variation so parts are easy to assemble while maintaining a close fit with no looseness. Compliance was also discussed in Chapter 8 as an enabler for proper constraint.

Robustness to dimensional variation is designed into the attachment system through proper constraint and constraint feature selection. Sometimes, however, this is not enough to ensure a close fit between parts. Compliance within a constraint pair is then used to supplement the systems performance.

Because they do not use clamp load (like threaded fasteners), a major design requirement for a snap-fit is that parts fit together tightly, with a line-to-line fit, for functional integrity and appearance. Another benefit when a line-to-line fit is maintained is that noise (generally squeaks and rattles) resulting from transient loads is eliminated. One way to get a line-to-line fit is by specifying very close tolerances on parts, but this can be expensive.

Noise in an interface results from energy inputs that cause parts to separate and snap back, resonate, or rub together. Preventing noise is a matter of holding parts tightly so separation and relative motion cannot occur under high transient loads or high frequency vibration conditions. The kinds of load cycles that cause noise in plastic are generally of very short duration whereas published material strength data is usually based on loads applied over a longer period. Because of the time-dependent behavior of polymers, the strains in the plastic parts under these loading conditions can sometimes exceed, without damage, the tested strain limits of the material for loads of longer duration. In any case, the compliance features must be designed for long-term effectiveness and resistance to cumulative damage over time. Long-term plastic creep and degradation of the material's properties must also be considered.

When considering how to effectively add compliance to the application, know the significant tolerances and stack-ups in the interface. Whenever possible, design so that tolerances can be taken up in a noncritical area and direction. Know where potential looseness or interference may occur due to differences in the mating materials' coefficients of thermal expansion. Design so looseness will not cause noise and so interference will not cause yield to the extent that looseness results.

See the discussion of critical directions and tolerances in Chapter 8. Deciding where to add compliance depends on alignment requirements and interface forces. In general, compliance is added *opposite* any fine-tuning enhancements and load-carrying constraint pairs.

Two ways to add compliance in the interface are *local yield* and *elasticity*. A third way involves adding additional pieces called *isolators* to the system.

9.3.3.1 Local Yield

Local yield involves using features like darts within a locator pair to create low levels of interference (through compressive stress). While creep to a lower stress may occur over time because of the compressive stress, as long as no significant additional loads or deflections are applied to cause further yield, a line-to-line fit will be maintained.

Designing for local yield may conflict with the need for good assembly feedback, and caution is required so feedback quality is not compromised. Local yield means that resistance to assembly forces will occur over a longer period and involve more than just the lock features.

Compliance through local yield should, ideally, occur *within* a constraint pair. Compliance *between* constraint pairs at opposing sites must be carefully managed because it can cause some level of over-constraint. While plastic yield is possible in tension, bending or compression, the only mode recommended for yield compliance is compression.

Because local yield requires strength in the features to force the compressive stress, it is rarely employed with locking features. Most of the time, local yield compliance will be found on locator features. Methods of obtaining interference through local yield include *darts, crush ribs*, and *tapered* features, see Fig. 9.16.

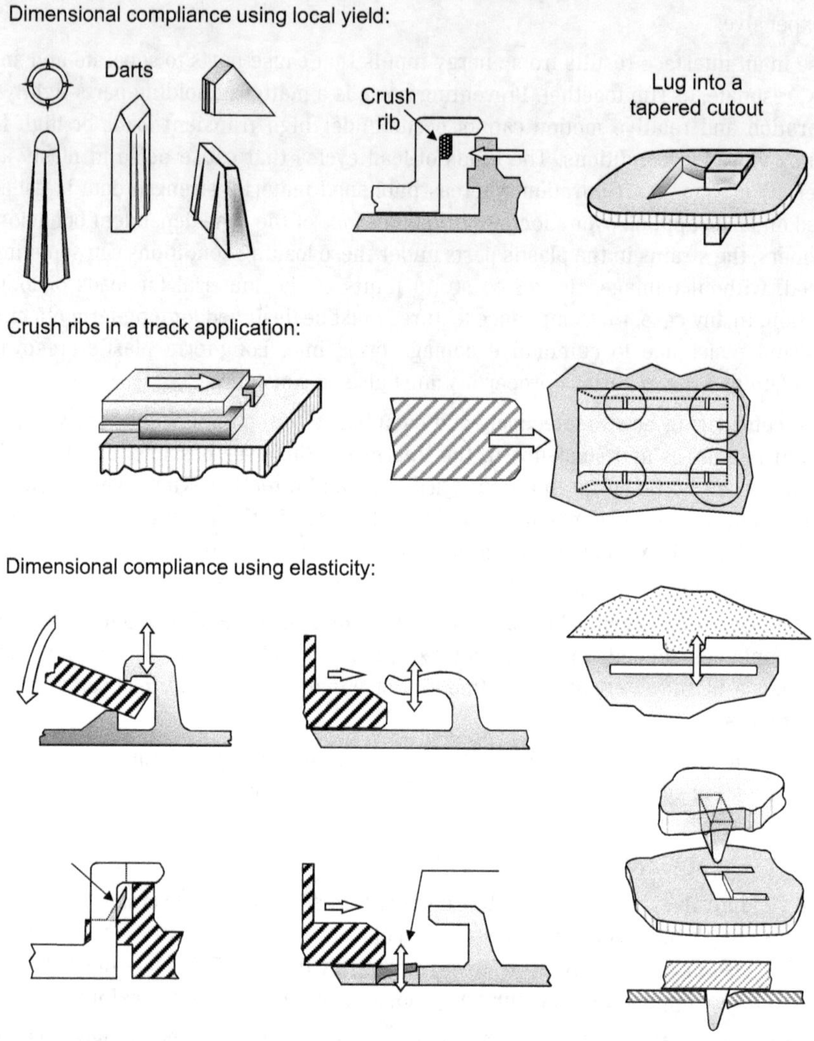

Figure 9.16 Compliance

Darts on pins, lugs, and wedges will embed into the edges of other locator features as the parts are pushed together. Darts, to be effective, should be placed on the harder of the two plastics in the interface. A dart-like rib on the softer material may become a crush-rib, which is not necessarily a bad thing. When the plastics are similar in hardness, a shallower included angle on the dart can ensure its effectiveness.

Crush ribs are exactly what the name implies. They are relatively thin ribs that are literally crushed or bent out of the way by the mating feature. The distorted rib fills the gap between the parts. In the track application, strategically placed crush ribs ensure that a bayonet type mount will remain tight in the mating track [5].

Another way to get local yield is by using lugs that engage tapered cutouts, tapered pins with holes, and tapered wedges with slots.

9.3.3.2 Elasticity

The inherent elasticity of plastic can also be used to establish and maintain a line-to-line fit between parts. If parts are structurally rigid, special molded-in features acting as springs can provide elasticity, see Fig. 9.16.

The slight distortion in some plastic parts, particularly panels, as they come out of the mold may provide sufficient elasticity for a close fit after one part is nested and locked in place to the other part.

Unlike local yield, which is best limited to locators, elasticity can be effectively used with both locator and lock features. Remember, however, that (most) lock features are weak. Use caution when taking up compliance in a lock pair and avoid lock bending as a means of getting compliance.

9.3.3.3 Isolators

As a last resort, because it will add cost and extra parts to the application, isolating materials can be added to the attachment to force a line-to-line fit. These can take many forms, including adhesive-backed foam or felt products, soft rubber or felt washers, and O-rings. Off-the-shelf O-rings may be slipped over a protrusion feature for a quick fix to a looseness problem. (Ensure there is no oil or other chemical on the isolator that could react with and degrade the plastic material and that the plastic is not reactive to the isolator material itself.) Also, ensure that any added materials do not create excessive stress or strain in lock or locator features.

9.3.4 Back-Up Features

Sometimes it may be wise to include contingency features in the design as a fastening alternative in the event the intended integral lock feature cannot provide reliable locking. Usually these are simply provisions in the mating and base-parts for threaded fasteners, push-in fasteners or metal clips should the integral lock feature fail.

The use of a back-up feature can be a strategic choice and does not necessarily indicate lack of confidence in the snap-fit.

If a snap-fit attachment is expected to fail then, of course, it should not be specified in the first place. But there can be gray areas:

- The cost saving potential of snap-fits often indicates their suitability as the mainstream attachment design for an application. However, technical and/or business issues may prevent their serious consideration. Back-up locks can help overcome some of those obstacles. When appropriate, a back-up lock can be made a part of the business case when evaluating an application's attachment alternatives. A significant factor in back-up lock decisions is the piece-cost of the part in question. A back-up lock may not be cost effective on a small inexpensive part but could be very desirable on a large or complex and expensive part.

- In some applications, a conservative approach to the snap-fit is desirable because the snap-fit may represent a technical reach. The potential benefits may be substantial but the risk of committing to a snap-fit may preclude its consideration unless a back-up fastening method can be designed into the interface at the same time. Should the snap-fit prove unreliable, the back-up lock allows the development program to continue with a reliable attachment for that application. Once the design is proven in testing and production, the back-up lock can be eliminated if it adds cost to the mold, or it can simply be ignored.
- While the design itself may not be a technical reach, incomplete data about the applied loads, material properties, or other application requirements may add uncertainty to the design. A back-up lock can allow the snap-fit design to proceed with the confidence that a reliable attachment is possible if the snap-fit does not work.
- The design may be such that the locking features of the snap-fit are susceptible to bending or breakage during shipping, handling, assembly, or disassembly. If the features cannot be protected by design (guards) and damage that would render the lock unworkable is possible, a back-up feature ensures the entire part will not be lost because of damage to one feature.
- If parts are intended for a new application and are also used on existing applications without snap-fits, allowing for both methods of attachment can accommodate both applications without creating a second set of parts.

Any fastening method may be a candidate as a back-up to a snap-fit and the design criteria should be appropriate to the technology. The same reliability considerations must be applied to the back-up lock as to the original snap-fit.

Back-up features need not be complex. Providing several clearance holes in a part and pilot holes, bosses, or clearance holes in the second part may be sufficient. Of course, if the back-up feature may become the mainstream design for production then all assembly and processing considerations must be included in the design. If necessary, clearance holes for threaded fasteners can be skinned over and punched, molded or drilled out if needed. Walls or ribs can be added on both parts in locations to accept and engage spring steel clips as back-up fasteners.

When a back-up feature is specified because of possible damage in disassembly for service or as a second attachment method on a service part, production assembly issues no longer apply. Give consideration instead to the tools and fastening methods required for service by the customer or service technician. Do not design a back-up feature that requires special fasteners or special tools.

Rules for back-up features include:

- Design to use fasteners like those already present in the product.
- Design for common or standard fasteners that repair facilities are likely to have.
- In consumer serviceable products, if high fastener strength is not an issue, design for fasteners that are available at hardware stores.
- Provide adaptable interfaces that permit several sizes, styles, or lengths of threaded fastener.

9.4 Manufacturing Enhancements

Manufacturing enhancements are design practices that support part and mold development as well as long-term manufacturing needs and part consistency. Many are documented in standard design and manufacturing practices for injection-molded parts and are already recognized as important factors in plastic part design. They fit neatly into the ALC as enhancements.

These enhancements make the part easier to manufacture and provide benefits in:

- Cost reduction
- Shape consistency
- Appearance
- Mold development
- Reliability
- Reduced internal stresses
- Process cycle time
- Performance consistency
- Fine-tuning for development
- Adjustment for variation and mold wear

This section is not a comprehensive guide to mold design and it will not make the reader an expert in the field. Because the part developer is most familiar with the application's requirements and is in the best position to ensure they are properly considered, a basic awareness of some processing concepts and practices is essential. The intention is to capture this aspect of snap-fit design as an enhancement and present a few of the more basic concepts that relate directly to snap-fits. The reader will learn enough to recognize design issues and then seek assistance from experts.

Remember that snap-fit features are subject to the same rules of good mold design as are the other features in an injection-molded part. For example, snap-fit features that protrude from a wall or surface should be designed according to the injection molding guidelines for protrusions. Be aware that the nature of some snap-fit features may require violating some guidelines. This is particularly true when features are tiny and/or close together. In these cases, discuss the requirements with the mold developer and the manufacturer.

Manufacturing enhancements fall into two groups. Those that improve part production are called *process-friendly* and are related to mold flow, mold and part cooling, and cycle times. Those that enable relatively easy dimensional changes to the mold for dimensional adjustments to parts are *fine-tuning* enhancements.

9.4.1 Process-Friendly Design

Required Enhancements:
Guides
Clearance
Feedback
Process-Friendly Design

Process-friendly design is simply following recommended and preferred plastic part design practices. Process-friendly parts are more robust to the molding process and are likely to be less expensive and more consistent in performance.

Part designs that violate recommended practices are likely to require special care during processing. For example, tiny features and very thin walls violate some of the general guidelines regarding section thickness. These features may not be as robust or process-friendly as larger features but they can be molded when processing accommodations are made and process variation is carefully controlled.

The information in this section was drawn from a number of publications. It represents *general design knowledge for a wide range of polymers* and can be found in multiple documents. Rather than cite numerous publications for each item presented, all the source publications are listed at the end of this chapter.

Process rules and guidelines can also change as materials and processing technology evolves and references can become outdated. The process-friendly guidelines given here are useful as a starting point but part developers must ensure their designs reflect current processing technology and best practices for their specific part material.

The single most important rule is to keep a design as simple as possible. Simple feature designs mean less costly molds and greater part consistency. Access for molding undercuts is always an issue in part design and snap-fits are no exception. Parts and features that can be produced without the added complexity and cost of slides and lifters (die-action) are always preferred.

Some general guidelines for process-friendly design are shown in Figs. 9.17 and 9.18.

In most of this book's illustrations, radii at all feature corners are not shown. However, the reader must know that a basic rule of plastic part design is to avoid sharp interior and exterior corners. This rule applies to snap-fit features where the feature meets the parent material as well as at all the angles and corners within the feature itself. Sharp internal corners create sites for stress concentrations. When at the base of a load-carrying constraint feature, sharp corners can cause feature failure.

Specify a radius for inside and outside corners. The idea is to maintain a constant wall thickness for smooth plastic flow through the mold; *the melt front does not like surprises.* Corners cause turbulence and are hard to fill. It is not enough to simply ask for fillets and radii in a general drawing note. Call out a fillet or radius dimension on the part drawing at every site where one is required.

Note how the protrusion height (H) limitation relative to wall thickness shown in Fig. 9.18 is frequently violated by cantilever beam lock features and pin locators, for example. This is acceptable if processing accommodations are made and the material permits it.

Treat every protruding feature (hooks, pins, tabs, lugs, etc.) as a rib and follow the guidelines for rib dimensions and rib spacing. Specify a wall thickness and protrusion thickness so that voids or residual stresses at the base of the feature do not occur.

If a part shows sink marks on the opposite side of a wall from a feature, this indicates that voids or residual internal stresses exist at the feature's base. These will weaken the feature and may result in failure.

Always include a draft angle. This allows the part to be easily removed from the mold. Start with the basic feature size then add the angle to each side. There are many sources of draft angle information, including [6].

Avoid thick sections and abrupt section changes for the same reasons sharp corners should be avoided. Another reason is the difficulty of cooling thick plastic sections. To properly cool a thick section results in significantly longer cycle times and higher cost.

Where die faces come together in shear, a shut-off angle is necessary. For instance, this will occur when access is required for molding undercuts in hooks or lugs.

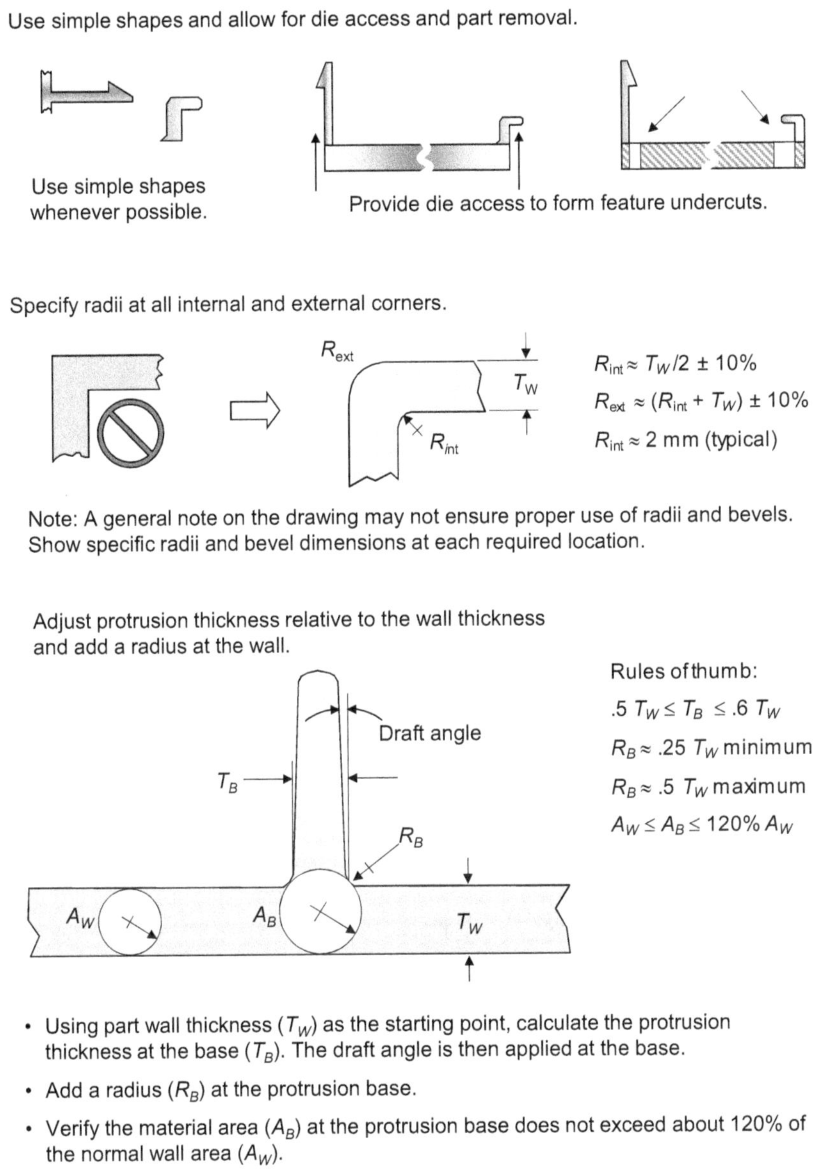

Figure 9.17 Common process-friendly design practices

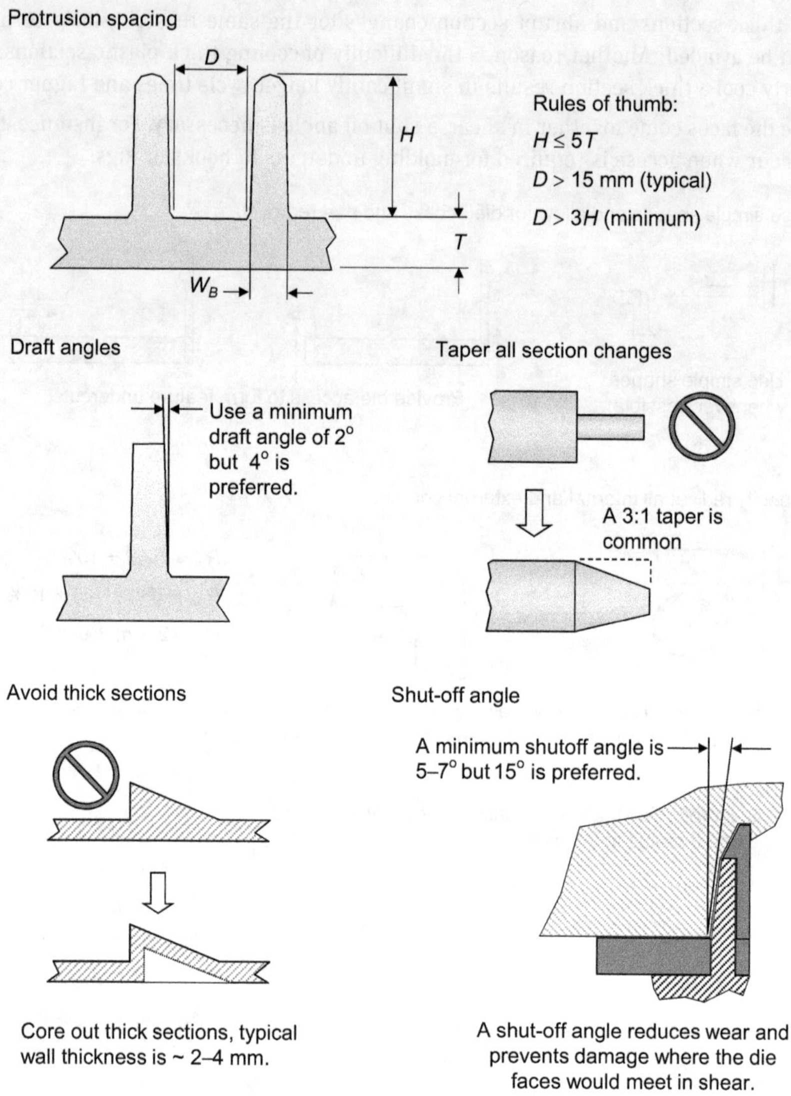

Figure 9.18 Process-friendly designs, continued

Pay attention to gates; they are areas where the plastic melt enters the mold cavity and gate style and location can affect snap-fit feature performance. Gates were discussed with respect to loop style locks in Chapter 6, Section 6.3.2.3.

Mold designers are not likely to know a part's critical areas and will put gates at locations they believe are the best sites for mold fabrication and the molding process unless the part designer indicates otherwise.

Gates should be located:

- Away from flexible features and impact areas.
- So that knit lines will not occur at high stress areas, including living hinges.

- In the heaviest/thickest sections so that flow is to the thinner, smaller areas.
- So flow is across (not parallel to) living hinges.
- So flow is directed toward a vent.
- In nonvisible areas.
- So that flow distance to critical features is not excessive.

Gate location can also affect part warpage. Be sure snap-fit features do not move out of position due to excessive part warpage. If they do, guide enhancements may be needed to bring the locks back into proper position for engagement.

Some of these process-friendly guidelines exist to help the manufacturer optimize the production process. Optimization includes minimizing cycle time. Some of the guidelines can be violated at the cost of higher cycle time. Very close communication between all stakeholders is required to ensure the required process parameters for quality parts are understood and maintained throughout the production life of the product. Beware that when a part design increases the cycle time, there may be a temptation to speed that time up once the part is in production.

Most importantly:

- Communicate directly with the material and part suppliers and mold maker to ensure all design requirements are understood and met. Section 13.3.1 in Chapter 13 describes how failure to communicate about draft angle requirements resulted in lock feature problems.
- Refer to current published rules and guidelines for mold design for the specified part material.
- Consider all protrusion features as ribs and follow rules for rib design and spacing.
- Always specify radii and smooth transitions between sections of different thickness.
- Pay special attention when, of necessity, a design falls outside of process-friendly guidelines.

9.4.2 Fine-Tuning Enablers

Fine-tuning capability makes initial mold adjustments easier. Despite continuous advances in materials, processing and part and mold-flow analysis, the nature of plastic means the first parts to come out of a mold are likely to require some fine-tuning.

Be aware of the relationship between compliance and fine-tuning sites.

Fine-tuning capability also accommodates long-term part and production variables. Once production begins, mold wear, variations or changes in raw materials, design changes, and variation in other parts may also require mold adjustments to maintain attachment integrity.

In anticipation of the need for initial and long-term adjustments, the developer should plan for mold tuning at strategic locations. The purpose is to avoid large-scale mold changes that would be expensive and time-consuming.

The first step in adding fine-tuning enhancements is identifying where compliance is possible relative to critical alignment and positioning requirements and the associated constraint sites, see Fig. 9.19. This process occurs in the design step of the snap-fit development process and was covered in Chapter 8, Section 8.6, in the constraint worksheet discussion.

Once these sites are identified, the developer can use *metal-safe* design (Fig. 9.20) or design for *adjustable inserts* (Fig. 9.21) at the corresponding areas of the mold.

Fine-tuning rules:

- Fine-tuning sites should be at critical positioning and alignment constraint pairs.
- Fine-tuning sites should be as close as possible to the critical dimensions that must be controlled.
- Compliance sites should not be fine-tuning sites.

Adding compliance to an application:

Application dimensional requirements:

- Edges at **A** must be flush to ± 0.1 mm
- Gap tolerance at **B** must be ± 0.2 mm
- **A** and **B** are critical alignment sites.

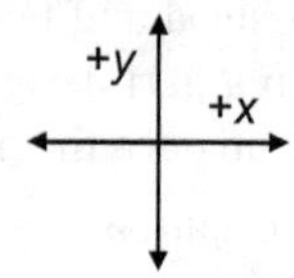

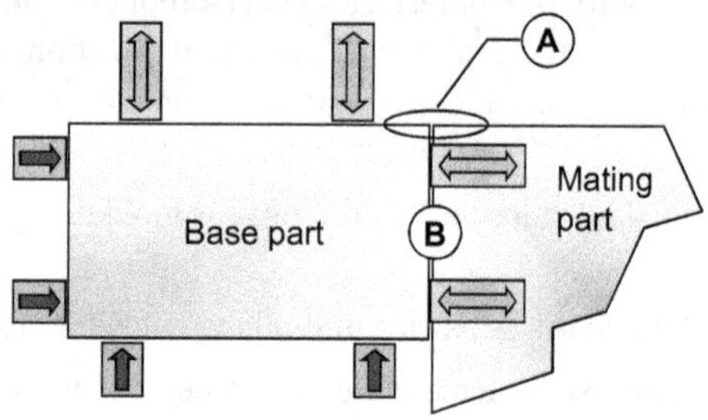

Locator pairs that must allow fine-tuning :

Locator pairs that must provide compliance:

Dimensional requirements must be satisfied using locator pairs that control the mating to base part positioning at **A** and **B**.

To control flushness at **A**, locator pairs at that site must position the mating part on the *y*-axis. These locator pairs must allow fine-tuning.

To control the gap at **B**, locator pairs at that site must position the mating part on the *x*-axis. These locator pairs must also allow fine-tuning.

Compliance for maintaining a line-to-line fit at the critical alignment sites on both axes is established at the locator pairs *opposite* the critical alignment sites.

Note: This example uses opposing constraint pairs to illustrate where compliance enhancements would be placed to ensure critical dimensional requirements are met. Recall the discussion in Chapter 8 about simplifying the attachment by *eliminating* opposing constraint pairs whenever possible.

Figure 9.19 Critical positioning requirements will determine fine-tuning sites

Metal-safe means to allow for part dimensional adjustment by removing rather than adding metal to the mold. Obviously, it is much easier to grind existing metal away in the mold than to first build up an area and then shape it by grinding metal away.

Once the critical alignment/positioning sites have been identified, select initial nominal dimensions and tolerances at or slightly beyond the minimum part material condition at those sites only, see Fig. 9.20. Be careful not to carry the idea of metal-safe design to such an extreme that the first parts out of the mold are not even close to design intent. This will render the parts useless for providing the dimensional information needed to guide fine-tuning.

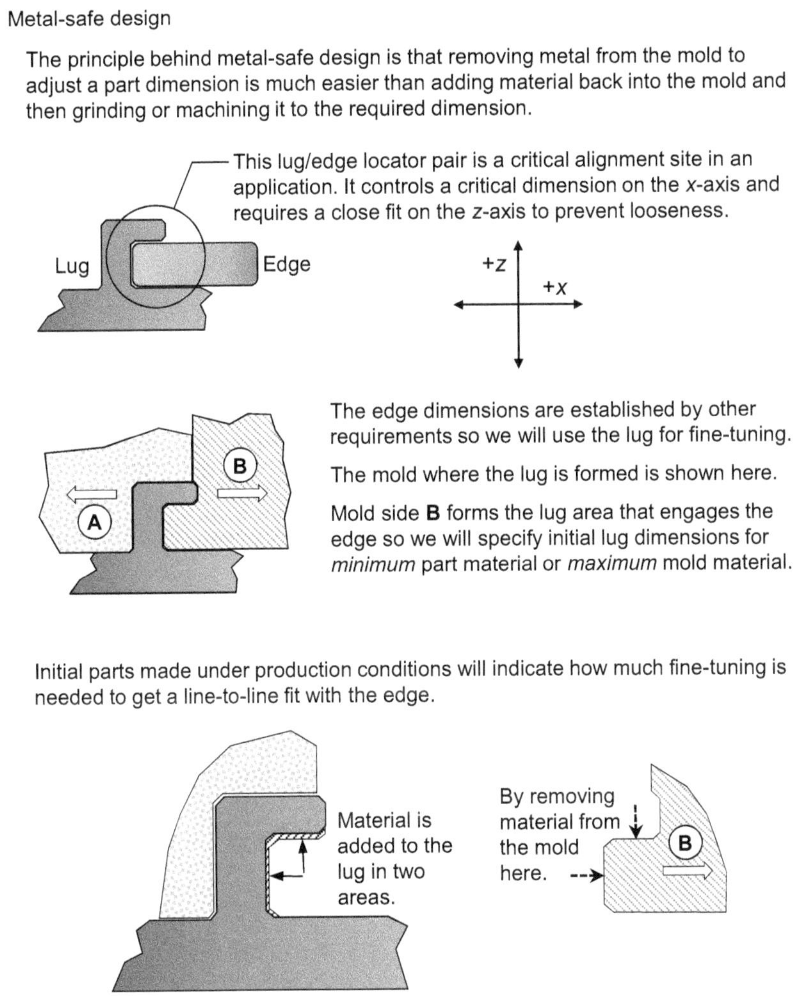

Figure 9.20 Metal-safe fine-tuning example

Metal-safe design allows easy adjustment of critical dimensions in one direction only, from a minimum (part) material condition toward a maximum material condition.

Adjustable inserts provide more flexibility than metal-safe design, allowing critical dimensions to be adjusted in two directions by either adding or removing material to the insert itself. They are separate pieces that are mounted in a receiver in the mold cavity and can be removed for modification or replaced, see Figs. 9.21 and 9.22.

They may be used anywhere tuning is needed, but their multidirectional capability can be particularly useful where (fine-tuning) lands are used in a constraint pair with two natural locators.

In Fig. 9.21, the edge-wall natural locators at all four sides of the panel-cavity interface make fine-tuning a close fit in the cavity very difficult. In the suggested design, tab locators at the panel perimeter can be easily tuned for a close fit by using adjustable inserts in the mold.

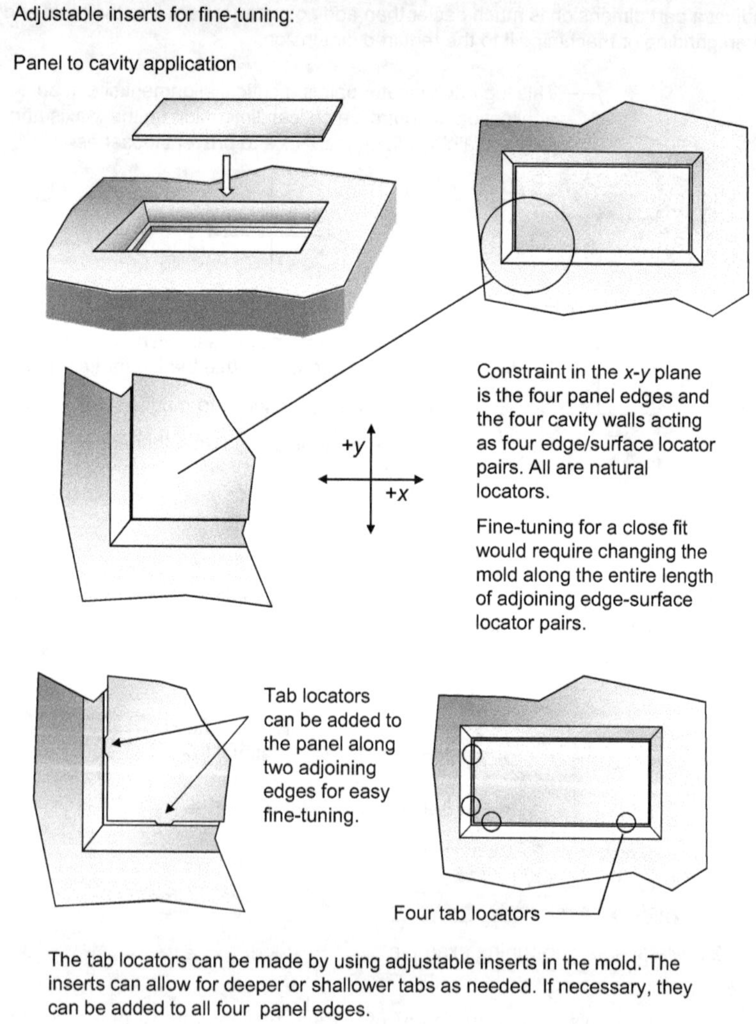

Figure 9.21 Tabs in this application are fine-tuned with adjustable inserts

In Fig. 9.22, a lock feature must have a close fit to prevent looseness. Adjustable inserts at the nearby catch/edge locator pair can control this fit.

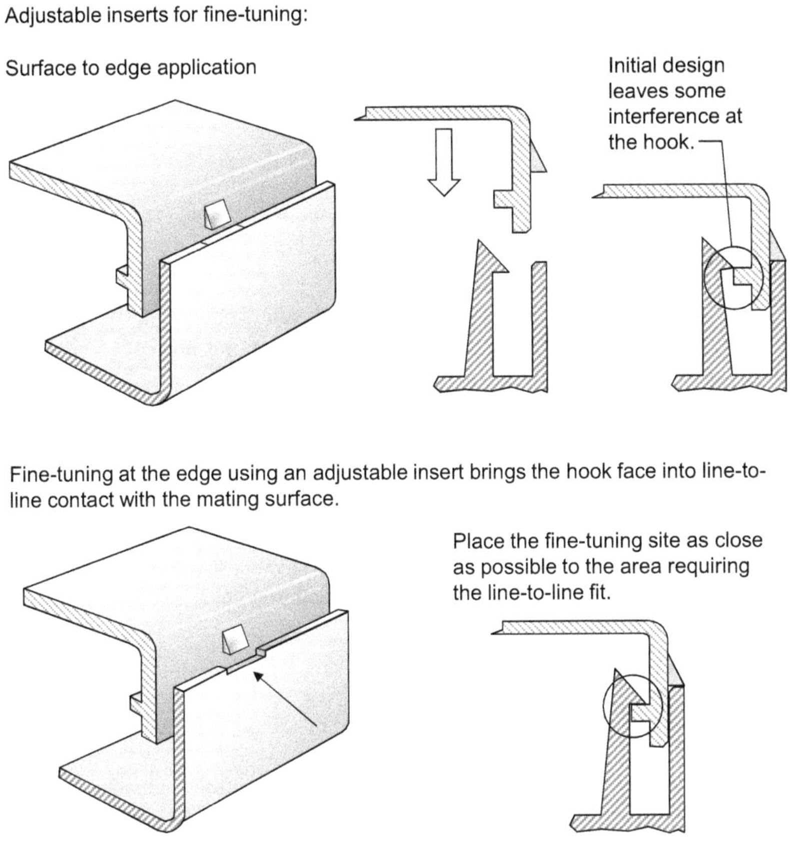

Figure 9.22 Adjustable inserts used for fine-tuning at an edge

9.5 Summary

Enhancements may be distinct physical features of an interface or attributes of other interface features. Enhancements improve the snap-fit's robustness to the variables and unknown conditions that can exist in manufacturing, assembly, and usage. Enhancements are often subtle details in a snap-fit application and they may not be obvious at first glance. The reader should study how enhancements are used in similar applications from different sources to see how effective use of enhancements can improve the product's quality.

As a customer: When soliciting bids on a snap-fit application, the required enhancements should be made a part of the business case and considered nonnegotiable. In short, they are essential to ensuring a high quality and successful snap-fit.

As a supplier: When bidding on an application, inclusion of enhancements may demonstrate the expertise and the attention to detail that wins the contract. Be prepared to provide technical reasons to explain extra cost, if any, incurred by the presence of enhancements.

Important points in Chapter 9:

- Knowing about the different kinds of enhancements will enable product developers to better understand snap-fit applications when they are studied during product benchmarking.
- During snap-fit development, include the *required* enhancements in the initial attachment concepts and in the first detailed parts made when possible.
- Desktop manufacturing methods can provide pre-prototype parts with enough detail that requirements for guides and assists can be identified.
- Other enhancements, assembly feedback and user-feel, for example, require parts made from prototype or pre-production molds and production materials to properly develop design details.

The need for retainers may not be apparent until parts undergo physical testing. Tables 9.2–9.4 summarize the enhancement discussion in Chapter 9.

Table 9.2 Enhancement Summary

Name	Why	What/How	Notes
Assembly			
Guides	Ease of assembly	Stabilize parts	No simultaneous engagement, use locators.
Clearance		No interference	Usually a feature attribute.
Pilots		Pilot-proper orientation	Use locator or guide as pilot if possible.
Feedback	Indicates good assembly	Tactile, audible, visual signals, and consistent behavior.	Maximize positive signals. Minimize system noise.
Activation and usage			
Visuals	Disassembly and operation Information	Text, arrows, symbols	Use standard symbols. Training and awareness required.
Assists	Enable disassembly or operation	Extensions for fingers, tool access	For nonreleasing locks. Possible feature damage, visuals, or guards may also be needed
User-feel	Perceived quality	Tactile, audible	Manually activated locks in moveable applications.

Table 9.2 Enhancement Summary *(Continuation)*

Name	Why	What/How	Notes
Performance and strength			
Guards	Protect weak or sensitive features	Prevent over-deflection, reduce strain	Cantilever hooks in particular may need protection.
Retainers	Improve retention performance	Strengthen or support the lock or stiffen the lock area	Cantilever hooks in particular may sometimes need retainers.
Compliance	Take up tolerances and prevent looseness	Elastic features or local yield	May interfere with feedback, use care.
Back-up lock	A back-up attaching system	Readily available fasteners, adaptable interfaces	For service and repair or as an alternative mainstream design.
Manufacturing			
Process-friendly	Efficient and consistent manufacturing process	Feature design and orientation	Follow mold and product design guidelines.
Fine-tuning	Development and manufacturing adjustments	Metal-safe design, adjustable inserts, local adjustments	Don't over-do, select sites carefully. Use only at locator sites controlling critical dimensions.

Some enhancements are required in every application; others depend on specific application needs, see Table 9.3.

Table 9.3 Enhancement Usage Recommendations

		Required in all applications	Required in some applications	Highly recommended in most applications
Assembly	Guides	✓		
	Pilots		✓	
	Clearance	✓		
	Feedback	✓		
Activation	Visuals		✓	
	Assists		✓	
	User-feel		✓	
Performance	Guards		✓	
	Retainers		✓	
	Compliance			✓
	Back-up lock		✓	
Manufacturing	Process-friendly	✓		
	Fine-tuning			✓

Table 9.4 shows the steps in the snap-fit development process where one is most likely to have enough information to add a particular enhancement. Of course, enhancements may also be added after the fact in response to product issues.

Table 9.4 Enhancement Decision Timing

8. Enhancement	Development stage				Comments
	R	C	D	T	
Guides		✓			Required: Combine with locators.
Pilots	✓				Required if a symmetric part can be improperly oriented.
Clearance		✓	*✓*		Required: Certain clearance features (lands) may be identified early. Details of clearance, bevels, and radii added during design.
Feedback		✓	*✓*	*✓*	Required: Details added in design. May require testing and evaluation.
Visuals	✓		*✓*		Need may be identified but implementation usually delayed until final parts.
Assists		✓			If a nonreleasing lock with limited access.
User-feel	✓		*✓*	*✓*	If a user activated lock in moveable application.
Guards		✓			Need may be identified early, part stacking or manual deflection, for example.
Retainers		✓	*✓*	*✓*	Sometimes predictable based on application concept (constraint features on thin walls). Sometimes identified in analysis or test.
Compliance		✓	*✓*	*✓*	Identify sites at concept development. Execute details during detailed design.
Back-up lock	✓		*✓*	*✓*	May be early or after testing indicates potential problem.
Process-friendly		✓	*✓*		Feature orientation decisions during concept. Details and dimensions added during design.
Fine-tuning			✓		Details and dimensions that support fine-tuning are added during design.

R - When establishing specific application requirements
C - While developing the attachment concept
D - Detailed design and analysis
T - Testing
✓ Need for enhancement is likely to be first identified
✓ Follow-up or secondary identification

References

[1] Private conversation with Rich Coppa, Senior Principle Engineer, Camera Division of the Polaroid Corporation, Boston MA (1994)

[2] The product redesign in the reflector application example was developed by Tom Froling and Tom Nistor (~1991)

[3] Bonenberger, P. R., The Role of Enhancement Features in High Quality Integral Attachments, Technical paper #294 at ANTEC, '95 Conf. Soc. Plast. Eng., Boston, MA, May (1995)

[4] SAE Recommended Practice J1344, Society of Automotive Engineers, Warrendale, PA, USA. http://standards.sae.org/

[5] From a shutter assembly on a Polaroid camera (model unknown)

[6] 5 Ways to Improve Part Moldability with Draft, Protomold: Design Tips for Rapid Injection Molding, (1/6/2016), https://www.protolabs.com/resources/injection-molding-design-tips/united-states/2016-01/

Bibliography

The following publications all provide highly useful information on plastics and designing for injection molding. They were used as general reference for this chapter. Also see the appendix for resources.

Beall, Glenn L., *Plastic Part Design for Economical Injection Molding*, Libertyville, IL (1998)

Dupont Polymers, *Dupont Engineering Polymers - Product Information Guide*, Dupont Polymers Dept., Wilmington, DE

GE Plastics, *GE Engineering Thermoplastics Injection Molding Processing Guide*, General Electric Company, Pittsfield, MA (1998)

Hoechst Technical Polymers, *Designing with Plastic - The Fundamentals*, Design Manual TDM-1, Ticona LLC, (Formerly Hoechst Celanese Corporation, now a division of Celanese AG) Summit, NJ (1996)

Malloy, Robert A., *Plastic Part Design for Injection Molding - an Introduction*, Hanser, Munich, Germany (1994)

Molders Division of The Society of the Plastics Industry Inc., *Standards and Practices of Plastics Molders*, Washington D.C. (1998)

Monsanto Company, *Monsanto Plastics Design Manual*, Monsanto Company, St. Louis, MO (1994)

Xerox Corporation, *Plastic Design Aid* (wall chart) (1987)

10 Applying the Snap-Fit Development Process

To provide some context for the elements and concepts discussed in Chapters 4 through 9, the Snap-Fit Development Process was introduced in Chapter 3.

This chapter explains in detail how those elements and concepts are used in the development process to create a snap-fit application, see Fig. 10.1.

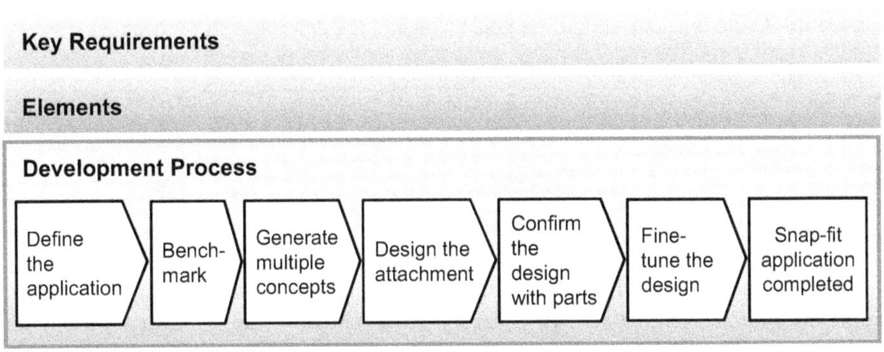

Figure 10.1 The snap-fit development process in the ALC

In Chapter 3, a preliminary *Step 0* was described in which the decision to use a snap-fit attachment was made. The discussion in this chapter assumes the choice to proceed with a snap-fit application has been made.

Recall from Chapter 3 how the development process begins with creating a good attachment *concept*. In the process, Step 3, - *Generate Multiple Concepts*, and Steps 1 and 2 that lead up to it, may appear to be a waste of time, but they are not because:

- Most of a product's cost is established during the concept development stage.
- Starting with a good concept will help ensure attachment reliability and quality.
- Issues that will require future correction are avoided and time-consuming development iterations are minimized.

The concept development stage may look difficult or time-consuming. Once the reader understands the process, it will become easy. It is primarily a thinking exercise with some product benchmarking. It does not involve detailed design - simple hand sketches of concepts and ideas are recommended.

Most snap-fit developers are not materials or processing experts. The snap-fit development process should include input from a polymers expert, preferably as early in the design process as possible. Input from processing experts is also recommended. If possible, also include the final part manufacturer(s) in the process.

Figure 10.2 repeats a figure from Section 3.5 showing where decisions about the spatial/descriptive and physical elements are made during the development process.

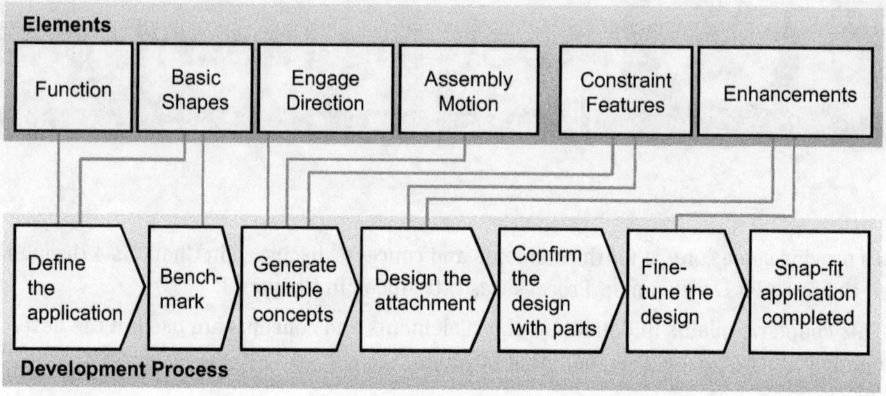

Figure 10.2 The relationship between elements and the development process

10.1 Step 1: Define the Application

The application is first defined using the descriptive elements *function* and *basic shape*. Function, summarized in Table 10.1, describes the nature of the locking requirements for the attachment. The purpose is to explicitly define what the lock feature(s) must do in the application so there can be no misunderstanding when decisions about lock feature selection are made later in the process. Refer back to Chapter 4 for details.

Table 10.1 Define the Lock Feature's Function in the Application

Action	Movable	Free movement or controlled movement
	or	
	Fixed	No movement once latched
Purpose	Temporary	Until final attachment is made
	or	
	Final	Snap-fit is the final attachment
Retention	Permanent	Not intended for release
	or	
	Nonpermanent	May be released
Release	Releasing	Releases with applied force on the mating-part
	or	
	Nonreleasing	Lock is manually deflected for release

Basic shapes are generic descriptions of the part's geometry, see Table 10.2. The common/rare designation is based on the author's observations. In any specific product field, the frequency of these combinations may be different and an appropriate frequency table can be developed.

Basic shape frequency is related to a business strategy of establishing a library of common/preferred basic shape combinations, which is discussed in Chapter 15.

Table 10.2 Likely Basic Shape Combinations

	Solid	Panel	Enclosure	Surface	Opening	Cavity
Mating-part	Common	Common	Common	Rare	Rare	Low
Base-part	Common	Rare	Rare	Common	Common	Common

Defining the application using these attachment level terms will help when design rules are applied later in the process. Their immediate value, however, is in helping the developer structure a search for ideas as they conduct technical benchmarking in the next step.

In addition to the general *key requirements* for snap-fits, each application will have specific performance requirements and in-service conditions which must be defined. Some of these need not be known at this stage of the process, but will be needed eventually. The sooner this information is collected, the better. Application-specific requirements and conditions include:

- Material properties
- Manufacturing limitations and capabilities
- Load-carrying and retention requirements
- Thermal history for the application
- Alignment and appearance requirements
- Environmental conditions such as chemical and ultra-violet exposure
- Product service conditions and requirements

At this time, the developer should begin rough hand-drawn sketches of the application in terms of its basic shapes. These *concept sketches* are used to capture ideas and alternatives throughout the concept development step. The developer should also begin thinking about how a crude model of the application can be constructed.

This is also the time to identify certain *red-flag* issues. These are not issues that would necessarily prevent use of a snap-fit, but they must be given extra attention because of their potential for special difficulties in attachment development.

Red-flag issues include:

- *Short grip length:* A lock feature having beam length less than ~5x its thickness. Cantilever hook locks typically do not work well in this situation. Use a lock style with higher decoupling capability.
- *Brittle or rigid material:* Will be much more sensitive to stress concentrations, small radii, assembly strain, over-deflection, and short grip lengths. This includes plated plastic parts.

- *Soft or flexible material:* Will be difficult to get reliable locking with cantilever hook style locks. Retainer enhancements may be required with cantilever hook locks or use a different lock style.
- *Impact forces:* Cantilever hook locks are particularly susceptible to release under impact. A concave retention face profile or a retainer enhancement may help absorb energy, but the preferred approach is to select a lock style with better decoupling capability.
- *Final material not yet known or subject to change:* A very conservative approach is indicated. Use lock features with higher decoupling capability (Level 3 is suggested) and design to a low maximum assembly strain (~1% is suggested). Verify behavior when the material is known.
- *Thin walls:* Features that protrude from thin walls (locks, locators, and guides) may create sink marks. Sink marks can be an appearance issue but they also indicate molded-in stresses, possible voids, and weakness at the feature's base. Wall flexibility may also reduce a protruding lock feature's effectiveness.

Chapter 10 is the *working* chapter of this book. Readers should use it during product development projects to guide themselves through the entire process. Table 10.3 and similar tables are provided throughout the chapter so users can list their own references.

Table 10.3 Cross-Reference Notes for Define the Application

10.2 Step 2: Benchmark

The term *benchmarking* has many meanings. In this development process, it is not marketing or customer or product feature benchmarking. It is *technical* benchmarking, and it is careful study of other snap-fit applications for understanding, learning, and ideas. It is not simply reverse engineering in order to copy other designs or ideas. Copying without understanding will lead to problems.

Benchmarking is a continuous process of learning and changing [1]. When a developer has a deep *attachment level* understanding of another product, aspects of that product's attachment can be adapted and improved for the new application. The idea of benchmarking is to stimulate creativity and ideas by becoming familiar with some of the available design options.

Products studied for benchmarking should include your own as well as competitors' products. Also study products that are unrelated to your own product or to the application under development. This is one reason why the concept of basic shapes is so important, the developer can find useful ideas in any products with similar basic shapes and is not limited only to applications similar to the one being developed; creative ideas become available everywhere.

Benchmark on basic shapes!

The checklist in Table 10.4 can be used as a reminder of snap-fit features and attributes to look for when parts are available to handle and assemble/disassemble for benchmarking.

Get real parts in your hands!

One of the simplest of snap-fit applications involves the panel to opening basic shapes. If the reader studies a number of panel-opening applications, they will discover a great variety of design interpretations. Some will be fundamentally better than others and some will be better for a given application. A developer can choose to invent a new panel-opening for an application, or they can study existing applications and apply the best ideas found and their own modifications to the new application.

Once an effective panel-opening design concept has been created, simply adapt it to the application at hand. A developer (or an organization) can build a library of good concepts for a variety of basic shape combinations and draw upon it as a reference for new design.

Why reinvent an attachment?

Table 10.4 Benchmarking Checklist for Hands-On Part Study

Application studied:	Comments
Is the application properly constrained?	
How is it constrained?	
Are the features effective?	
Any evidence of damage?	
Stress marks?	
Damage to edges or corners?	
How does it feel?	
Assemble it, is it easy?	
Shake it, any noises or movement?	
Drop it, what happens?	
Take it apart.	
Is disassembly obvious and easy?	
Any damage?	
Are tools required for disassembly?	
Look for all enhancements, especially the required ones.	
Guides Clearance	
Feedback Process-Friendly Design	
Is the design efficient?	
Locators, guides, and pilots functions are combined.	
Efficient use of constraint pairs.	

Table 10.4 Benchmarking Checklist for Hands-On Part Study *(Continuation)*

Application studied:	Comments
Are there any special issues that need enhancements?	
Is it easy to make? No die action required Simple features Process-friendly design	
How would you rate the application if **you** had to assemble it 8 hours a day?	Poor Fair Good Excellent
How would you rate the application if **you** were a service technician?	Poor Fair Good Excellent
How would you rate the application if **you** had to manufacture it?	Poor Fair Good Excellent
How would you rate the application if **you** were a customer?	Poor Fair Good Excellent

Important benchmarking guidelines include:

- Benchmark on basic shapes and do not limit the search to just one type of product or application. Many plastic products are available for study including toys, electronics, small appliances, etc. Most of the time, ideas drawn from several applications will influence the final design.

Enhancements can be clues to a problem someone else has already had to fix.

- Make it a point to look for enhancements. They are often added after initial design and testing or after production begins because a problem was discovered. When one can recognize enhancement features and understand what they do, they can either be included in the attachment design from the start or the condition that made the enhancement necessary can be avoided.
- Study the constraint features and how constraint and compliance issues are resolved. Understand why the locks and locators were selected and arranged as they are. Study how they behave and interact as the parts are brought together through the required assembly motion.
- Assemble and disassemble the parts. How do they feel? Shake the parts. Do they squeak and rattle? Are the parts stiff enough? Look at the distribution of constraint features on the parts. Are there enough to compensate for part flexibility? Flexibility is of particular concern with large panels as they are usually weak in bending.
- Study the parts for witness marks indicating over-stress or assembly problems. A lighter color or whitened area at the base of a lock or locator indicates damage. Broken or damaged edges or corners on parts indicate interference and difficult assembly. These are clues to strength requirements as well as enhancements that may be needed.

Table 10.5 is another checklist that can be used during benchmarking. It presents a more detailed list of specific design criteria for an application.

Also use this checklist when evaluating concept proposals and final part designs later in the development process.

Table 10.5 Benchmarking and Proposal Screening Checklist

		Execution				Why				
	Use this checklist for screening attachment proposals. The objective is to understand and compare each design's potential and trade-offs required for creating a reliable, durable, and cost-effective attachment.	Good	Fair	Poor	N/A	Ease of Assembly	Piece cost	Serviceability	Reliability/Durability	Product Quality
Required in all Applications	Constraint in 12 DOM (less if a moveable application)					✓			✓	✓
	Sufficiently strong lock and locator features								✓	✓
	Feature, basic shape and assembly motion compatibility					✓		✓	✓	✓
	Guide and clearance enhancements					✓				
	Operator feedback					✓				✓
	Process-friendly								✓	✓
	Proper radii called out at all critical locations								✓	
Preferred in all Applications	Fine-tuning enhancements						✓			✓
	Normal (commercial) rather than fine/close tolerances						✓			
	Design is robust to potential material changes						✓		✓	✓
Application Specific	Other enhancements as needed					✓	✓	✓	✓	✓
	High (customer) perceived quality*									✓

* Perceived quality is different than (real) quality of appearance and performance. Perceived quality is what the customer believes about the application. It is particularly important in customer-activated applications.

Table 10.6 Cross-Reference Notes for Benchmarking

Benchmarking is easy to ignore, but it is extremely important for enabling creativity. As familiarity with snap-fit technology increases, attachment level benchmarking will occur naturally when products are studied.

■ 10.3 Step 3: Generate Multiple Concepts

Figure 10.2 showed how the elements of a snap-fit map to the development process. In that diagram, note that four of the six elements are brought together in this step. Step 3 is the most critical stage of the process because it is where important decisions about the attachment concept are made and because it helps to enable creativity during application development.

This is really important.

By identifying combinations of allowable engage directions and assembly motions, the developer can create several *fundamentally different attachment concepts* rather than mentally locking themselves into only one idea or just variations on one basic theme.

Developing concept alternatives is an important enabler for creativity [2]. Constraint features and some enhancements are then added to each alternative, the concepts are evaluated and one is selected for analysis and design.

Step 3 need not be long or difficult and it can be conducted as a personal or as a group brainstorming session. Knowledge gained during application definition in Step 1 and benchmarking in Step 2, *with attachment level understanding* is now applied as applications concepts are generated.

Important This step is recommended as an activity during design-for-assembly workshops. The use of different assembly motions to force generation of fundamentally different attachment solutions lends itself quite well to a group effort in a workshop.

Step 3 consists of five sub-steps, see Fig 10.3, where *ED* is engage direction and *AM* is assembly motion.

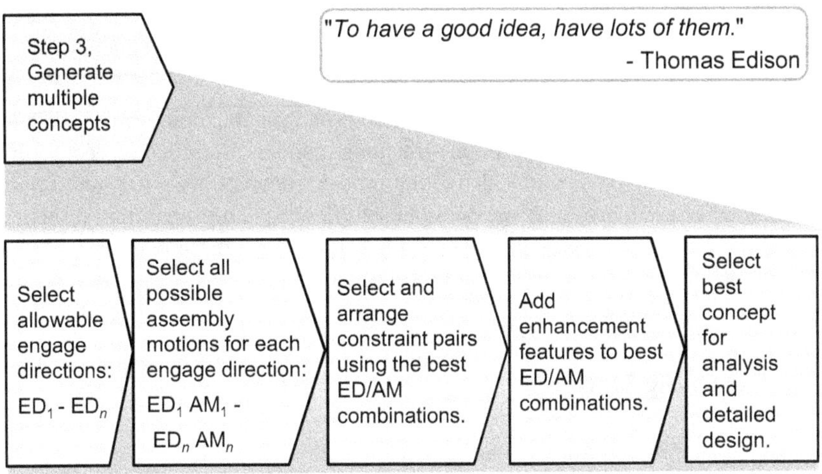

Figure 10.3 Details of Step 3

10.3.1 Engage Direction

Once the application's general requirements and shapes are defined, engage direction is the first decision made in the development process.

Remember: *Engage direction* is not the same as *assembly motion.*

Careful selection of the engage direction is important because it is associated with a separation direction that, in turn, determines lock feature orientation. In most snap-fit applications, the separation direction is opposite the engage direction and the lock-pair(s) disengage or separate in the opposite direction from which they engage. The basic rules for selecting allowable engage directions are:

- Engage directions and the associated separation directions must be compatible with the basic shapes.
- They must be compatible with access for assembly, service, and usage. Also consider if other parts added later would interfere with access or service.
- They must meet ergonomic requirements if the parts will be assembled by human operators.
- If intended for automatic assembly, consider the impact of access and motion complexity on capital equipment costs.
- If possible, avoid significant forces acting in the separation direction. If they cannot be avoided, retainer enhancements are available or a lock style with a higher level of decoupling should be used. It is up to the developer and the materials expert to determine if a force is significant. The determination must consider magnitude, duration, and frequency of the force itself and environmental conditions, of which temperature is usually the most significant.

Early in the development process, information on the magnitude of forces across the snap-fit interface may not be available or information may be based on estimates with more accurate data to follow. At this point, it is not necessary to know the exact magnitudes of these forces. It is more important to know the direction and relative magnitudes of each force. Most of the time, engage direction decisions can be made with this information.

Some applications will have only one allowable engage direction, others will have more than one. Possible engage and separation directions can be represented by vectors on a convenient coordinate system and added to the concept sketches begun in Step 1. Some applications will have more than one possible set of engage and separation directions and all *feasible* directions can be identified. In Fig. 10.3, these directions are identified as ED_1, ED_2, etc.

10.3.2 Assembly Motions

Recall that five simple assembly motions: push, slide, tip, spin, and pivot describe final mating-part motion as the lock(s) engage. Each assembly motion will allow certain constraint feature configurations and preclude others. In Fig. 10.3, assembly motions are identified as AM_1, AM_2, etc.

Given the engage directions from Step 3.1 and the basic shapes, the developer will find that only some of the five assembly motions are possible. Identify these assembly motions for each engage direction. Certain conditions may render some of these combinations undesirable although they may be feasible. Eliminate those combinations from consideration. By the end of this step, a few *best* combinations will have been identified, (typically 2 or 3, but sometimes only one).

The application conditions that influence assembly motion decisions include many of the same conditions that influence engage direction decisions:

- Assembly motions must be compatible with the engage direction.
- They must be compatible with the basic shapes.
- They must be compatible with access for assembly, service and usage. Also consider if any other parts added later would interfere with access.
- They must be ergonomically friendly if the parts will be assembled by human operators.
- If intended for automatic assembly, consider the impact of access and motion complexity on capital equipment costs.

Figure 10.4 shows examples of how different assembly motions will force *fundamentally different* interface concepts for a solid to surface application with a separation force in the $+y$ direction:

- With a push motion, all engagement, thus all resistance to separation, must occur only with lock features.
- With a tip assembly motion, the strong lug(s) must be engaged first before the tip motion can begin. The lug(s) are then available to help resist separation.
- With a slide motion, the mating-part is placed against the surface so the lugs are aligned with the edges they will engage. When the mating-part slides to engage the

lugs, a trap locks it in place. The lugs are available to resist separation and no locks are involved.

The only constraint features shown in these illustrations are those directly involved with assembly motion, feature selection, and the resulting separation resistance. Other constraint features required in these attachments are not shown.

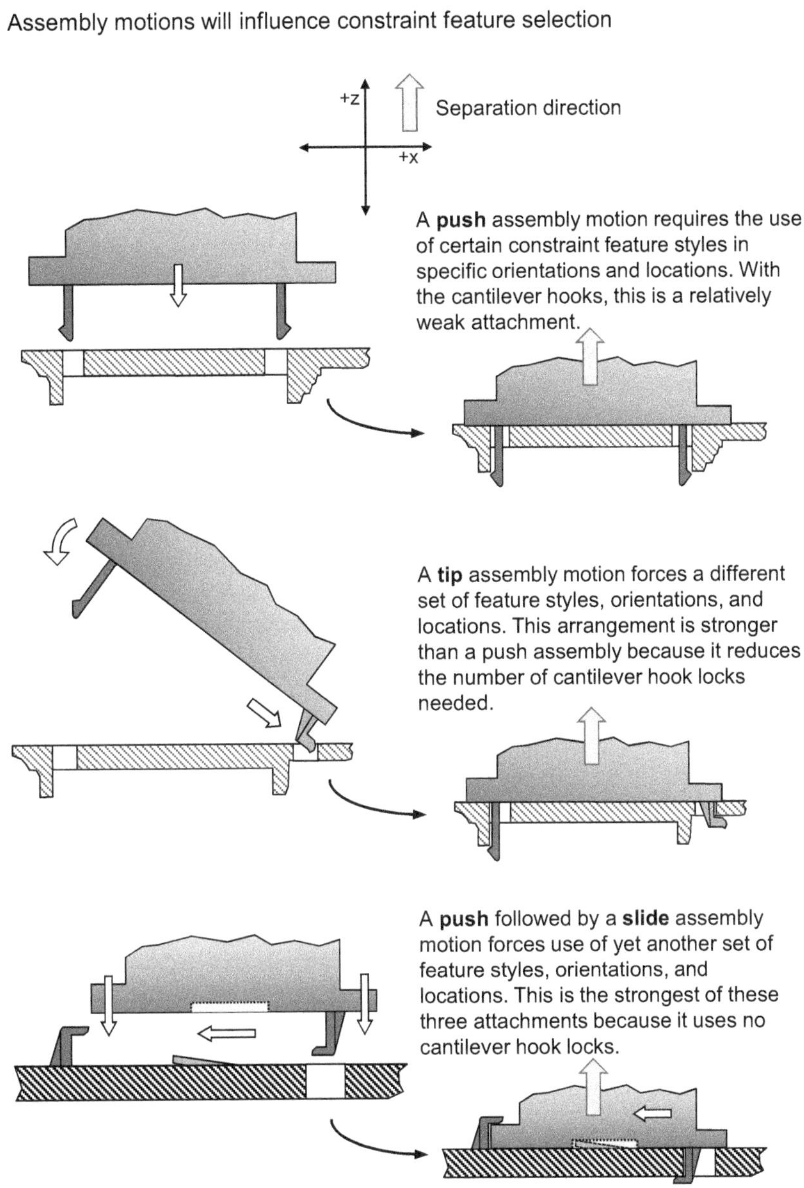

Figure 10.4 Different assembly motions will force fundamentally different attachment options

Rules for selecting an assembly motion are:

- In general, the push assembly motion will result in a weaker attachment because more degrees-of-motion must be removed by the lock features. Other assembly motions allow more DOM to be removed by (stronger) locators and are generally preferred.
- The tip motion has certain advantages over some of the others. The first locator-pair, once engaged, stabilizes the mating-part relative to the base-part for easier engagement of the remaining constraint pairs. The tip motion also minimizes potential for simultaneous engagement of constraint features.
- A disadvantage of the tip motion is that the rotational movement may require more space than is available for assembly.
- In general, the tip and slide assembly motions are preferred over the push motion. However, at this stage of the process, the intention is to use assembly motion alternatives to generate ideas. All feasible assembly motions should be considered at this time.

It is desirable to have several ED/AM combinations at the end of this step although in some applications this will not be possible. Constraint pairs and eventually enhancements can now be sketched onto each available ED/AM combination.

Engage Directions, Assembly Motions, and Worker Ergonomics

With respect to assembly ergonomics, developers should seek out ergonomic guidelines specific to the application and the assembly process to ensure their final design is *assembly-friendly*. For snap-fit applications, maximum allowable assembly forces for hands, arms, and fingers, assembly direction, work area height, cycle times or frequency, and operator motions are typically important.

Table 10.7 Cross-Reference Notes for Engage Direction and Assembly Motion

Some general rules for ergonomic design include [3–5]:

- Products with short cycle times repeated over extended periods should have low assembly forces.
- The operator should not be required to strike or pound (as with the palm of the hand) the mating-part to locate or lock it the base-part. Any impact is undesirable.
- Avoid designs that force the assembler into an unnatural body position or force unnatural arm, shoulder, wrist, or hand positions while applying assembly force.
- The areas of a part where the operator must apply assembly force should distribute pressure over a sufficient area of the finger or hand. Do not require pushing against edges, corners, or points.

- Part assembly while wearing gloves might interfere with feedback and process efficiency. If gloves are a factor in assembly, the part design should reflect it.
- Avoid part designs that favor right-handed over left-handed workers.

Specific work-related musculoskeletal disorders (WMSDs) [5] related to snap-fit assembly include carpal tunnel syndrome, epicondylitis (tennis elbow), neck tension syndrome, shoulder tendonitis, tendonitis, ulnar artery aneurysm, ulnar nerve entrapment, and De Quervain's syndrome.

10.3.3 Identify Constraint Pairs

Constraint features are lock and locator-pairs that prevent relative movement between parts. Ideas gained during benchmarking will now help the developer select the best constraint features for the application. The strategy of identifying several engage direction/assembly motion combinations will now pay off. As constraint features are selected and arranged, basic shape and assembly motion interactions will force the use of different constraint feature styles for each possible ED/AM combination. This drives creativity by forcing development of fundamentally different attachment concepts rather than just variations on one theme.

Up to this point, the developer should have been working with *hand-drawn* concept sketches (leave that computer and the design programs alone!). It is now time to create a 3-D model of the application. Again, as with the concept sketch, the purpose is to invoke spatial reasoning skills and creativity. The model at this stage of the process need not (in fact, it cannot be) very detailed or even accurate. The most important thing is to get something in three dimensions that can be held and manipulated in space. While making constraint features decisions, use the model(s) to visualize the interactions of the mating and base-parts under the different ED/AM combinations. Also use the model to visualize how the interface will react to input forces.

Models can be very useful as a visual device when explaining or justifying a snap-fit concept to others. The highly spatial and sometimes complex nature of snap-fits can make them difficult to explain with words or drawings. The developer willing to provide a model has a better chance of getting their point across.

Rapid-prototyping technology makes it possible to produce detailed models early in the development process. With some applications, this may be desirable and helpful. In other cases, however, the effort and expense of producing these models is better left until later in the process.

The use of only rapid-prototype parts also takes away the learning and creative advantages of making models by hand. Even when rapid-prototype models are indicated at this stage of the process, some crude hand-made models should also be built.

For creativity, get your hands involved!

Model-making possibilities include:

- Modify existing parts
- Cut, glue or tape cardboard and heavy card stock shapes
- Carve styrofoam shapes

- Cut and shape wood pieces
- Use craft materials that can be shaped and hardened
- Plaster-of-paris can be molded and cut, and then filed and sanded to shape
- Available objects like a box, book, or coffee cup may serve as a model
- A table top may represent a surface
- A piece of card stock can be a panel
- An open box can be a cavity or an enclosure.
- Constraint features can be cut from card stock and glued onto models

Adding Constraint Pairs

Consider the first ED/AM combination and begin selecting and arranging constraint pairs. Note how a single, wider constraint pair could also divide into several constraint pairs or multiple pairs combined into one larger pair. This can be a very subjective decision but there is no functional difference between identical constraint pairs that act in parallel and in the same sense, with respect to constraint, Fig 10.5. Molding or other issues may, however, affect this decision. (This figure also appeared in Chapter 8, Section 8.4.)

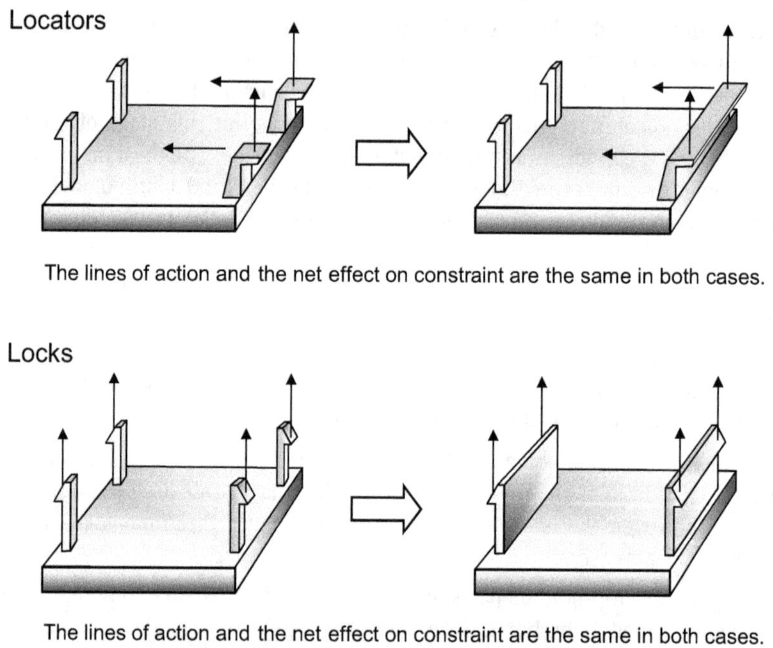

Figure 10.5 Constraint feature variations with equivalent effects

It is easiest and logical to add constraint pairs to a concept in the order in which they will engage during assembly. Ideally, the first constraint pair(s) engaged should be the most constraining locator-pair possible and must, of course, be compatible with the engage direction and assembly motion (ED/AM). Add it to the concept sketch and note all the constraint vectors associated with this pair.

The remaining constraint pairs, again considered in order of assembly engagement, will be locator or lock pairs with the lock pairs always added last. They too, must be compatible with the first constraint pair, the ED/AM, and with each other. These constraint pairs will likely be less constraining than the initial pair(s) and none of their constraint vectors should be coincident with those of the initial pair(s).

Repeat this process for all remaining ED/AM combinations.

Some developers have an intuitive feeling for constraint and will quickly understand the process. For others, some practice is required in order to reach an understanding of constraint. Use the constraint worksheet discussed in Chapter 8, Sections 8.5 and 8.6, as an aid to arranging and evaluating constraint pair usage.

Decisions about placing a feature, particularly lock features, on the base-part or the mating-part are influenced by many factors. Manufacturing options, material strength, basic shape and assembly motion are the most common. Another is the relative value of the parts and the chances of feature (particularly lock) breakage. Design the application so if a feature does break, the part carrying that feature is easily replaced, inexpensive, easily serviced, or repaired.

Rules for Selecting and Arranging Constraint Pairs

Many of these rules were introduced in Chapters 5, 6, and 8. Refer back to those chapters for details.

Refer to the constraint discussion in Chapter 8.

- Lock-pairs should constrain in as few degrees-of-motion (DOM) as possible and locator-pairs in as many DOM as possible.
- Lock-pairs should constrain only in the DOM associated with part separation.
- Generally, the more DOM removed with locators, the stronger the attachment.
- A tip or slide assembly motion is preferred over the push motion because more DOM are removed with locators and because of ease of assembly.
- The application should not be over-constrained due to redundant constraint pairs.
- Over-constraint due to opposing constraint pairs is undesirable but sometimes necessary.
- Where constraint pairs oppose each other (two constraint pairs with collinear strength vectors of opposite sense), placing the pairs as close to each other as possible will minimize tolerance effects and the potential for opposing internal forces within the constraint system.
- Where constraint pairs create a couple (parallel strength vectors of opposite sense), they should be placed as far apart as possible for mechanical advantage and reduced sensitivity to dimensional variation.
- Where constraint pairs have parallel strength vectors (of the same sense) they should be placed as far apart as possible for reduced sensitivity to dimensional variation.
- Applications should not be under-constrained. Under-constraint occurs when no constraint pairs provide strength in one or more translational DOM or when a constraint couple is ineffective in removing rotational constraint.

- In a fixed application, the mating-part must be constrained to the base-part in exactly 12 DOM. An exception is certain functional attachments where free movement is allowed in some DOM.
- Locking features should not carry high forces or sustained forces.
- Compliance should generally occur within a constraint pair rather than between pairs.
- The lock retention face can be used to take up some tolerance. A slight angle on the retention face of a (90° nonreleasing) hook will absorb tolerance without affecting retention. A contoured face can ensure maximum retention angle at any level of hook deflection.
- Select and orient constraint pairs whenever possible to avoid a die-lock condition, which will require die-action. But die-action considerations cannot be allowed to affect feature performance.
- Interface efficiency is important to minimize cost and interface complexity. Combine locator, guide, and pilot functions into one feature when possible. Use locator-pairs that constrain in multiple DOM. Minimize locator-pairs that constrain in only one DOM.

After adding constraint features to the concept sketches, the developer may also wish to build them into the 3-D models. Rapid prototype models may again be considered. Including the locator features on models is useful because their presence allows one to evaluate ease of assembly and some aspects of constraint. If the material to be used in the rapid-prototype models is brittle, there is no value in including (flexible) locking features on the models as they will soon be broken off. If the prototype material is flexible, then lock features can be included. It is also possible, if necessary, to make locking features out of flexible plastics and attach them to the models with screws or adhesives for visualizing assembly motions and feature behavior.

Table 10.8 Cross-Reference Notes for Select Constraint Pairs

[illegible]	[illegible]
[illegible]	[illegible]
[illegible]	[illegible]
[illegible]	[illegible]

Interface Efficiency

Fewer features in the snap-fit interface make it cheaper and easier to manufacture consistently. Combining locating, guide, and guard functions is one way to increase efficiency. Another is strategic use of locator pairs to remove multiple DOM within one pair. (This is influenced by the choice of assembly motion and engage direction.)

The author cannot judge whether the (mating) part shown in Fig. 10.5 is or is not truly inefficient because the base-part is not available, but we'll use it as the basis for a discussion of efficiency.

Interface efficiency

This 60 x 60 mm panel has ten interface features and is the mating part in a snap-fit application. The author has not seen the base part and can not make any judgments about the actual application's efficiency. It is possible that all ten features are necessary.

For the sake of this discussion, we will make some assumptions about the application in order to explain the concept of interface efficiency.

The assembly motion is a push.

Feature identification:

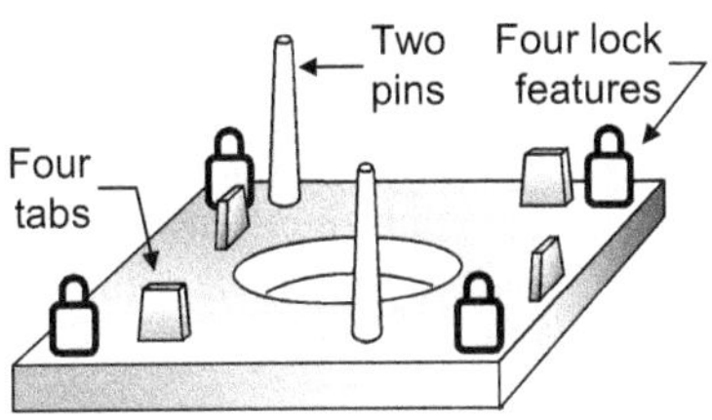

Assumptions:

- The pins are guide (enhancement) features.
- The tabs are locators. They may be fine-tuning lands but, given their shape, this is not likely.
- Because orientation of this part is important, either the nonsymmetry of the pins or one of the tabs may be serving as a pilot (an enhancement).

More efficient design:

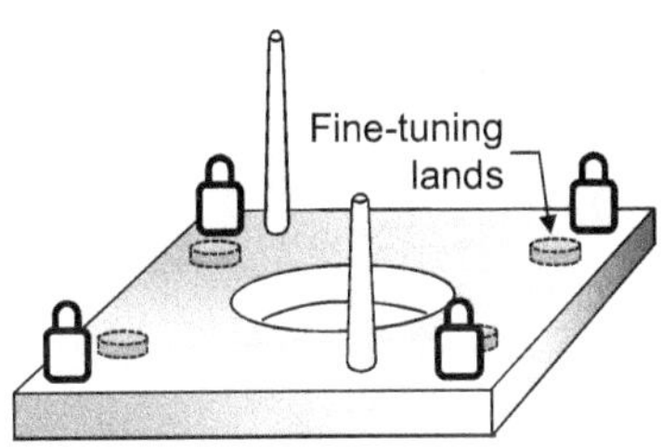

Two pins can serve as guides, locators, and a pilot.

- The pins are offset relative to the part shape to provide the pilot function.
- One pin is longer than the other for initial engagement. It will engage a hole. The shorter pin will engage a slot.
- Both pins engage before any lock features make contact.
- If the mating and base part surfaces are both natural locators and fine-tuning is required, fine-tuning lands can be added.

Figure 10.6 Combining feature functions for interface efficiency

If fine-tuning is not required, the number of interface features is reduced from ten to six.

The poorly designed lock features on this part require comment. Two fundamental issues are apparent, high assembly force and lock feature damage. Clues indicating these issues are:

- The cantilever hook locks have a very low L/T ratio.
- The ribs on the lock beams were likely added to the original design because the beams were damaged or breaking at the base during assembly, rendering the locks ineffective.

- The ribs will not prevent some damage, but they can help the hooks survive assembly.
- The ribs further shorten the beam and increase the assembly force.
- The steep insertion face angles also increase the assembly force.
- The deep undercuts on the hooks indicate possible excessive assembly deflection.

Recall the discussion in Chapter 8, Section 8.1, about perfect constraint and how three points define a plane. An observant reader may question the use of four fine-tuning lands as shown in the suggested design in Fig. 10.6. We'll address that question here:

- Use of only three fine-tuning lands in the example would more closely satisfy the three-point/plane condition. But this application is an appearance panel surrounding a functional element in the central hole. Fingers can push anywhere on the panel surface and only three lands would not be sufficient to prevent rocking. Placing four lands as far apart as possible and close to the lock features at each corner is required in this case.
- It is possible that, for the reasons mentioned above, the four tabs in the original design are actually being used as fine-tuning lands and are necessary. But their shape and positioning suggest otherwise.
- If lands are needed, the author would recommend placing them on the mating or base-part surface where they would not interfere with constraint feature positioning for function or molding.

10.3.4 Add Some Enhancements

For each concept alternative, decide which enhancements are needed. At this step in the development process, one can often predict the need for some enhancements depending on the nature of the application. Guides are required enhancements and should be added now. Pilots, visuals, assists, and guards, if needed, can usually be added to the attachment concept now. The remaining enhancements are normally added later when the detailed design is created.

Table 10.9 Cross-Reference Notes for Add Some Enhancements

10.3.5 Select a Concept for Analysis

To this point, snap-fit development has involved creativity within a structured process. The result is several fundamentally different and technically sound snap-fit attachment concepts. Each concept has constraint features arranged to provide proper mating to base-part constraint. Some enhancement features have also been added to each concept.

Each concept should now be reviewed by the appropriate stakeholders. Likely stakeholders include the product engineer/developer(s) for both the mating and base-parts, as well as cost analysts, purchasing agents, materials experts, part manufacturers and manufacturing, assembly, and process engineers.

The best concept is selected to be carried forward into design and recommendations for improvements may also be made. The other concepts, if judged feasible, can be ranked in order of preference and kept available should the selected design become unacceptable as the program proceeds. The models and sketches created to this point should be available for this concept review and can be valuable tools for explaining the details of each design. Table 10.10 is a worksheet for comparing alternatives when selecting the best concept.

Table 10.10 Worksheet for Select the Best Concept

	Concept Alternatives		
Criteria	#1	#2	#3
Constraint execution			
Efficient use of features			
Meets minimum requirements for enhancements			
Ease of assembly			
Estimated piece cost			
Supports the business case			
Ease of manufacturing			
Meets ergonomic requirements			

Blank rows are for additional criteria.

Whenever possible, plan for the prototype supplier to also be the production supplier. Plastic part tooling and processing require a thorough understanding of the application. Supplier input during the initial design can help ensure a functional design that can be reliably produced. Single sourcing of parts may be desirable for the same reasons. These kinds of decisions are often made based on purchasing and organizational procedures or policies, but the developer may be able to present a solid business case for a single knowledgeable supplier.

This concludes the concept development phase of the process. We exit Step 3 with a fundamentally sound attachment *concept* and move into the more familiar and traditional area of feature analysis and detailed design.

10.4 Step 4: Design and Analyze Features

In this chapter we are discussing the entire development process. So the flow of this discussion is not interrupted by an extensive digression into material, feature design, and analysis topics; we will save that detailed discussion for Chapters 11–13.

This section will briefly discuss some general issues associated with feature design and analysis.

To this point, we have been working with concepts and ideas, not dimensions or details. Step 4 is the traditional feature level step involving material property considerations and lock performance analysis. The objective is to determine dimensions for:

- Acceptable assembly and separation forces.
- Separation (retention strength).
- Acceptable strain levels during deflection for assembly and separation.
- Acceptable stress and strain levels under applied loads.
- Squeak and rattle resistance.

Note that analysis of any kind is of limited value unless the snap-fit interface is properly constrained.

In Step 4, the selected constraint features are designed and analyzed for performance. Typically, such an analysis is required only for lock features but, in some applications, locators may also require analysis.

Sometimes feature design is carried out based on experience and analysis is used only if testing indicates a need for it. In other cases, analytical methods are applied immediately to evaluate feature performance and determine feature dimensions.

Analysis may indicate that the selected features can be designed to meet all application requirements. Analysis may also indicate the need for additional constraint features for increased strength or retainer enhancement features to improve retention beyond the inherent strength of the lock-pair(s) in question. The results may also indicate that the preferred concept cannot be designed to meet the required objectives and that one of the alternative snap-fit concepts should be tried.

Recall that the purpose of the first three steps of the process is to create a fundamentally sound attachment concept. A concept can be sound and yet fail to meet one or more of the objectives because of a combination of material performance limitations and conflicting performance requirements. For example, a high retention strength requirement can be in conflict with a low insertion force requirement (an issue that *should* be resolved by proper lock style selection) or part material properties will not support the required assembly or retention performance. Why then should we spend our time creating a fundamentally sound attachment if it might not work? Because the fundamentally sound concept has the best chances of working as a snap-fit and because, as we will see in the next section, we have created an optimized interface that can also accept loose fasteners.

A properly constrained interface concept with locating features and enhancements now exists. When we recall that lock features are the last constraint features added to the design and the last features engaged during assembly, we see that, for many applications, replacing an integral lock with an alternative locking feature can be relatively simple and with relatively minor changes, the interface may be modified to accept loose fasteners.

10.4.1 Lock Alternatives

This development process begins with a rational decision that a snap-fit attachment is feasible for the application. But it is possible that we may find that any snap-fit lock feature simply cannot meet the performance requirements. In that case, an alternative fastening method is necessary.

Because constraint principles have been followed, the number of loose fasteners needed in the attachment is minimized.

A decision at this point to use *another* fastening method instead of a snap-fit is not the same as including a back-up fastening method as discussed in Chapter 9, Section 9.3.4.

Details of the fastening methods available as alternatives to snap-fit locks are beyond the scope of this book. The purpose here to introduce a few alternatives and give the reader a starting point for further investigation.

Some of the lock alternatives described here lend themselves to automatic assembly to the parent component. In that case, they can be installed before final assembly and, as far as assembly operators are concerned, they are integral locks and the assembly labor savings apply just as with integral locks. Even threaded fasteners can sometimes be preassembled to plastic parts so the operator does not need to handle loose fasteners. The mechanism used to capture the screw in the part can be a snap-fit feature.

Any fastening method using loose fasteners will add some cost to the attachment but by following the snap-fit development process and designing for proper constraint using locators, the number of loose fasteners is minimized as is the associated cost impact. Three common lock alternatives, screws designed for use in plastic, push-in fasteners, and spring clips are described here. Other attachment methods like hook-loop fasteners and double-back tape may also be alternatives to an integral lock feature.

10.4.1.1 Threaded Fasteners

In place of an integral lock, a loose threaded fastener, a screw, can be used. If locators are used and the attachment is properly constrained, the number of loose fasteners required will be minimized and the design will be optimized for design-for-assembly.

Do not use sheet metal or machine thread screws for tightening directly into plastic. Use special fasteners designed for that purpose. These screws use various combinations of a special thread form, thread pitch or cross-section shape to reduce the stresses produced as plastic is displaced to form the internal threads. Some styles also have flutes that cut plastic away to form the thread.

If a screw is to be used, some options are:

- An internal thread formed by a specialized screw in a pilot hole. The pilot hole is usually molded into a boss added to the part. Boss design guidelines are available from the resin suppliers or from the specialty screw manufacturers.
- When the feasibility of threading directly into plastic is in question, metal inserts with machine threads can be molded in-place or pressed into a hole in the part after molding. These inserts add cost to the process and to the parts but provide higher thread strength and allow the use of a machine thread screw. They also provide a solution where the screw must run into a wall and the extra length of thread engagement available with a boss is not possible.
- Add-on clips (spring-steel or other) carrying single or multiple internal threads. Sometimes called *J-nuts* or *U-nuts*, like inserts, they eliminates issues associated with threading directly into plastic bosses.

When choosing a thread-forming or a thread-cutting screw, consider the application's tolerance for the chips created when thread cutting screws are used and the properties (hardness/toughness are most important) of the plastic material in which threads are to be created.

Considerations when using screws in plastics include:

- A boss provided for a pilot hole may cause a sink mark on the opposite side of the wall on which it sits. Sink marks indicate high residual stresses and/or voids that will weaken the area. Design that area to eliminate the conditions causing sink marks and residual stress.
- Follow design guidelines for depth of core pin, pilot hole diameter, and wall thickness carefully.
- Special screws are required for use in plastics. Often the screw manufacturer will provide recommended plastic boss dimensions for these screws.
- If the screw enters the free end of the boss rather than at the base, allow for a stress relief area at the end of the boss by recessing the pilot hole in a slightly larger diameter area. The pin removal feature at the base of the pilot hole core pin in the mold can be used to form the recess.
- Screws with countersunk heads should not be used in countersunk holes in plastic. The wedging action of the screw head will crack the material unless the tightening process and clamp-load are very carefully controlled.
- Screws coated with oil or having an oil-based finish should not be used in plastic because some plastics degrade in the presence of oil.
- If necessary, especially over slots, distribute the pressure under the head of the screw over sufficient area by using a captive washer on the screw.
- If the assembly method is high speed tightening of the screw, a captive washer is recommended for screws tightened against plastic. Heat build-up due to friction under the head of the screw may melt the plastic if a washer is not present.

- High speed tightening may also cause heat build-up in the plastic threads during forming or cutting. This can degrade plastic properties in the threads themselves making them weak. Limiting the tightening speed may be necessary. Tightening at 1000 rpm is likely much too fast; tightening at ~300 rpm may be acceptable. Verify with testing.
- Screws will develop clamp load. Plastics tend to creep and a high clamp load may result in long-term cracking of the plastic under the fastener and/or in the boss.
- Over-tightening and stripping of the plastic threads is possible. This is a hidden condition and it may leave the assembly failure undetected.
- If screws are to be removed and reassembled into the plastic multiple times for repair, cumulative damage to the plastic threads is a possibility. Hand removal and retightening may be required.
- If the application does not permit perfect alignment of the screw to the pilot hole for assembly, then cracking of the boss during rundown is likely.
- The driving impression style on the screw head must be selected for screw stability during rundown. The driver bit must be selected for compatibility with the driving impression.

Sometimes loose fasteners can be captured in the plastic parts prior to final assembly. This eliminates operator handling and saves time. Some methods for capturing screws in parts are shown in Fig. 10.6.

References [6] and [7] provide discussions of many of the issues associated with loose fasteners in plastics.

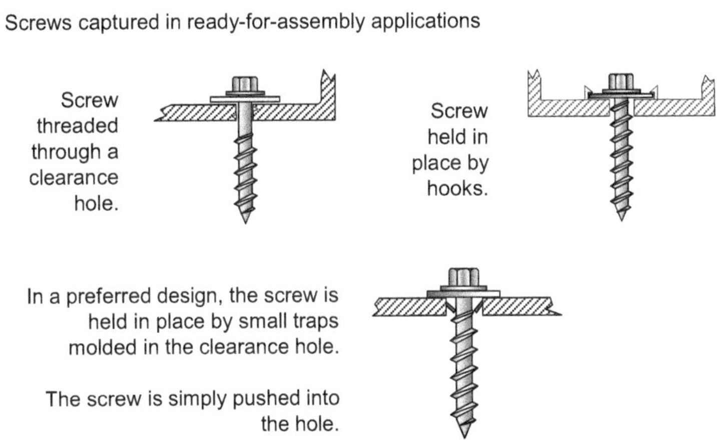

Figure 10.7 Threaded fasteners captured in plastic parts

10.4.1.2 Plastic Push-In Fasteners

We do not tend to think of push-in fasteners as snap-fits, but in many ways they are. Push-in fasteners are usually spring steel or plastic (Fig. 10.7) and they involve integral feature deflection and return for interference, just like a snap-fit lock. Like snap-fits, push-in fasteners do not generate any significant clamp load. The common one-piece style push-in style fasteners do not decouple assembly and retention performance but the two-piece fasteners do perform as level 4 decoupling devices, see Chapter 7.

Sometimes, push-in fasteners can be installed automatically in the part before reaching final assembly. In this case, as far as the assembly operator is concerned, they are integral snap-fits. As a developer, keep in mind that all the rules for good snap-fit design will apply to these applications. Other push-in fasteners are installed by the operator and will add cost as a separate part in the assembly process. Ergonomic limits on push-in forces apply to these fasteners when they are hand-installed. Removal for service can also be an issue. The popular ribbed plastic fasteners can be difficult to remove and may be damaged during removal. Some one-piece styles will also damage the mating-parts during removal unless the parts are designed to be quite strong in the area of the push-in fastener. The two-piece push-in styles, because they are a form of level 4 decoupling, can be very easy to remove when a feature is provided to assist in removal.

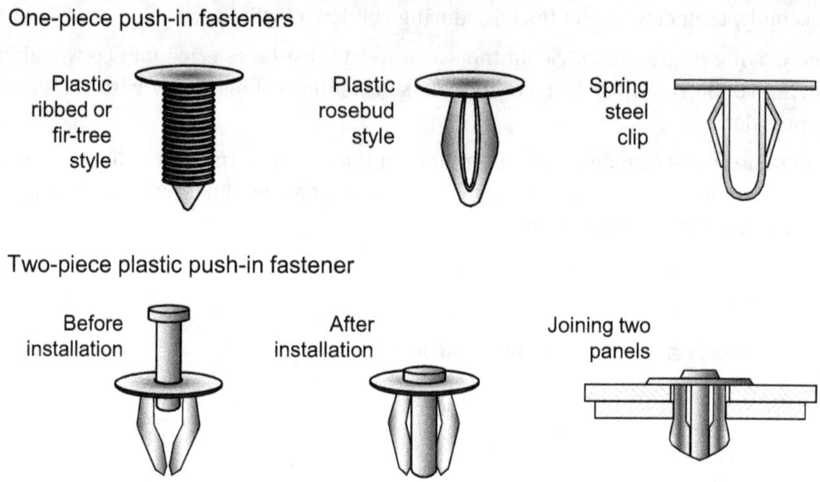

Figure 10.8 Push-in fastener alternatives to integral lock features

For some applications that use screws, simply replacing screws with push-in fasteners should be considered as an intermediate step for proof-of-concept in conversion to integral snap-fits. With little or no change to the parts, holes already provided for screws can be used as attachment sites for push-in fasteners.

10.4.1.3 Spring-Steel Clips

These fasteners, an example is shown in Fig. 10.8, are designed to grip features on the plastic parts to hold them together or as already discussed, to provide an internal thread for threaded fastening. Some styles enter final assembly already attached to a part. As with preinstalled push-in fasteners, the application is a snap-fit as far as the operator is concerned and, once again, all the rules of snap-fit design apply. These fasteners are frequently designed with sharp barbs intended to dig into and hold the parts together. The barbs can be effective, but they will cut grooves into the plastic when removed. This makes them less effective in retention when reassembled. Other clips, like the one in

Fig. 10.8, grip with a spring-like action and are friendlier to disassembly and reassembly.

Table 10.11 Cross-Reference Notes for Feature Analysis and Design

[illegible]

The appendix lists suppliers of plastic and steel fasteners that could be used as loose fastener alternatives to snap-fits. The list is far from all-inclusive, a more general search on the fastener style will locate more sources.

10.5 Step 5: Confirm Design with Parts

In this step, the first parts are produced and evaluated. Results of this initial evaluation will likely indicate the need for modifications to the design. Specific application performance requirements will determine how these parts are evaluated, but it is essential that an evaluation include evaluation/verification of:

See Tables 10.4 and 10.5, in Section 10.2, "Benchmarking," for additional review lists.

- Ease of assembly, including access, motions, assembly force, and operator feedback.
- User-feel if the customer will be operating the snap-fit frequently.
- Process-friendly design.
- Serviceability and/or maintenance, if expected.
- Proper constraint.
- Strength and resistance to squeak and rattle.

Evaluation and testing will indicate whether or not changes must be made to the design. Thorough understanding of the application as a result of following the snap-fit development process should make it easy to identify the changes needed. Chapter 14 describes a diagnostic process for investigating problems and recommending changes to snap-fit applications.

The attachment level development process is now completed. As a final review, the evaluation worksheet in Table 10.12 can be used to confirm that all aspects of attachment level snap-fit development have been considered.

There are also other checklists in this chapter, each with a different approach or emphasis on the product and the development process, and there is some redundancy between these lists. The reader can modify and use these list(s) in a manner that works best for them.

Table 10.12 Final Snap-Fit Evaluation (Page 1)

Basic Shapes	Mating-part is	Solid Panel Enclosure	
	Base-part is	Solid Surface Cavity Opening	
Function	Action is	Fixed or Moveable (free or controlled)	
	Attachment type is	Temporary or Final	
	Retention is	Permanent or Nonpermanent	
	Lock type is	Releasing or Nonreleasing	
Assembly motion	Preferred	Push Tip Slide Twist Pivot	
Strength and forces	Requirements and directions identified	**YES***	No
Alignment	Requirements and directions identified	**YES**	No
Packaging	Assembly & service/removal access	**EASY**	Difficult
	Clearance for part movements	**YES**	No
Material	Final material properties identified	**YES**	No
Geometry	Dimensions and tolerances identified	**YES**	No
Separation direction	Any significant forces	Yes	**NO**
If significant forces in separation direction	Retainer enhancements added	**YES**	No
	Trap locks and/or twist assembly motion	**YES**	No
	Decoupling considered	**YES**	No
Locators	Used as guides	**YES**	No
	Used as pilots (if needed)	**YES**	No
	Usage is maximized	**YES**	No
	Have clearance enhancements	**YES**	No
	Well distributed on panels to prevent squeak and rattle	**YES**	No
	Critical pair selected as datum for other constraint pairs	**YES**	No
	Simultaneous engagement of multiple locator features	Yes	**NO**
Locks	Usage is minimized	**YES**	No
	Carrying high or sustained forces	Yes	**NO**
	Well distributed on panels to prevent squeak and rattle	**YES**	No
	Simultaneous engagement of multiple lock features	Yes	**NO**
	Simple shapes	**YES**	No
	Engage strong locators	**YES**	No
	Constrain only in separation direction	**YES**	No

Table 10.12 Final Snap-Fit Evaluation (Page 1) *(Continuation)*

Benchmarking	Carried out on	Basic shapes Similar applications **BOTH**	
Constraint	Proper constraint verified	**YES**	No
	Under-constrained	Add constraint features where needed.	
	Over-constrained in opposition	Combine opposing constraint pairs.	
	Over-constrained by redundant features	Eliminate one constraint pair.	

* Where applicable, preferred choices are shown in **BOLD UPPERCASE** font.

Table 10.12 Final Snap-Fit Evaluation (Page 2)

Compatibility	Between basic shapes & assembly motion	**YES***	No
	Between locator-pairs & assembly motion	**YES**	No
	Between lock-pairs and assembly motion	**YES**	No
	Assembly & disassembly motions are the same	**YES**	No
Required enhancements	Guides	**YES**	No
	Feedback	**YES**	No
	Clearance	**YES**	No
	Process-friendly	**YES**	No
Recommended enhancements	Compliance	**YES**	No
	Allowance for fine-tuning	**YES**	No
Other enhance-ments as needed	Pilots Guards User-feel Visuals Assists Retainers Back-up locks	The need for these enhancements depends on specific application requi-rements.	
Feedback	Preferred	**TACTILE** Audible Visual **MULTIPLE**	
General design practices	Sharp internal/external corners	Yes	**NO**
	Thick sections	Yes	**NO**
	Sudden section changes	Yes	**NO**
	Gate location near constraint features	Yes	**NO**

Table 10.12 Final Snap-Fit Evaluation (Page 2) *(Continuation)*

Evaluate first parts from the mold	Assembly interference	Yes	**NO**
	Acceptable assembly force	**YES**	No
	Feature damage during assembly	Yes	**NO**
	Operator feedback	**YES**	No
	Compatibility & constraint	**YES**	No
	Attachment durability	**YES**	No
	Good perceived quality & user-feel	**YES**	No
	Part and feature consistency	**YES**	No
	Internal and external radii as specified	**YES**	No
	Any sink marks created by constraint features	Yes	**NO**
Fine-tuning	Establish location and close fits	Mold adjustments at critical alignment sites.	
Compliance	Take up tolerance	Opposite critical load bearing sites.	
	No looseness	Opposite critical alignment sites	

* Where applicable, preferred choices are indicated in **BOLD UPPERCASE** font.

10.6 Step 6: Fine-Tune the Design

In this step, changes indicated by the testing and evaluation in Step 5 are made. Any required mold changes will be easier if fine-tuning enhancements, Chapter 9, Section 9.4.2, have been included in the design.

Testing may indicate the need for lock feature improvements or the need for retainers. Enhancements like assembly guides, clearance effectiveness, assembly feedback, and user-feel that cannot be fully evaluated until actual parts are available can now be fine-tuned.

Multiple cycles of design confirmation with parts and fine-tuning may be necessary before the attachments are acceptable, but a major objective of this development process is to *minimize* the number of iterations required to get good parts.

Table 10.13 Cross-Reference Notes for Fine-Tune the Design

10.7 Step 7: Snap-Fit Application Completed

The application is now finished and verified. It is highly recommended that the application be classified according to basic shapes and assembly motions and captured in a searchable knowledge base. Lessons learned once the application is in production and any required design changes should also be captured in that knowledge base. See Chapter 15 for more information about how such a knowledge base or library of proven good snap-fit applications can provide valuable design information for future products.

10.8 Summary

This chapter explained the snap-fit development process in detail. By design, the snap-fit development process is conceptual and creative in the early steps before it is analytic. As the process is followed, it becomes second nature as users grow in their understanding of snap-fits.

Important points in Chapter 10:

- The snap-fit development process is compatible with and supports design-for-assembly (DFA) principles. Specific aspects of this process should be included in the DFA thought process and in DFA workshops. The concepts of generic basic shapes and assembly motions as well as the use of assembly motion to drive interface design alternatives are particularly important and applicable to DFA [9].
- By using basic shapes and combinations of assembly motions and engage directions, the development process encourages and enables improved spatial reasoning and creativity by developers when developing snap-fit attachment concepts.
- With only minor changes, the attachment level development process for snap-fits is applicable to other attachment development and design situations. Simply treat the lock selection step as allowing any mechanical fastener as a locking option rather than limiting the selection to only integral or snap-fit locks.
- The process encourages snap-fit developers to generate truly different concepts, not just variations on one theme. Take advantage of this opportunity.
- If a snap-fit is indicated, play it safe (within the parameters of the business case) and over-design the attachment if necessary. The snap-fit attachment will still save money over more conventional fastening methods.
- Developers must get their hands and their spatial reasoning involved in the creative process by making sketches, handling parts, and building models.
- Get others involved in the development process; many mechanically-minded people find that snap-fit applications are very interesting and fun, especially if they have parts to play with.

References

The appendix contains updated information for some of these references.

[1] Meeker, D.G., *Benchmarking, Its Role in Product Development*, Proc. 1994 Int. Forum Des. Manuf. Assem., Boothroyd Dewhurst Inc., Newport, RI (1994)

[2] Michalko, M., *Thinking Like a Genius*, THE FUTURIST, May (1998) pp. 21–25

[3] Woodsen, W.E., *Human Factors Design Handbook*, McGraw-Hill Book Company, New York (1981)

[4] Marras, W.S., *Biomechanics of the Human Body*, Handbook of Human Factors and Ergonomics, Salvendy, G., (Ed.), John Wiley & Sons, New York (1997) pp. 233–265

[5] Karwowski, W., Marras, W.S., *Work-related Musculoskeletal Disorders of the Upper Extremities*, Handbook of Human Factors and Ergonomics, Salvendy, G., (Ed.), John Wiley & Sons, New York (1997) pp. 1124–1172

[6] Malloy, Robert A., *Plastic Part Design for Injection Molding*, Hanser, Munich, Germany (1994)

[7] *Designing with Plastic – The Fundamentals, Design Manual TDM-1*, Ticona LLC (formerly Hoechst Celanese), Summit, NJ (1996)

[8] Tres, P., *Designing Plastic Parts for Assembly*, Hanser, Munich, Germany (2000)

[9] Bonenberger, Paul R., *An Attachment Level Design Process for Snap-Fit Applications*, ASME Int. Mech. Eng. Congr. Expo., Anaheim, CA, DE-Vol. 99, MED-Vol. 7 (1998)

11 Feature Development: Material Properties

This chapter introduces the plastic material properties used in feature performance calculations. Actual analysis will be discussed in Chapter 13. Four material properties normally appear in these calculations: stress (σ), strain (ε), modulus of elasticity (E), and coefficient of friction (μ).

Most of the time, snap-fit performance is not a consideration when a material is chosen for an application. The snap-fit must be made to work with the given material. A developer may, however, be able to influence selection of a particular formulation within a chosen material family if the case can be made for improved snap-fit performance.

With a basic understanding of properties and some analysis, the developer can identify potentially difficult situations early in the development process and know the questions to ask of a polymers expert.

Some polymer knowledge is relatively constant but the technology is always evolving. In particular, consider the various data tables in this chapter as useful examples and guidelines for the discussion, but always seek out the most current polymer material property data and design recommendations from industry sources, current literature, and the internet.

Refer to the appendix for material property data resources.

11.1 Sources of Material Property Data

Material property data, primarily stress-strain and strength information, is available in several forms; some more useful than others.

Product brochures contain information appropriate only for general product comparison or for initial screening of products for an application. It should not be used for part design or for snap-fit feature analysis.

Material data sheets represent actual laboratory test data. Data sheets are more detailed and useful than brochure information but are, of necessity, based on general use assumptions and specific test conditions. They only provide data at specific points and the data presented is subject to normal differences of data interpretation. If used for analysis, ensure the data represents the information needed (a supplier's terminology may not be the same as yours) and that you fully understand the conditions under which the data was generated. Be aware that test and sample preparation procedures may differ between suppliers. Data should be generated using standard testing methods, ISO Standards are one example.

There are also other online material databases. See the appendix.

Materials encyclopedias, supplier databases, and universal databases contain information similar to that in the materials data sheets. However, different test and sample preparation methods may make direct comparisons difficult. An exception is the CAMPUS® database [1]. CAMPUS® stands for Computer Aided Material Preselection by Uniform Standards. The CAMPUS® database contains data from many plastics producers and includes stress-strain curves and other mechanical, thermal, and electrical properties. One of its primary attractions is that the data is based on uniform testing to ISO Standards, making direct comparisons of material properties possible. The database is available to qualified customers of the member companies [3].

Some online databases also contain stress-strain curves.

Stress-strain curves generated directly from laboratory testing are the preferred form of data for snap-fit feature design and analysis. (These may be the same curves that someone else will interpret to generate material data sheets.) The actual material stress-strain curves allow the developer and the materials expert to interpret the data as they see fit for a particular application. The developer must still verify that the conditions under which the data was generated represent the application and if not, make appropriate allowances. Sometimes, stress-strain curves need to be generated for a particular set of conditions.

Much of the published stress-strain information is based on tensile testing. Tensile test data is desirable for tensile loading conditions and acceptable for other conditions when no other is available, but data generated in tests that represent actual loading conditions is preferred. For most snap-fit features, *bending* is the primary mode of deflection so stress-strain curves generated by flexural testing are preferred if they are available. For shear conditions, data generated by shear testing is desirable.

Part developers must recognize the nature of plastic and the many variables involved. The following quote is from one resin supplier's design guide [2] but it is applicable to material property data from all sources.

Values shown are based on testing of laboratory test specimens and represent data that fall within the standard range of properties for natural material ... These values are not intended for use in establishing maximum, minimum, or ranges of values for specification purposes. [The user] must assure themselves that the material as subsequently processed meets the needs of their particular product or use.

This means no matter how good the material property data is, it is the result of laboratory testing under standard conditions. These conditions cannot represent all the variables and conditions associated with any particular application. End use testing of production parts is required to verify performance.

11.2 Material Property Assumptions

Analysis calculations for plastic, unless otherwise noted, are based on three assumptions about the material. Plastics in production parts do not meet these assumptions without qualification, although some plastics come closer than others.

- The plastic is *linearly elastic.* The stress-strain curve is linear in the area of analysis. In reality, most plastics are not linear over the useful area of their stress-strain curve. We compensate for this by assuming a linear stress-strain relationship (the secant modulus) for the range of stress and strain in which we are doing analysis. The opposite of *elasticity* is *plasticity.*
- *The plastic is homogeneous.* The material's composition is consistent throughout the part and a small area of the part will have the same physical properties as the whole part. In reality, part composition depends on many factors, including raw material mixing, mold flow, and cooling. Proper mold and part design can help ensure that material properties in the areas of analysis are reasonably close to the predicted properties. Safety factors and conservative calculations also compensate for variation. The opposite of *homogeneity* is *heterogeneity.*
- *The plastic is isotropic.* The physical properties at any point in the material are the same regardless of the direction in which the sample is tested. In reality, filled and glass reinforced materials in particular do not exhibit isotropic behavior and the material data must indicate the direction of testing. Sometimes the data will only reflect the maximum performance direction. Proper part and mold design should ensure that the high performance properties are oriented in the correct directions in the final part. Values used in analysis should reflect anisotropic behavior if it exists. The opposite of *isotropy* is *anisotropy.*

These assumptions are necessary if we are to apply relatively simple calculations using traditional equations of structural analysis and the results are reasonable for most snap-fit analysis. One reason the results are acceptable is that predictive analysis of snap-fit behavior is not an exact science in any case. The effects of these assumptions on analysis accuracy are not as significant as are the effects of many other variables on the calculations. Some of these other effects are discussed in this chapter.

11.3 The Stress-Strain Curve

The most important information needed for analysis is the material's stress-strain relationship. This relationship is best represented in a stress-strain curve, a graph of stress vs. strain for a material under a given set of laboratory test conditions, see Fig. 11.1. The initial modulus is the slope of the stress-strain curve at relatively low stresses and strains. It starts at the origin and is immediately tangent to the curve. If the plastic exhibits some linear behavior, the initial modulus will be the slope up to the proportional limit. Or, if the initial portion is nonlinear, the initial modulus may be reported as a secant modulus at a given percentage of strain offset from the origin.

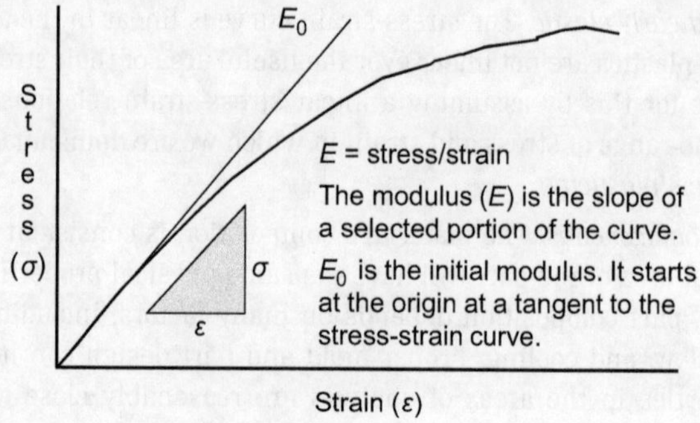

Figure 11.1 The basic stress-strain curve

Unlike a stress-strain curve for steel, where the curve's shape is characteristic of all steels, stress-strain curves for plastics may be quite different from material to material. This is another reason why it is good to get the material's actual stress-strain curves when doing an analysis. Some typical stress-strain curves for plastics are shown in Fig. 11.2. Note that some points do not appear on every curve.

Figure 11.2 and the definitions of terms that follow are from *Designing with Plastic – The Fundamentals, Design Manual TDM-1*, courtesy of Ticona LLC [2]. In the author's opinion, this document is an excellent blend of material and design information for the snap-fit developer who is not a polymer expert. It is highly recommended.

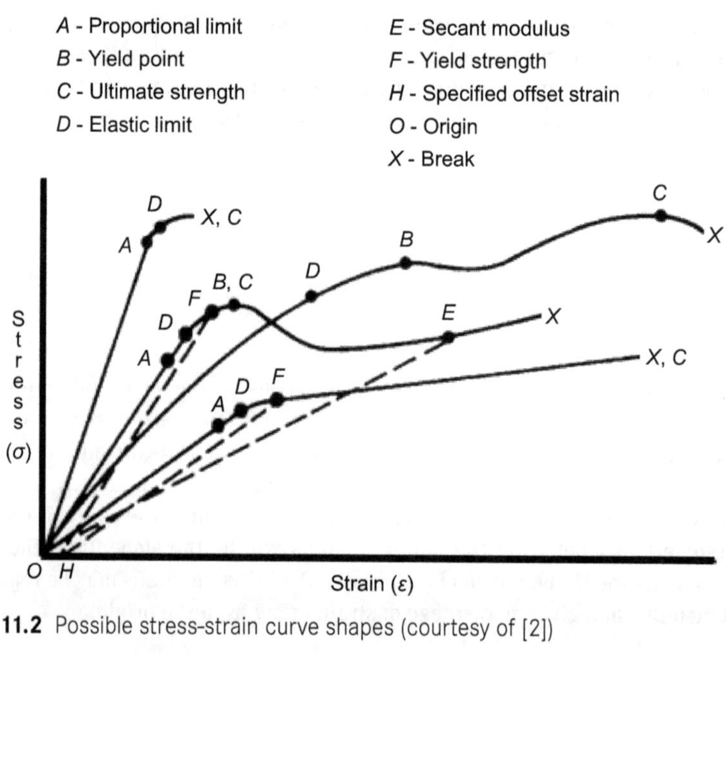

Figure 11.2 Possible stress-strain curve shapes (courtesy of [2])

A - Proportional limit: With most materials, a point exists on the stress-strain curve where the slope begins to change and linearity ends. The proportional limit is the greatest stress at which a material is capable of sustaining the applied load without deviating from constant proportionality of stress to strain. This limit is expressed as a pressure in MPa (or in psi). Note that some materials maintain this proportionality for large measures of stress and strain, while others show little or no proportionality, as previously discussed.

B - Yield point: The yield point is the first point on the stress-strain curve where an increase in strain occurs without an increase in stress. The slope of the curve is zero at this point. Note that some materials may not have a yield point.

C - Ultimate strength: The ultimate strength is the maximum stress a material withstands when subjected to an applied load. This is also a pressure expressed in MPa (or psi).

D - Elastic limit: Many materials may be loaded beyond their proportional limit and still return to zero strain when the load is removed. Other materials, particularly some plastics, have no proportional limit in that no region exists where the stress is proportional to strain. However, these materials may also sustain significant loads and still return to zero strain when the load is removed. In either case, the elastic limit on the stress-strain curve is the point beyond which the material is permanently deformed (will not return to zero strain) when the deflecting load is removed.

E - Secant modulus: The secant modulus is the ratio of stress to corresponding strain at any point on the stress-strain curve. In Fig. 11.2 the secant modulus at Point *E* is the slope of the line *OE*. Also, recall from the section introduction that, if the stress-strain curve is non-linear from the origin, the initial modulus may be reported as a secant modulus at some offset expressed as a percentage of strain.

F - Yield strength: Some materials do not exhibit a yield point. For such materials, it is desirable to establish a yield strength by picking a stress level beyond the elastic limit. Although developed for materials that do not exhibit a yield point, this value is often used for plastics that have a very high strain at the yield point to provide a more realistic yield strength. The yield strength is generally established by constructing a line parallel to *OA* at a specified offset strain: Point *H*. The stress where this line intersects the stress-strain curve at Point *F* is the yield strength at *H* offset. For instance, if Point *H* were at 2 % strain, then point *F* is the *yield strength at a 2 % strain offset.*

Three basic types of plastic stress-strain behavior are shown in Fig. 11.3. Toughness is a measure of a material's resistance to impact loads and is represented by the area under the stress-strain curve. Thus, the rigid (brittle) and flexible materials represented here will have lower toughness than the ductile material. Tough plastics are the preferred materials for snap-fits. Snap-fits in brittle plastics require very careful design and analysis with particular caution if impact loads are present in the application. Flexible materials normally do not lend themselves to snap-fits.

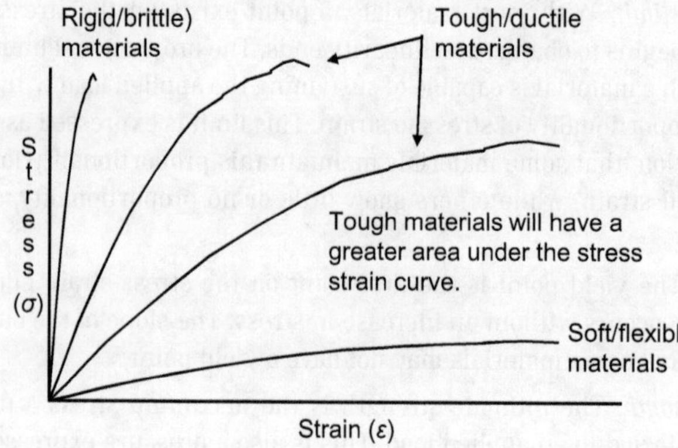

Figure 11.3 Plastic toughness, brittleness, and flexibility, adapted from [2] and [3]

The stress-strain curve is an important source of information for feature analysis. If one is not available, a reasonable representation can sometimes be constructed [3] from the information provided on the material data sheet, see Fig. 11.4. From this constructed curve, other values needed for analysis can be estimated. Of course, the accuracy of this curve (or any stress-strain curve) must be taken into account when interpreting analysis results. If a stress-strain curve must be constructed, it is useful to find representative curves for similar materials from the same family. These curves will provide a good idea of the general shape of the curve for the material in which you are interested. Refer to Fig. 11.2 for an idea of how various curves might be shaped depending on which data points are available.

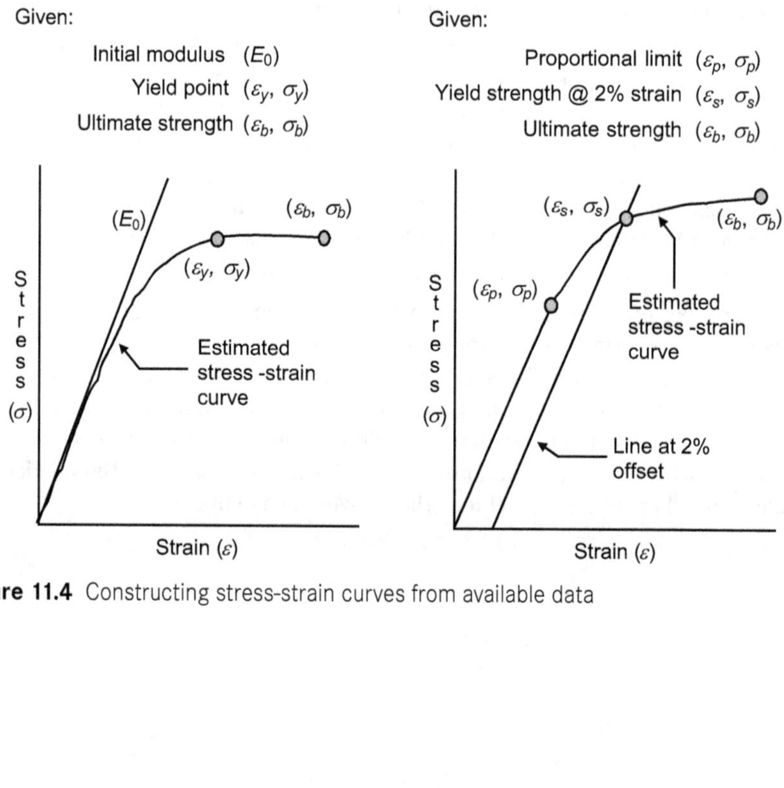

Figure 11.4 Constructing stress-strain curves from available data

11.4 Determining a Design Point

For setting preliminary design targets, the strength values on standard product data sheets can be multiplied by the percentages shown in Table 11.1. [3]. Remember, this is for *early* analysis and product screening and short-term or intermittent loading only, not for setting final design requirements.

Reference [3] also provides a deeper discussion of safety factors and the introduction of other factors reflecting materials and processing effects into determining the design point.

Table 11.1 Maximum Strength Estimates for Preliminary Part Design [3]

	When feature failure is not critical	When feature failure is considered critical
For intermittent loads (not repetitive loads)	25-50%	10-25%
For constant loads	10-25%	5-10%

For final analysis and design decisions, establishing the design point from a stress-strain curve is recommended. The design point represents the maximum stress and strain allowed in the feature being analyzed. The design point also establishes the secant modulus. It may be necessary to determine several design points using several stress-strain curves, each one representing a different condition under which the snap-fit is expected to perform. It may also be necessary to ask the supplier to generate curves representing specific conditions for the application.

Conditions for which a design point should be established may include both a short and long-term timeframe if the material can degrade over time due to its properties or as a result of specific environmental conditions.

Typical short-term conditions may be a new/fresh material at room temperature. This would be generally appropriate for evaluating initial assembly behavior unless initial assembly involves temperature extremes or aged material. Typical long-term conditions could include load history, expected number of assembly/disassembly cycles, thermal, ultraviolet and chemical aging effects, material creep properties, and ambient temperature effects.

Once stress-strain curves have been obtained, the following guidelines can be used to establish an initial design point for each curve. The graphs in Figs. 11.5-11.7 were created using information found in [4].

11.4.1 Applications with Fixed Strain

These are applications in which a feature is deflected during assembly then remains at some level of deflection for the life of the product. This is a long-term load/deflection condition where the fixed load/strain may be anywhere from low to high.

- For ductile and high-elongation plastics set the maximum permissible strain at 20% of the yield point or yield strain, whichever is lower, and
- For brittle and low-elongation plastics that don't exhibit yield set the maximum permissible strain at 20% of the strain at break, see Fig. 11.5.

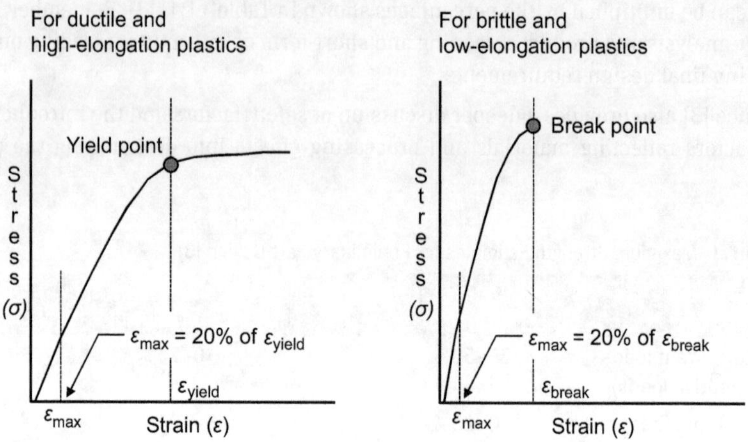

Figure 11.5 Design points for fixed strain applications

11.4.2 Applications with Variable Strain

The assembly process itself involves a strain increase. When deflections occur very rapidly, as in assembly or impact loading, feature analysis should be based on *dynamic strain*, not on stress or static strain. Because of the time-dependence of plastic behavior, it is very possible for calculated stress to exceed the stress at yield without causing damage when deflection occurs rapidly. When loads or deflections occur rapidly, as they do during assembly, use the dynamic strain limit when determining the design point. When loads or deflections occur more slowly, as during disassembly or sustained loading, maximum allowable stress or static strain values can be used in the calculations.

Some suppliers may recommend a maximum working stress level in their material design information or use stress for evaluating long-term loading conditions.

In the analysis procedures presented in this book we will use strain as our target or design value. Other analyses sometimes use stress. A polymers expert will be the best judge of which one to use.

Two material situations apply in applications with variable strain:

- *Materials with a definite yield point*:
  - For a low number of assembly/disassembly cycles (1 ~ 10 cycles), set the maximum permissible strain at 70% of the strain at yield, see Fig. 11.6.
  - For higher assembly/disassembly cycles (> 10 cycles), set the maximum permissible strain at 40% of the strain at yield, see Fig. 11.6.

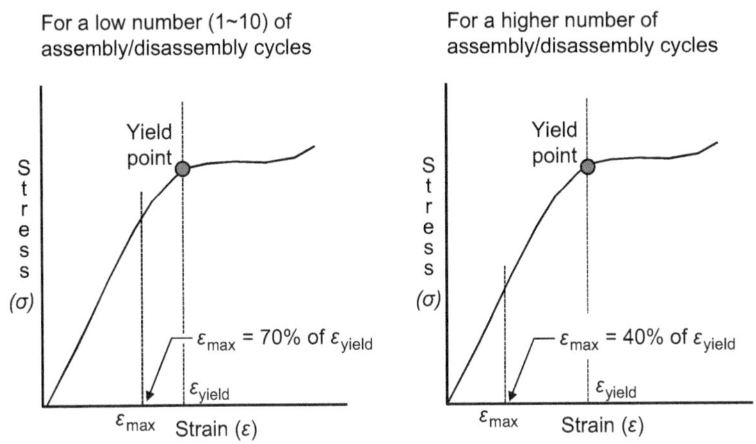

Figure 11.6 Design points for variable strain applications and materials with a definite yield point

- *Materials without a definite yield point*:
  - For a low number of assembly/disassembly cycles (1 ~ 10 cycles), set the maximum permissible strain at 50% of the strain at break, see Fig. 11.7.
  - For higher assembly/disassembly cycles (> 10 cycles), set the maximum permissible strain at 30% of the strain at break, see Fig. 11.7.

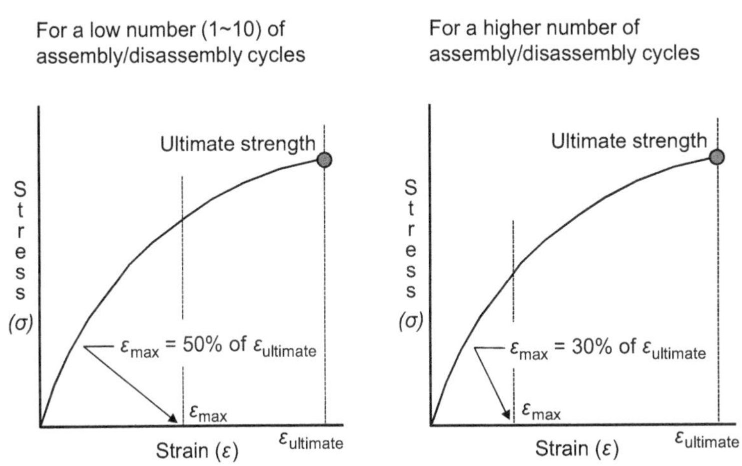

Figure 11.7 Design points for variable strain applications and materials without a definite yield point

11.4.3 The Secant Modulus

Once a design point is established, the secant modulus (E_s) is the slope of a line from the origin through the design point, see Fig. 11.8. The secant modulus will be used in the analysis calculations where we assume the plastic material behavior follows the line of the secant modulus rather than the actual stress-strain curve. (Use of the secant modulus reflects our assumption of elastic behavior in the application's working range.)

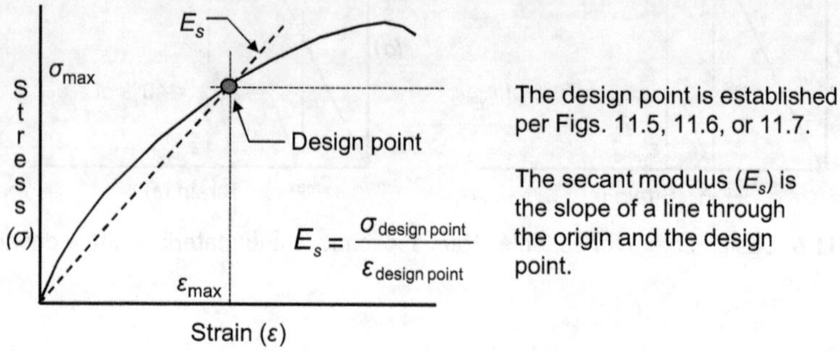

Figure 11.8 Calculating the secant modulus from the design point

11.4.4 Maximum Permissible Strain Data

Maximum permissible strain values for some material families are shown in Table 11.2. They can be useful for initial performance calculations, but always verify actual allowable strain values for the expected usage environment with the supplier or a polymer expert.

Table 11.2 Maximum Permissible Strain

Material	Typical ε_{max}	Source
Most plastics fall within	1–10%	X
Glass filled plastics tend to fall within	1–2%	X
Polypropylene PP	8–10%	X
Polycarbonate (30% glass-fiber reinforced) PC	1.8%	X
Polyphenylene sulfide (40% glass-fiber reinforced) PPS	1%	X
High heat polycarbonate PC	4%	B
Polycarbonate/ABS blend	2.5%	B
Acrylonitrile-styrene-acrylate ASA	1.9%	B
Polycarbonate blends	3.5%	B
Polycarbonate PC	4%	B
Polyamide (conditioned) PA	6%	B
Polyamide (dry) PA	4%	B
Polyamide/ABS	3.4%	B

Table 11.2 Maximum Permissible Strain *(Continuation)*

Material	Typical ε_{max}	Source
Acrylonitrile-butadiene-styrene ABS	1.8%	B
Polycarbonate (10% glass reinforced)	2.2%	B
Polyamide/ABS (15% glass reinforced)	2.2%	B
Polycarbonate (20% glass reinforced)	2%	B
Polyamide conditioned (30% glass reinforced)	2%	B
Polyamide dry (30% glass reinforced)	1.5%	B
Polyetherimide PEI	9.8%	A
Polycarbonate PC	4-9.2%	A
Acetal	1.5%	A
Nylon 6 (dry)	8%	A
Nylon 6 (30% glass reinforced)	2.1%	A
Polybutylene terephthalate PBT	8.8%	A
Polycarbonate/Polyethylene terephthalate PC/PET	5.8%	A
Acrylonitrile-butadiene-styrene ABS	6-7%	A
Polyethylene terephthalate PET (30% glass reinforced)	1.5%	A

B - *Snap-fit Joints for Plastics a design guide*, Polymers Division, Bayer Corp., 1998
A - *Modulus Snap-Fit Design Manual*, Allied Signal Plastics, 1997
X - Unidentified

- Find more current information from the sources listed in the appendix.
- Materials in the table are unreinforced unless noted otherwise.
- These values are for short-term strain and low cycle or single cycle operation. For multiple cycles, use 60% of the values shown.
- The strain data is at room temperature.
- *Conditioned* refers to standard test conditions of 50% relative humidity and 20 °C, unless other specific humidity/temperature conditions are noted.
- *Dry* means low or no moisture content. Often it is *dry as molded.*

This concludes the discussion of stress and strain. Important points to remember include:

- Published data in brochures is acceptable for initial screening but material data sheets and preferably actual stress-strain curves should be used to establish the design points for final analysis.
- Use stress-strain data that represents actual application conditions.
- Maximum permissible strain tends to be higher for ductile and lower for brittle polymers.
- Recognize that many conditions may affect the actual maximum permissible strain and that end-use testing is necessary to verify predicted performance.

■ 11.5 Coefficient of Friction

Coefficient of friction is related to a material's lubricity. It is a measure of one material's ability to slide across another without galling or other surface damage. Materials with good lubricity will tend to have lower coefficients of friction. Those with poor lubricity will tend to have higher coefficients of friction. Information about a material's lubricity can sometimes be found in its data sheet.

In lock feature calculations, the coefficient of friction relates the normal force on the retaining member's assembly or retention face to the force required to slide a mating feature on that face. It is used when calculating assembly and separation forces.

Some published coefficient of friction values are shown in Table 11.3. However, be aware that published values are based on specific tests and materials that may have little or no relation to a specific application or to the common snap-fit condition of an edge sliding over a retention feature surface. The best source of friction data is testing under actual conditions, but this is rare. Use the published data along with lubricity information and your own judgment to determine a coefficient of friction. From the data shown, one can see that values of μ range from 0.2 to 0.7. For initial analysis, unless other information is available, values of 0.2 for low friction materials and 0.4 for high friction materials are reasonable estimates. Coefficient of friction variability will affect the reliability and accuracy of assembly and separation calculations.

The data identified as from source *T* in Table 11.3 is associated with information on spin-welding and was most likely developed with that technology in mind. However, it is useful in that it shows the kind of variation that can occur depending on the test. For example, note the difference between steel vs. polypropylene and polypropylene vs. steel. Unless obtained through actual testing of the application, coefficient of friction data should be considered by the developer as information that will allow for an *educated estimate* of the friction value(s) for use in analysis.

It may be desirable to have estimates of the coefficient of friction under both static and dynamic conditions. Again, unless application specific tests are run to generate this data, an educated adjustment to the available published data will be necessary. As we'll discuss in Chapter 13, the variability and uncertainty of published friction data can often make the distinction between static and dynamic friction meaningless.

Table 11.3 Published Coefficients of Friction

Material	μ	Source	Notes
Polyetherimide PEI	.20 - .25	A	*
Polycarbonate PC	.25 - .30	A	*
Acetal	.20 - .35	A	*
Nylon 6	.17 - .26	A	*
Polybutylene terephthalate PBT	.35 - .40	A	*
Polycarbonate/Polyethylene terephthalate PC/PET	.40 - .50	A	*

Table 11.3 Published Coefficients of Friction *(Continuation)*

Material	μ	Source	Notes
Acrylonitrile-butadiene-styrene ABS	.50 - .60	A	*
Polyethylene terephthalate PET	.18 - .25	A	*
Polytetrafluoroethylene PTFE	.12 - .22	B	**
Polyethylene PE rigid	.20 - .25 (2.0)	B	**
Polypropylene PP	.25 - .30 (1.5)	B	**
Polyoxymethelene; Polyformaldehyde POM	.20 - .35 (1.5)	B	**
Polyamide PA	.30 - .40 (1.5)	B	**
Polybutylene terephthalate PBT	.35 - .40	B	**
Polystyrene PS	.40 - .50 (1.2)	B	**
Styrene acrylonitrile SAN	.45 - .55	B	**
Polycarbonate PC	.45 - .55 (1.2)	B	**
Polymethyl methacrylate PMMA	.50 - .60 (1.2)	B	**
Acrylonitrile-butadiene-styrene ABS	.50 - .65 (1.2)	B	**
Polyethylene PE flexible	.55 - .60 (1.2)	B	**
Polyvinyl chloride PVC	.55 - .60 (1.0)	B	**
Slider specimen vs. plate specimen	Tested at 10.6 mm/sec		
Polypropylene (as molded) vs. Polypropylene (as molded)	.71	T	***
Nylon (as molded) vs. Nylon (as molded)	.65	T	***
Polypropylene (abraded) vs. Polypropylene (abraded)	.27	T	***
Nylon (machined) vs. Nylon (machined)	.47	T	***
Mild Steel vs. Polypropylene (abraded)	.31	T	***
Mild Steel vs. Nylon (machined)	.30	T	***
Polypropylene (abraded) vs. Mild Steel	.38	T	***
Nylon (machined) vs. Mild Steel	.40	T	***

A - *Modulus Snap-Fit Design Manual*, Allied Signal Plastics, 1997

B - *Snap-fit Joints for Plastics - a design guide*, Polymers Division, Bayer Corp., 1998 and [4]

T - *Plastic Process Engineering*, James L. Throne, Marcel Dekker Inc., 1979 [5]

* The values are for the given material tested against itself.

** Values are for the material tested against steel. Friction between different plastics will be equal to or slightly lower than these values. Friction between the same materials will generally be higher; a multiplier is shown in parenthesis if it is known.

*** Unlubricated tests, dynamic coefficient of friction.

■ 11.6 Other Effects on Material Properties

Plastic materials have many other properties that, while they do not appear in the calculations, can influence analysis because of their effect on stress and strain behavior. Some will also affect the dimensional stability of the parts.

Additives are chemicals added to enhance certain functional or processing capabilities of a plastic. Because additives may adversely affect mechanical properties, they can affect snap-fit feature performance. Examples of additives include impact modifiers, UV stabilizers, coloring agents, and flame-retardants.

Plastics will exhibit *accelerated aging* at elevated temperatures. All plastics will experience degradation of mechanical properties at elevated temperatures over the long term. A comparison of thermal stability values will indicate the severity of the degradation. Sometimes stress-strain curves are generated to show performance at elevated temperatures.

Creep is a relatively long-term increase in strain (i. e., deflection) under a sustained load. The rate of creep for a material depends on the applied stress, temperature, and time. Stress-strain curves showing the effects of long-term creep are required for long-term performance analysis. From these curves, a creep modulus can be determined and used in the calculations.

Plastic properties are sensitive to *temperature effects*. In general, materials become softer and more ductile and the modulus decreases with increasing temperature. The deflection temperature under load (DTUL), also called the heat deflection temperature or HDT, is a single point measurement that may be useful for quality control or for initial screening of materials for short-term heat resistance. However, the DTUL value should not be used as design data.

Fatigue endurance For applications subjected to cyclic loads, SN (stress vs. number of cycles to failure) curves can be generated. Cyclic loading, particularly reversing loads, can significantly reduce the life of a plastic part.

Notch sensitivity is the ease with which a crack propagates through a material from a notch, initial crack, or a corner. A stress concentration factor related to the effect of sharp corners on local stress should be included in all calculations.

Chemical and ultra-violet effects may degrade mechanical properties. In general, as temperature and/or stress level increases, the plastic's resistance to these other effects will decrease.

Mold design and part processing can affect feature performance. Thick sections and improper cooling can cause voids or internal stresses. Mold flow patterns, knit lines, and placement of gates can adversely affect feature strength. Identical features in different areas of a part may have different strength and strain capabilities.

Plastic behavior is *rate dependent*. This means it is affected by the speed of the applied load. Stress-strain tests are conducted at a standard speed and may not represent the actual rate of loading in an application. For a given plastic, a high load rate will typically result in behavior similar to that at a low temperature: more rigid and brittle. A slow load rate results in behaviors similar to high temperature behavior (more ductile and flexible), see Fig. 11.9.

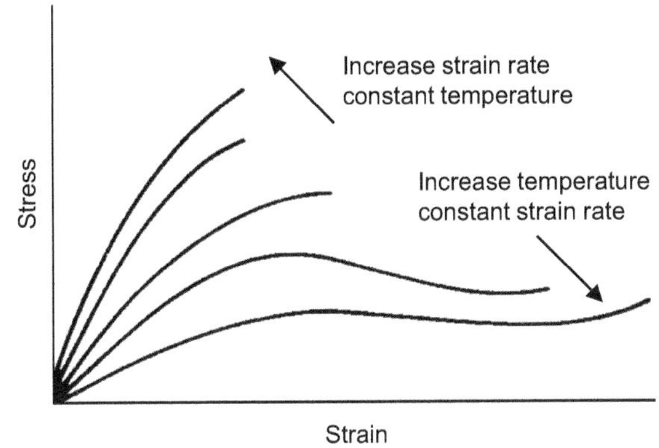

Figure 11.9 Effects of temperature and strain rate on stress-strain behavior (from [2])

The amount of recycled content or *regrind*, as well as the effectiveness of the material mixing process (for uniformity) prior to molding, can affect mechanical properties and part-to-part consistency.

Stress relaxation is a relatively long-term decrease in stress under a constant strain. (Creep involves constant stress while stress relaxation involves constant strain.) Data similar to creep data can be generated and a relaxation modulus determined, but relaxation data are not as available as creep data. The creep modulus can be used as an approximation of the relaxation modulus.

Toughness is the ability to absorb mechanical energy (impact) through elastic or plastic deformation without fracturing. Material toughness is measured by the area under the stress-strain curve. Tests for impact resistance under specific conditions include the Izod and Charpy tests of notched specimens, the tensile impact test, and the falling dart impact test.

Moisture content in some polymers, due to both ambient humidity and moisture in the environment can affect mechanical properties (especially stiffness), electrical conductivity, and dimensional stability. Materials with low water absorption will have better dimensional stability. For susceptible polymers, mechanical property data is often given at two humidity conditions: Dry as molded (DAM) and 50% relative humidity. Nylon is particularly susceptible to moisture content; use *impact modified nylon* to minimize moisture sensitivity.

Coefficient of Linear Thermal Expansion (CLTE) is a measure of the material's linear (not volumetric) dimensional change under temperature changes. The lower the materials' CLTE values, the greater the dimensional stability and the effects of thermal expansion and contraction will be minimized. Table 11.4 shows CLTE values for some plastics and, for comparison, some common metals.

In general, linear thermal expansion will be less of an issue when CLTE values are similar for both mating and base parts.

Use CLTE to estimate compliance requirements in the interface, particularly when the distance between opposing part features is high. When these conditions exist, it also becomes even more important to avoid over-constraint caused by those opposing features.

Mold shrinkage is the percentage of shrinkage as a part cools from the actual mold shape and it will affect final part dimensions. In general, amorphous plastics have lower shrinkage than crystalline and glass-filled are lower than unfilled or *neat* plastics. Reference [6] is an excellent source of tolerance data for a wide variety of polymers.

Table 11.4 Published Coefficients of Linear Thermal Expansion (CLTE)

Material	in./in. / °F 10^{-5}	cm/cm / °C 10^{-5}
Liquid Crystal (GR*)	0.3	0.6
Glass	0.4	0.7
Steel	0.6	1.1
Concrete	0.8	1.4
Copper	0.9	1.6
Bronze	1.0	1.8
Brass	1.0	1.8
Aluminum	1.2	2.2
Polycarbonate (GR)	1.2	2.2
Nylon (GR)	1.3	2.3
TP Polyester (GR)	1.4	2.5
Magnesium	1.4	2.5
Zinc	1.7	3.1
ABS (GR)	1.7	3.1
Polypropylene (GR)	1.8	3.2
Epoxy (GR)	2.0	3.6
Polyphenylene Sulfide	2.0	3.6
Acetal (GR)	2.2	4.0
Epoxy	3.0	5.4
Polycarbonate	3.6	6.5
Acrylic	3.8	6.8
ABS	4.0	7.2
Nylon	4.5	8.1
Acetal	4.8	8.5
Polypropylene	4.8	8.6
TP Polyester	6.9	12.4
Polyethylene	7.2	13.0

Courtesy of Ticona LLC, *Designing with Plastic – the Fundamentals*. Also see [7] for additional CLTE data.
* GR indicates a glass-reinforced material.

11.7 Summary

This chapter provided an introduction to important material properties related to feature strength and analysis. Methods for determining a design point were given. The details of feature analysis will be discussed in Chapter 13.

Use the tabulated material property information in this book as guidance, but consult a polymers expert and refer to the many online resources and current printed material property information when researching material properties for application development.

Important points in Chapter 11:

- Use material property data from product brochures and sales literature only for initial screening and rough estimates of performance.
- Material data sheets can provide more application specific data and more complete data than brochures. Use this data from for more accurate calculations for initial evaluation and design. Recognize that many conditions may affect the actual maximum permissible strain and that end-use testing is necessary to verify predicted performance.
- Actual stress-strain curves are the preferred source for stress-strain data.
- Use stress-strain data that represents actual application conditions whenever possible, but no matter how representative the data are with respect to the application, end-use testing is the only way to verify feature performance.
- Maximum allowable strain tends to be higher for ductile polymers and lower for brittle materials.

References

The author relied heavily on references [2, 3, and 4] in creating this chapter on plastic materials. Reference [3] is a very detailed and in-depth book and Chapter 3, Section 3.4, was particularly helpful. All are highly recommended resources for anyone responsible for plastic part design who may not be a plastics expert. There are many other excellent sources of information on the subject, these are the ones I have used and liked.

The appendix contains updated reference material information.

[1] CAMPUS, [Computer Aided Material Preselection by Uniform Standards] is a registered trademark of Chemie Wirtschaftsforderungs-Gesellschaft (CWFG).

[2] *Designing with Plastics - The Fundamentals, Design Manual TDM-1*, Ticona LLC (Formerly Hoechst Celanese Corporation, now a division of Celanese AG) Summit, NJ (2000) www.ticona.com

[3] Malloy, Robert A., *Plastic Part Design for Injection Molding*, Hanser, Munich, Germany (1994)

[4] *Snap Joints and Springs in Plastics*, Bayer Plastics Business Group, Bayer AG KU-Europe (2000) www.plastics.bayer.com

[5] Throne, James, L., *Plastics Process Engineering*, Marcel Dekker Inc., New York (1979)

[6] *Standards and Practices of Plastics Molders*, 1998 ed., Molders Division of The Society of the Plastics Industry Inc. Washington, D. C. (1998)

[7] Tres, Paul A., *Designing Plastic Parts for Assembly*, Hanser, Munich, Germany (2000)

Bibliography

Trantina, G. G., Minnicbelli, M. D., *Automated Program for Designing Snap-Fits,* GE Plastics, Plast. Eng., Pittsfield, MA, Aug. (1987)

Rackowitz, David R., *Beyond the Data Sheet – Developer's guide to the interpretation of data sheet properties*, BASF Plastic Materials, Wyandotte, MI

Designing Cantilever Snap-Fit Latches for Functionality, Technical Publication #SR-402, Borg-Warner Chemicals

Smith, Zan, *It's a SNAP!*, Hoechst Celanese Corporation, Assembly Magazine, Summit, NJ, Oct. (1994)

Roy, Dhirendra C., *Snap-Finger Design Analytics and Its Element Stiffness Matrices*, United Technologies Automotive, SAE Technical Paper Series (SP-1012), Int. Congr. Expo. (1994)

Standard Test Method for Kinetic Coefficient of Friction of Plastic Solids, ASTM Standard D 3028, ASTM Committee D-20 on Plastics

Sopka, D., Fletcher M., Smith Z., *The Give and Take of Plastic Springs*, Mach. Des., Nov. (1997) pp. 69–72

Noller R., *Understanding Tight-Tolerance Design*, Plast. Des. Forum, March/April (1990) pp. 61–72

12 Lock Feature Development: Rules-of-Thumb

Lock design is often an iterative process and rules for establishing feature dimensions can provide a reasonable starting point for lock analysis. These rules are not hard and fast, and with some experience and awareness of brittle vs. ductile behavior and material properties, the developer can bias the rules in the right direction when setting initial feature dimensions.

Look for gross violations of these rules.

The rules are also useful in other ways:

- A developer can quickly eliminate from consideration lock features or applications with dimensional requirements that fall *far* outside of the rules. For example, if an application calls for a cantilever hook feature with a 2:1 length to thickness ratio, calculations are not needed to know there will be problems with that feature. If the L/T ratio is 4:1, the developer *could* decide to proceed with caution.
- When diagnosing feature failures, root causes and possible fixes can often be quickly identified.
- Knowledge of these rules can help make product benchmarking more meaningful.

Some of these guidelines are related to processing capabilities and following them can help avoid marginal processing situations that may cause inconsistent feature performance. As always, feature performance, especially on critical applications, must be verified by analysis and end-use testing.

The L/T ratio mentioned throughout this chapter refers to the L_b and T_b dimensions shown in Fig. 12.1.

Most lock features are based on cantilever beams as the deflecting member while the catch configuration is a common retaining member. The catch as a retaining member also appears on other deflecting members. The primary focus of this chapter is on the beam and the catch.

In cases where the application is low demand, the lock feature(s) are simple beams and the material falls in the *tough* range of the stress-strain curve, these rules-of-thumb may be sufficient, without further analysis, to provide final feature dimensions for parts suitable for end-use testing.

12.1 Beam-Based Locks

In Chapter 6, Section 6.8, we identified two major areas of a lock feature: the deflecting member and the retaining member. In Fig. 12.1, the lock feature, a cantilever hook, consists of a catch at the end of a beam. Refer to Chapter 6 for examples of the many deflecting/retaining member combinations available with beam-based locks.

During engagement and separation, the retaining member applies a deflection force and a moment at the beam's end. Rotation due to the moment is ignored in most calculations, including those in this book. If the rotation is considered significant and important, it can be included in the performance calculations.

The example we will present first is the common cantilever hook lock configuration; a catch at the end of a beam.

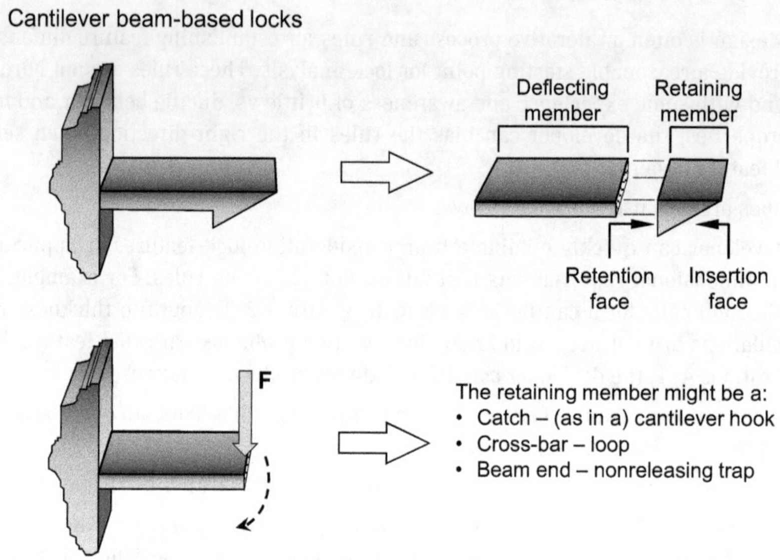

Figure 12.1 Deflecting and retaining members of a lock feature

We'll use the deflecting member as the starting point for applying these rules-of-thumb. Applicable beam dimensions are shown in Fig. 12.2.

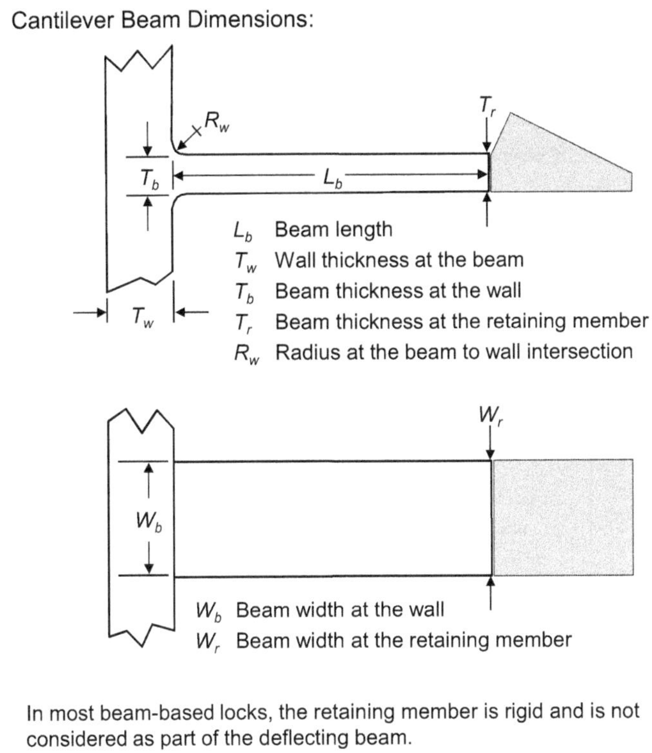

Figure 12.2 Beam dimensions

12.1.1 Beam Thickness at the Base

The dimensions and geometry of the lock's mounting area on a part are usually predetermined and they are the first limiting parameters on lock design. So we will start there, where the beam meets the parent material.

A beam may extend from a wall or surface in many ways, the most common are a 90° protrusion or in-plane at 180°, see Fig. 12.3.

If the beam protrudes from a wall, then the beam's thickness at the base (T_b) should be about 50 to 60% of the wall thickness. Beams thinner than 50% may have filling and flow problems. Beams thicker than 60% may have cooling problems at the base because of the thick section. This can lead to high residual stresses and voids which will weaken the feature at its point of highest stress. Sink marks, which may appear on the opposite side of the wall, also indicate residual internal stress at the beam's base. Sink marks are also unacceptable on appearance surfaces.

These requirements also appeared in the process-friendly discussion in Chapter 9, Section 9.4.1.

If the beam is an extension of a wall, then (T_b) should be equal to the wall thickness. If the beam thickness must be less than the (in-plane) wall thickness, a gradual change in thickness at a 1:3 ratio from the wall's edge to the desired beam thickness should be used to avoid stress concentrations and mold filling problems.

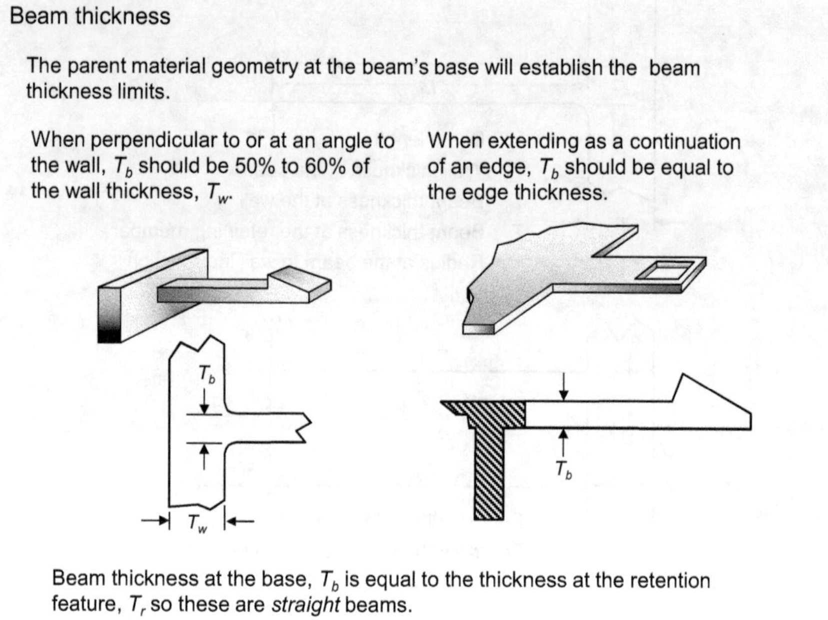

Figure 12.3 Beam thickness at the parent material

Beam draft angle warning

See Fig. 12.4. Most of the time, the beam's long axis will be in the same direction as mold separation or, if present, the die-action. Design the lock so the required draft angle does *not* result in lower thickness at the beam's base than at the end. Consider changing the lock style or, if there are no other options, minimize the draft angle. An angle as low as 0.5° is possible, but die wear will increase and part extraction may become more difficult. Draft angles of 1.5–3.0° are frequently recommended [1].

See more information about draft angles in [1] (p. 59, Table 8.02) and in [2].

Use caution with draft angles on beams.

Mold separation or die-action may also occur along the two axes orthogonal to the beam's long axis. There are two possibilities:

1. The draft angle may be *across the beam's short axis* (and 90° to the long axis) and the deflection strain will no longer be evenly distributed along the beam's base as shown in Fig. 12.4. Strain levels will be higher at the thicker side of the base. The beam thickness rules-of-thumb or a strain calculation using nominal or minimum beam thickness values will underestimate the maximum strain at the thick end of the base. Calculations using the thicker dimension will be more accurate, but they will also, incorrectly, assume an even distribution of the strain along the base.

   The thicker end of the base is a stress-concentration and beam failure can initiate at that point. This must be considered in the lock design and watched for during testing. See Chapter 13, Section 13.3.1 for an example.

2. Mold separation or die-action is 90° to the beam's surface. The effect is less dramatic, however, because strain will be evenly distributed across the base. The design should put the beam's wider surface on the tensile stress side of the beam if at all possible and the maximum allowable strain adjusted to a lower value.

   In all cases, unless the die action is coincident with the beam's long axis and the beam is thinner at the retention feature than at the base, the draft angle should be minimized and its effect carefully evaluated.

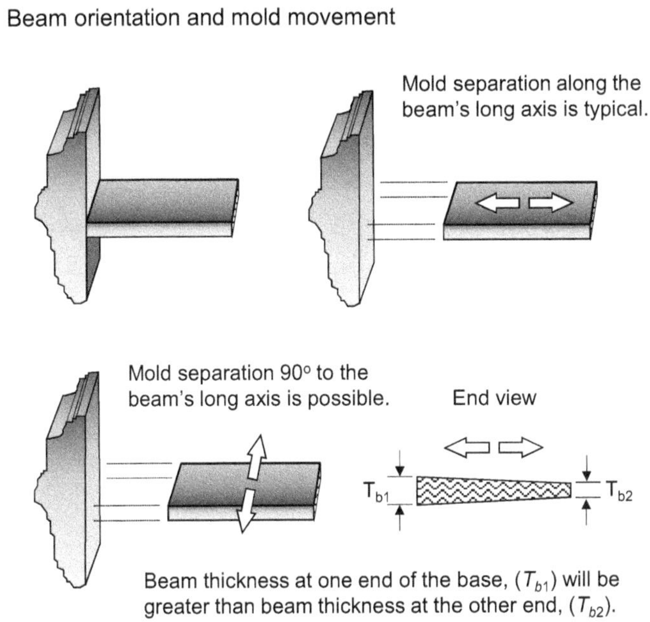

Figure 12.4 Die taper effects on beam thickness

12.1.2 Beam Length

The *total* lock feature length (L_t) includes beam length (L_b) and retention feature length (L_r), see Fig. 12.4. These two are considered separately because when bending is calculated, only the flexible beam portion of the hook is included. (Loops with an opening at the end of the beam are a special case.)

IMPORTANT: These rules assume beam deflection is *equal to or less than* beam thickness at its base.

Retention feature length (L_r) is not yet known, but we can establish the beam length (L_b) now. Later, we will add (L_r) and (L_b) to find (L_t). Ideally, we want to be free to select a beam length without any restrictions, but it may be limited by available space or mating part dimensions.

Beam length (L_b) should be at least 5x beam thickness ($5 \times T_b$) but closer to 10x thickness ($10 \times T_b$) is preferred. Beams can be longer than 10x thickness, but warpage and filling may become problems. Check the design against the material's *spiral flow curves* to ensure the long beam is properly filled.

Beam length

Once beam thickness is established, a length can be determined. A general rule of thumb when the beam assembly deflection is roughly equal to the beam thickness (T_b) is:

- For unplated materials, beam length (L_b) should be at least 5 × T_b, a length of 10 × T_b is preferred.

When the plastic is plated, the plating process will degrade the plastic surface and the plating itself is brittle.

- For plated beams, the L/T ratios should be greater, at least 10 × T_b.

More flexible materials will better tolerate lower L/T ratios while more rigid materials will require a higher L/T ratio.

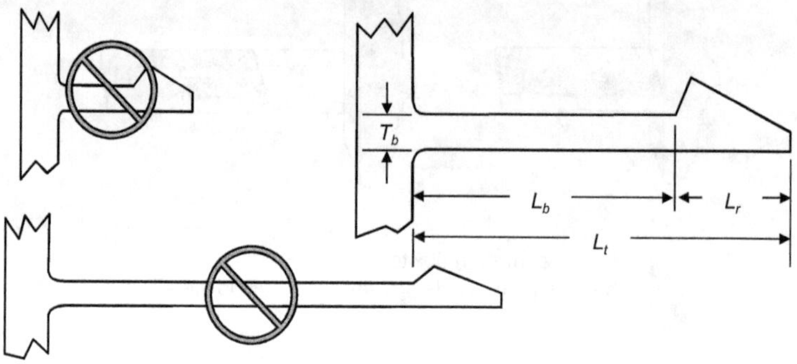

Figure 12.5 Beam length dimensions

Beams shorter than 5 × T_b will experience significant shear effects as well as bending at the base. Not only does this increase likelihood of damage during assembly, it renders the analytical calculations (based on beam theory) much less accurate. Shorter beams are much less flexible and create higher strains at the base. Longer beams are more flexible for assembly but also weaker for retention.

Higher length to thickness ratios are recommended for plastics that are harder and more brittle.

If clearance restricts *total* lock length and forces the beam length to violate the L/T rule, a different lock style better suited to a short grip length should be considered.

12.1.3 Beam Thickness at the Retention Feature

Often the beam thickness at the retention face (T_r) is equal to the beam's thickness at its base (T_b), as in Fig. 12.5. However, when strains at the base are high, tapering the beam over its length will more evenly distribute strain through the beam and reduce the chances of over-strain at the base, see Fig. 12.6. Common taper ratios ($T_b : T_r$) range from 1.25:1 up to 2:1. In shorter beams, tapering can reduce strains at the base by as much as 60%. However, tapering will also reduce the retention strength. Tapering is one possible solution to high strains when design constraints force a beam to violate the 5:1 minimum length to thickness rule.

Tapering will also result in a thinner beam at the end and rotation at the retaining member can become more important. Never make a beam so thin at the end that damage occurs at the beam/retaining member interface.

Do not taper a cantilever beam from the retention face back to the base. This moves virtually all the deflection strain to the beam's base and damage is just about guaranteed.

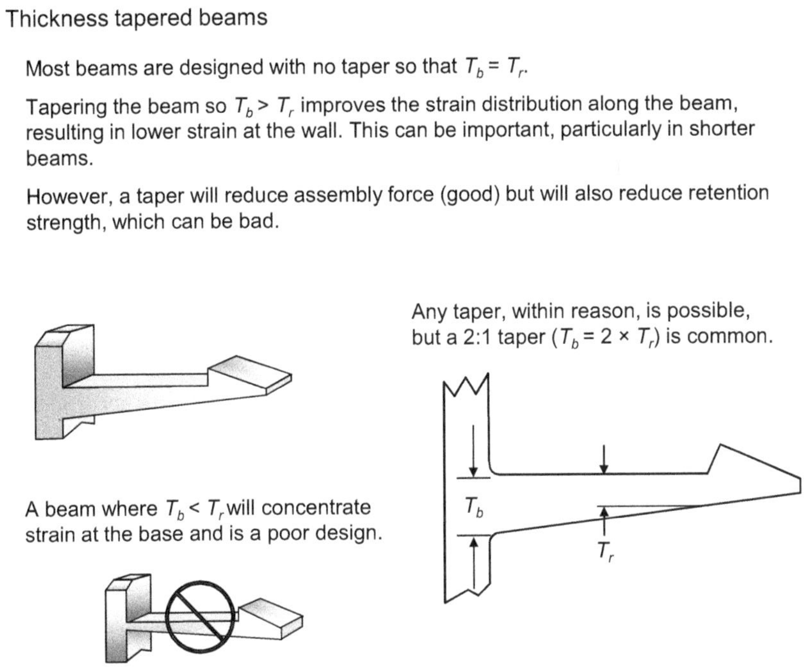

Figure 12.6 Thickness tapered beam

If necessary, a beam weakened by tapering in thickness can be made wider to compensate for that loss in retention strength.

12.1.4 Beam Width

Get more strength by making a beam wider!

Beam width does not appear in the strain equation. This means the beam's bending strength can be changed by changing its width without affecting the maximum assembly strain, see Fig. 12.7. If more retention strength is needed, simply make the beam wider. *Always* consider this option before making a beam stronger by making it thicker.

Width, unlike thickness, is not a squared term in the beam bending equation, so increasing the width is not as effective in improving strength as is increasing thickness.

Of course, increasing either thickness or width for more retention strength will also increase assembly force.

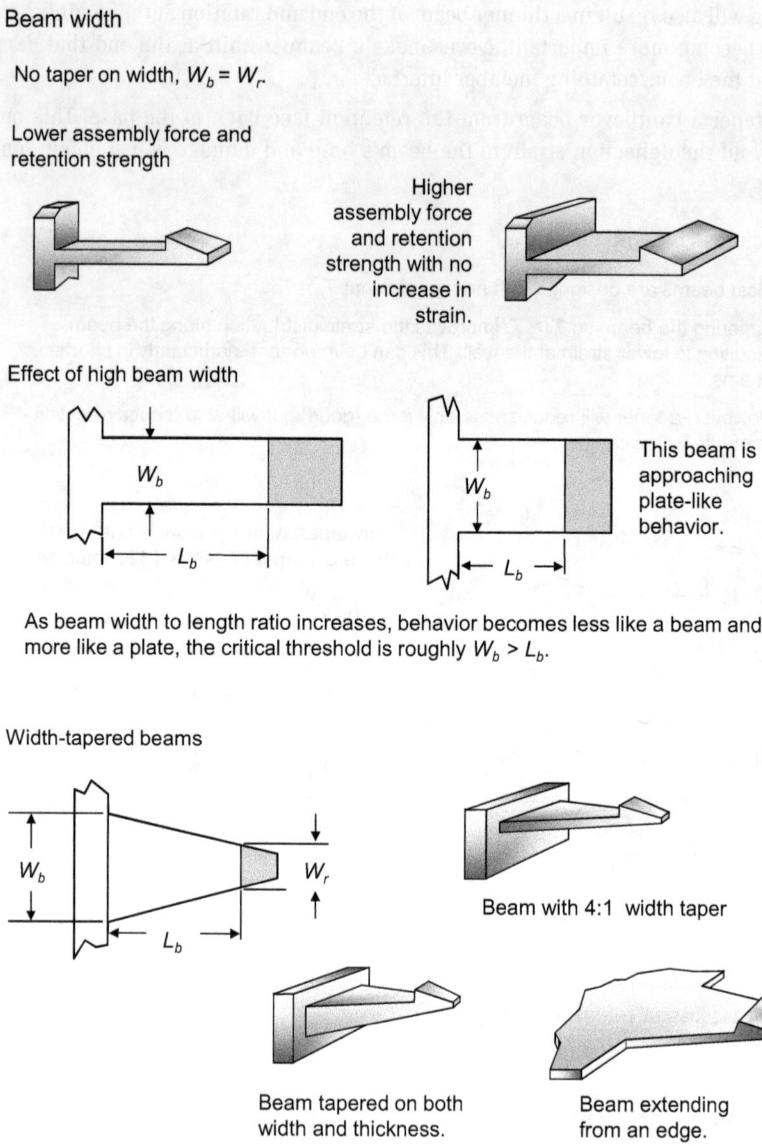

Figure 12.7 Beam width

An initial value for beam width is very much a judgment call. To minimize distortion during deflection, the beam's width should certainly be greater than its thickness. It must also, with the thickness, provide sufficient cross-sectional area for plastic to fill the entire lock feature protrusion during molding. Unless there is a compelling reason to make the width a particular value, the author suggests beginning with a beam width around 3–5x the beam's thickness. Beam length and material may also be factors in selecting the initial width value. A longer beam may require more cross-sectional area for filling. Mold flow issues like this should be discussed with a processing expert.

For beam theory to apply, the beam's width should be less than or equal to its length. As the width becomes greater than 50 % of the length, the feature *begins* to behave more like a plate than a beam. When width becomes greater than the length, the beam should be considered a plate. However, given the many other variables and the assumptions made to permit these calculations, relatively minor inaccuracies at higher beam widths can generally be ignored and cantilever beam calculations used.

Where a beam extends in-plane from an already thin wall, increasing the width for more strength may be the only option. Beams can also be *tapered* in width as shown.

12.2 Retaining Member: Catch

Now we'll consider the lock feature's retaining member. A catch locator is often used with a beam to create a cantilever hook. However, catches can also be used with traps, another beam-based lock, as well as with torsional locks and planar locks. This section focuses on the catch only, not the deflecting member on which it is mounted.

For a catch, we are concerned with the insertion and retention face angles and the retention face height.

12.2.1 The Insertion Face

The insertion face angle affects assembly force. The steeper the angle, the higher the force required to deflect and engage the lock.

Ideally, the insertion face angle should be as low as possible for low assembly force, see Fig. 12.8.

- An angle of 25–35° is normally reasonable. Shorter beams require a lower angle than longer beams.
- Angles of 45° or greater are difficult to assemble and should be avoided.

For a cantilever hook, the initial insertion face angle will increase during insertion; this is another good reason to start out with that angle as low as possible. This change in insertion face angle is greater when the lock's L/T ratio is low. The subject is discussed in detail in Chapter 13.

For example:

- For a lock with an L/T ratio of 5:1, a lower value of 25° or even 20° would be more appropriate.
- For a lock with an L/T ratio of 10:1, a higher value of 35° would be acceptable.

A profile on the insertion face can improve assembly performance and is also discussed in Chapter 13.

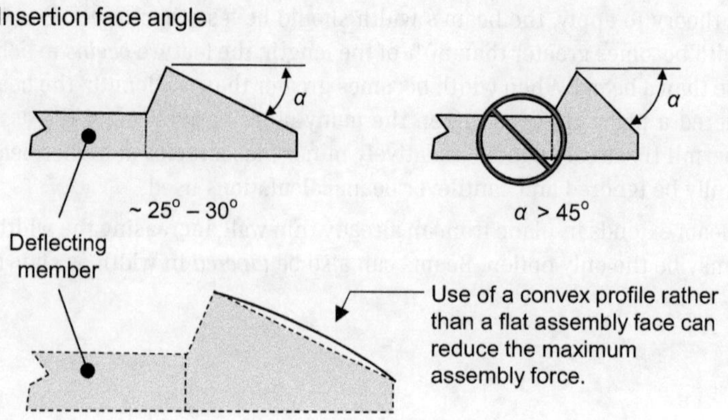

Figure 12.8 The catch insertion face

12.2.2 The Retention Face

For any lock, the full retention face depth should be used for engaging the mating locator. Thus retention face depth will equal deflection ($Y = \delta$) which, in non-tapered beams, should be no greater than the beam thickness. This helps ensure that separation forces on the catch occur as close as possible to the neutral axis of the beam and minimizes rotational forces at the end of the beam that would contribute to unintended release.

The retention face depth (Y), sometimes called the *undercut*, determines how much the beam will deflect during engagement and separation. (*Separation* can mean either unintended release due to an external force or intentional release for disassembly.)

For a beam length (L_b) to thickness (T_b) ratio in the range of 5 : 1, the initial retention face depth (Y) should be less than T_b.

For a L_b/T_b ratio closer to 10 : 1, the initial retention face depth can be equal to T_b. Harder and more rigid plastics (higher modulus) will tolerate less deflection for a given length than will tougher or more ductile plastics.

Some mitigating factors related to allowable deflection are discussed in Chapter 13.

When analysis calculations are based on a known strain limit for the material, a maximum allowable deflection can be determined. The maximum retention face depth is then set near to the maximum allowable deflection.

The retention face angle will affect separation behavior. The steeper the angle, the higher the separation force and retention strength, see Fig. 12.9.

For a releasing lock where no separation forces are acting on the mating part (aside from an intentional manually applied separation force), a retention face angle of about 35° is generally acceptable, see Fig. 12.9. The exact angle will depend on the friction coefficient between the materials and the actual stiffness of the lock material. If the application has a high number of usage cycles, as with a moveable snap-fit, then a lower angle is preferred to reduce cyclic loading on both the lock and the mating feature. If the lock will be released only a limited number of times, a higher angle may be possible.

If some relatively low external separation forces are expected, then a retention face angle of about 45° is a reasonable starting point. Again, consider friction and hook stiffness. These locks may still be a releasing style, but manual separation forces will be high and a high number of removal cycles is not recommended.

If the lock must resist high external separation forces, then a releasing lock is not recommended and a permanent or nonreleasing lock (manual deflection needed for disassembly) should be designed. The retention face angle should be close to 90°. A retention face angle of exactly 90° is usually not necessary. Because of frictional effects, any angle above a limiting value called the *threshold angle*, discussed next, will behave like a 90° angle.

Like the assembly face angle, deflection will change the retention face angle. However, the retention face angle will *decrease* with deflection, tending to reduce its retention capability. This is also discussed in Chapter 13.

Adding a concave profile to the retention face can sometimes improve its performance by maintaining the original retention face angle throughout lock deflection.

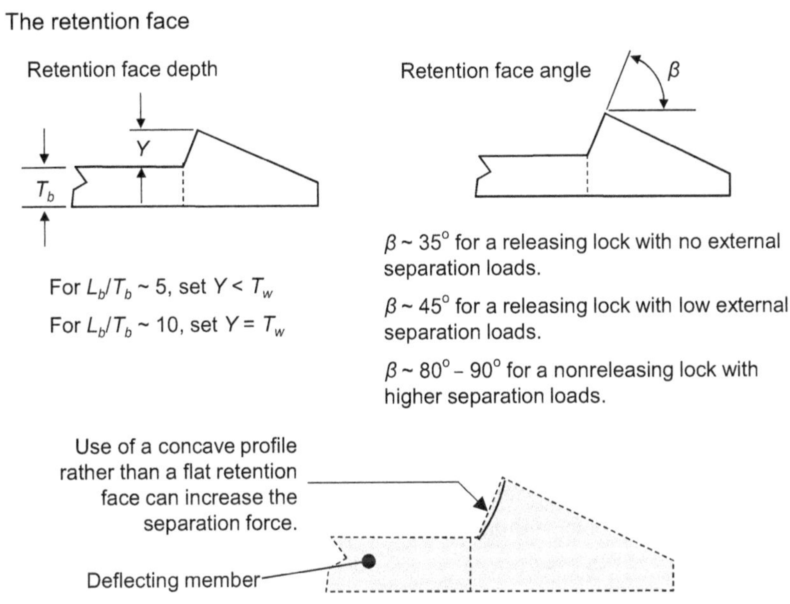

Figure 12.9 The catch retention face

Because of friction between the retention face and the mating surface, a retention face angle less than 90° can still *behave* like a 90° angle, see Fig. 12.10.

The threshold angle can be calculated if the friction coefficient is known or estimated. At a friction coefficient of 0.3, this angle is approximately 80°, meaning that any angle above 80° will behave like a 90° angle for initial resistance to separation.

$$\beta_{\text{threshold}} = \text{Tan}^{-1}\left(\frac{1}{\mu}\right) \tag{12.1}$$

Using a retention face angle close to the threshold angle may sometimes be useful because it will provide slightly more dimensional compliance and robustness than a 90° angle.

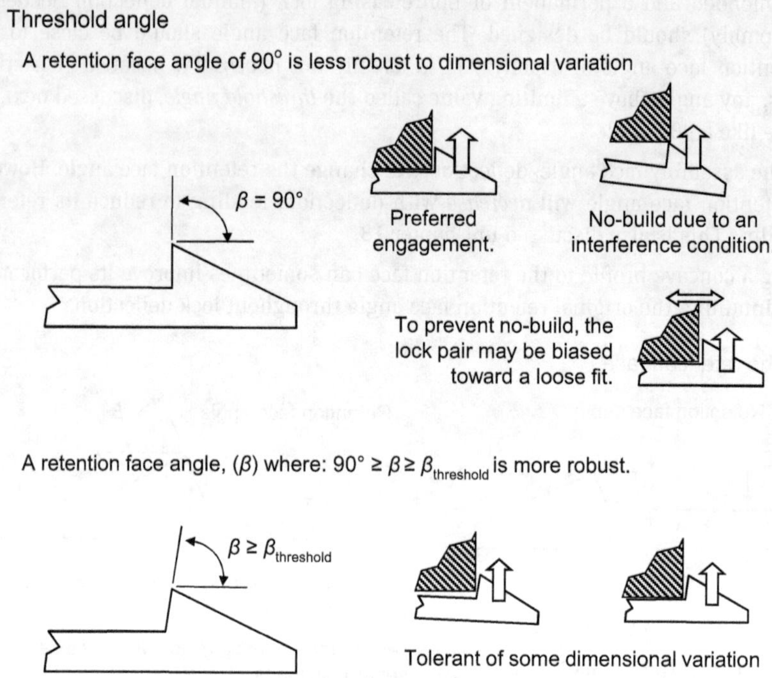

Figure 12.10 Threshold angle vs. a 90° angle

This concludes the discussion of rules of thumb for the most common deflecting and retaining mechanisms: a beam and a catch. They appear together very frequently as a cantilever hook style lock. They also appear together or separately in other lock styles and some of the above rules-of-thumb can be applied in those cases as well.

12.3 Loops

Loops may consist of two parallel deflecting members connecting the loop's base to the cross-bar. Or one beam will extend from the base to an opening with two parallel beams extending to the cross-bar.

For the deflecting member

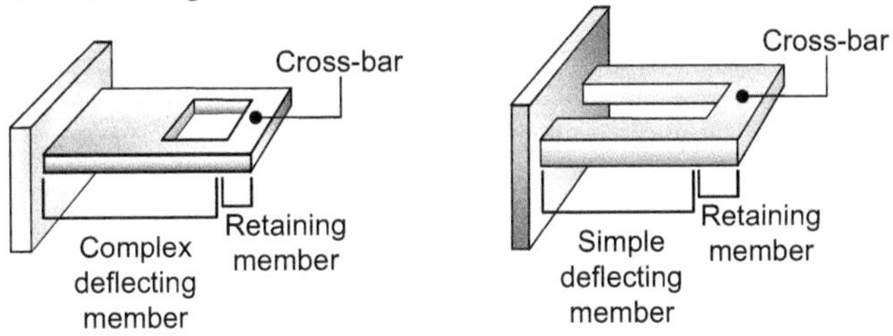

The beam rules-of-thumb presented in Section 12.1 apply here.

For the retaining member

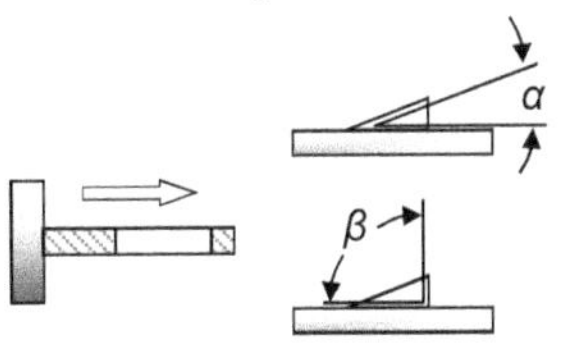

The retaining member, a cross-bar, usually engages a catch which carries the insertion and retention faces.

Because the catch is stationary, the insertion and retention face angles do not change as the loop itself deflects.

The catch design rules presented in Section 12.2 apply here but changes in the angles will no longer be a consideration.

All insertion and retention face profile options are available for the catch.

Figure 12.11 Loop locks

12.4 Traps

Traps are significantly different from the cantilever hook in that engagement and separation directions with respect to the beam's base are opposite those of the hook.

This difference is reflected in some changes to the rules for initial assembly and retention face angles, see Fig. 12.12.

For the deflecting member and engagement

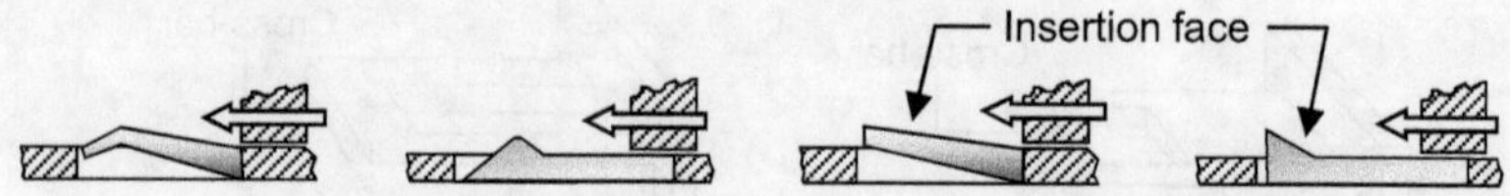

Whether the trap is releasing or nonreleasing, the beam rules-of-thumb in Section 12.1 will apply.

The insertion face rules in Section 12.2 can be adjusted to permit a slightly higher insertion face angle because for a trap, as deflection occurs, the initial insertion face angle will *decrease*.

For the retaining member in a releasing trap

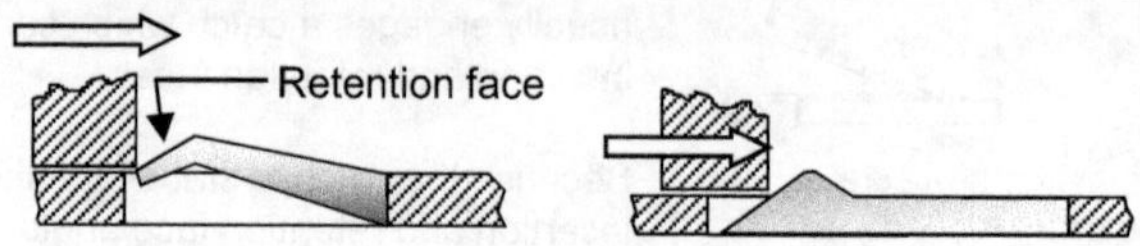

The insertion face rules in Section 12.2 can be adjusted to a slightly lower retention face angle because for a trap, as deflection occurs, the initial retention face angle will *increase*.

For the retaining member in a nonreleasing trap

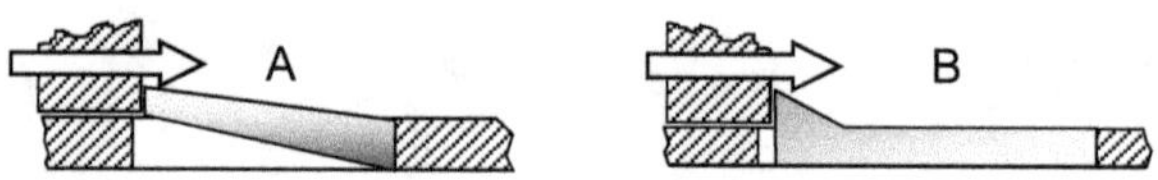

The retention face rules in Section 12.2 will not apply.

For Case A above, the entire trap becomes the retaining member and acts as a column loaded in compression, although it is not *pure* compression. A maximum 10:1 L/T ratio *seems* to be a reasonable maximum for achieving high retention strength without buckling, and is within the recommended beam L/T ratio limits. Remember, the free end of the trap must be restrained from slipping on the mating surface.

For Case B, the face of the catch becomes the retaining member and creates an off-center compression load and a moment on the column, again not pure compression. A length to thickness ratio lower than 10:1 is indicated here.

Figure 12.12 Trap locks

12.5 Other Lock Styles

The primary deflecting members in these lock styles are not beams, so different equations are used to calculate the deflection force. However, beam deflection may *sometimes* be a secondary deflection that should be considered.

Rules-of-thumb for the deflecting members in these styles are not common like they are for beams.

Catches as retaining members are common in these other styles and once the deflection force is known, assembly and separation forces can be calculated using catch behavior. The rules-of-thumb discussed above for initial catch dimensions still apply.

12.5.1 Torsional Locks

In torsional locks, the primary deflecting member will act as a torsional spring. There may also be beam-like areas of the lock feature, see Fig. 12.13.

For the deflecting member

Deflecting member rules for a beam-based lock will apply to torsional locks only with respect to process-friendly design of the beam portion of the lock.

An exception would be if, as is possible in configuration A below, the beam's L/T ratio is high enough to contribute to catch deflection. In this case, the rules related to beam bending would apply to the beam portion only and deflection due to torsional action would also be taken into account.

For the retaining member

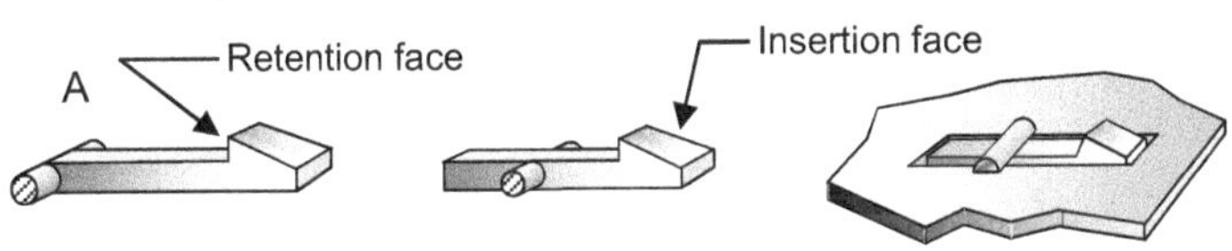

When the retaining member is a catch, all the rules in Section 12.2 for the insertion and retention face angles will apply, including consideration of the changes in these angles due to deflection.

Figure 12.13 Torsional locks

12.5.2 Planar Locks

Planar locks involve deflecting members where the width and length are significantly greater than the member's thickness, see Fig. 12.14.

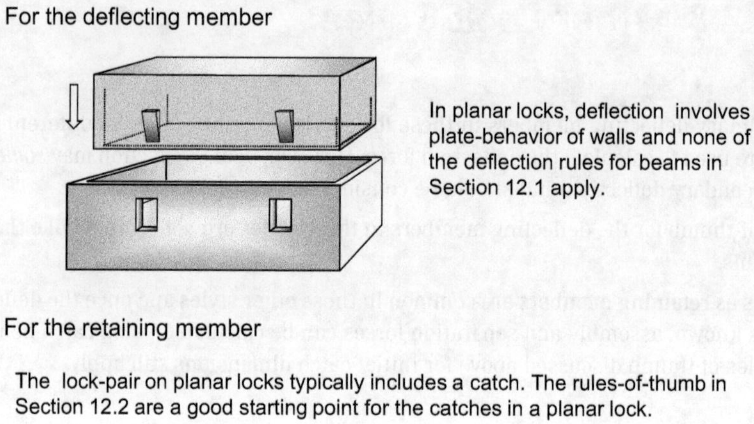

Figure 12.14 Planar locks

12.5.3 More Lock Styles

Figure 12.15 shows some other beam-based lock styles. These are not common, although, in the author's opinion, the lock style shown in the first example should probably be used more often than it is. It certainly is applicable to some short grip-length applications. This lock is essentially two side-action locks joined at the center.

The beam/prong style in the second example can be very strong and also useful in tamper-proof or tamper-resistant applications.

The recommendations here also assume that beam deflection is equal to or less than the beam thickness.

For the deflecting member:

No explicit rules-of-thumb have been found for this situation.

Rules related to process-friendly design in Section 12.1 will apply.

Issues of knit line weakness, as discussed in Chapter 6 for loops, will apply.

The author has seen a *limited* number of applications with this lock configuration. An L/T ratio *around* 10:1 seems to be common, so a 10:1 ratio is suggested as a reasonable starting point. Note the exceptions below.

For the retaining member:

We assume that beam deflection is equal to or less than the beam thickness.

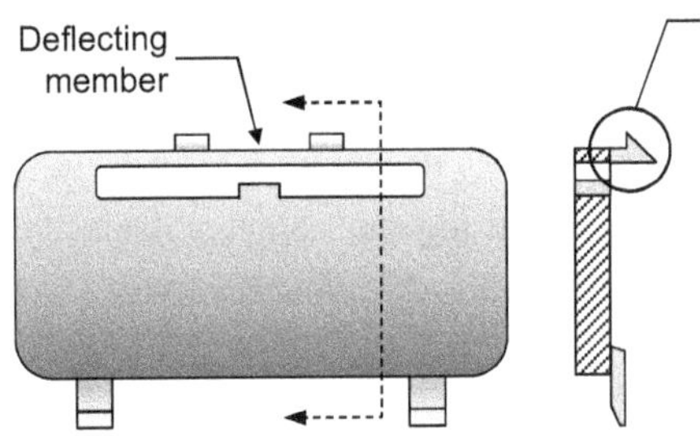

In this example, the beam is carrying two catches. The catch feature rules-of-thumb from Section 12.2 apply.

If beam twisting can occur during deflection, changes in the insertion and retention face angles may need to be taken into account.

The 10:1 *L/T* ratio suggested above for the beam assumes *one* deflection contact point at the beam's center.

In this example, there are *two catches* located away from the beam's center point so an adjustment to the beam length may be needed.

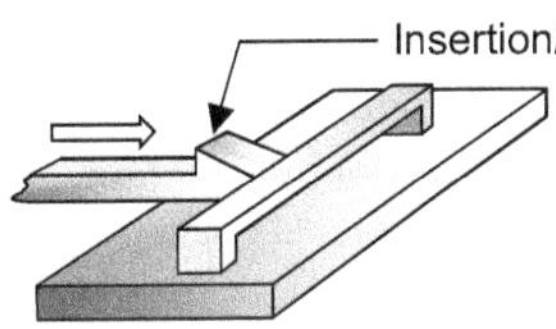

In this example, the deflecting member is a beam fixed to a surface and engaging a prong. Recall from Chapter 6 that a prong can *look* like a cantilever hook, but it does not deflect and is a locator.

The above rules for a deflecting member fixed at both ends will apply.

We assume the prong engages at the beam's center and the prong is not wide. If the prong is wide, the suggested *L/T* ratio of 10:1 may need to be increased.

To reduce the effect of prong width, the prong's retention face can be contoured across the insertion/retention face juncture to match the beam's curvature at full deflection.

The catch feature rules-of-thumb from Section 12.2 apply to the catch on the prong.

Figure 12.15 Other beam-based lock styles

12.6 Summary

In this chapter, rules-of-thumb for establishing initial feature dimensions were given. Following these rules should provide reasonable lock feature dimensions as a starting point for analysis. Some of these rules, particularly those for the cantilever beam deflecting member are common in the literature. Others are based on the author's judgment as a subject-matter-expert.

Again, these rules are not intended to provide final feature dimensions. They are a starting point for development and analysis. End-use testing should be the final proof of product and snap-fit capability.

Important points in Chapter 12:

- Use these rules-of-thumb as a *starting point* for lock design prior to analysis.
- Keep them in mind during product benchmarking and application problem diagnosis.
- Tapering a beam's thickness can significantly reduce strain at the base. A 2:1 taper is common.
- Beams can also be tapered in width.
- Strain is independent of beam width. This means retention strength can be increased by increasing the beam's width with no increase in strain. This change will, however, increase the insertion force.
- Be aware of draft angle direction and magnitude and their effects on beam behavior.

References

The appendix contains additional and updated reference material information.

[1] *Designing with Plastics - The Fundamentals, Design Manual TDM-1*, Ticona LLC (Formerly Hoechst Celanese Corporation, now a division of Celanese AG) Summit, NJ (2000) www.ticona.com

[2] *5 Ways to Improve Part Moldability with Draft, Protomold: Design Tips for Rapid Injection Molding*, (1/6/2016); https://www.protolabs.com/resources/injection-molding-design-tips/united-states/2016-01/

Bibliography

Hoechst Technical Polymers, *Designing With Plastic - The Fundamentals*, Design Manual TDM-1, Ticona LLC, (Formerly Hoechst Celanese Corporation, now a division of Celanese AG) Summit, NJ (1996)

Malloy, Robert A., *Plastic Part Design for Injection Molding - an Introduction*, Hanser, Munich, Germany (1994)

Modulus Snap-Fit Design Manual, Allied Signal Plastics, Morristown, NJ (1997)

New Snap-Fit Design Guide, Allied Signal Plastics, Society of Plastics Engineers ANTEC (1987)

Snap-Fit Joints for Plastics - a Design Guide, Polymers Division of the Bayer Corporation, Pittsburgh, PA (1998)

Tres, P., *Designing Plastic Parts for Assembly*, Hanser, Munich, Germany (2000)

13 Lock Feature Development: Calculations

Feature analysis is possible and appropriate only after a fundamentally sound attachment concept has been created and proper constraint ensures that forces in the attachment are statically determinate.

These calculations are the traditional *feature level* snap-fit technology.

Because most lock features are based on cantilever beams, the cantilever beam-based locks are the primary focus of this chapter. However, sources of equations for some other lock configurations will be provided.

Likewise, the catch locator feature is most commonly used as the retaining feature at the end of a beam so our focus in this chapter will be the behavior of cantilever beam deflecting members and catches in various combinations.

This chapter presents the traditional equations for cantilever beam bending and discusses how they apply to all beam-based locks. With these equations, and given any two of the three variables: beam deflection, strain or deflection force, we can calculate the third. We will be primarily concerned with the maximum values of these variables although there will be times when calculating intermediate values will be useful. The ultimate goal is to develop a beam *solution* that meets all performance requirements without exceeding the maximum allowable strain. This is typically an iterative process because conflicting requirements must be satisfied and multiple cycles through the calculations may be necessary.

Adjustments to the traditional equations are required for snap-fits.

Bending equations for cantilever beams are readily available in the literature and an internet search will also turn up some tools for doing those calculations directly online. Having access to tools for performing these calculations will make the iterative process much easier.

However, *adjustments* to snap-fit calculations that use the classic beam equations are required to reflect the unique behavior of lock features. These adjustments are specific to snap-fits and do not appear in the general cantilever beam equations nor do they appear in most of the published snap-fit calculations. These adjustments should be included when possible because their effect on the calculation results can sometimes be significant. If online or other beam analysis tools are used, the adjustments can be made by hand.

Locator features, in most cases, require little analytical attention because they are strong and inflexible. Locator strength calculations, if required, normally involve simple shear, tensile, possibly combined stresses, or compression strength calculations. For this reason, locator calculations are not discussed in this book.

Normally we are interested in calculating:

- Assembly force
- Assembly strain
- Separation force (or retention strength)

Technically, *separation force* refers to intentional separation for part disassembly while *retention strength* emphasizes the attachment's resistance to unintended separation due to applied forces. In calculations, they are equivalent. For simplicity, we will use *separation force* to refer to both.

13.1 Assumptions and Allowances

To allow calculations, certain assumptions and allowances are required for both simple and complex beam shapes and other lock styles. They are necessary because some information is simply unknowable (the future effects of processing variables, for example) or, at the time of analysis, information is not available. Addressing some of these issues would introduce extreme complexity into the analysis with little confidence it would bring significantly more accuracy to the results.

Material Properties

If you have not yet read Chapter 11, read it now.

In preparation for feature analysis, developers must have access to the best material property data available, see Chapter 11. They must know the expected loads, deflection cycles, and environmental factors (for example, temperature and chemical) in order to make adjustments to the target or maximum allowable strain. Input from a polymers expert may be required. Sometimes property data for both fresh as well as aged material may be needed.

Four material properties are normally used in feature analysis calculations: stress (σ), strain (ε), modulus of elasticity (E), and coefficient of friction (μ). The earlier in the development process the developer has this information, the better.

Lock feature calculations use strain (rather than stress) as the limiting parameter for feature performance because polymer material property data typically uses strain values.

Some assumptions about material properties are required for lock feature calculations. They are discussed in Chapter 11.

Feature Geometry

We must also make assumptions about the feature itself so classic beam equations can be applied:

- The beam material is homogeneous with the same modulus of elasticity in tension as in compression.
- The beam is straight or is curved in the plane of bending with a radius of curvature at least 10 times the beam depth.
- The beam cross-section is uniform.
- The beam has at least one longitudinal plane of symmetry.

- All loads and reactions are perpendicular to the beam's axis, and lie in the same plane, which is the longitudinal axis of symmetry.
- The beam is long in proportion to its depth.
- The beam is not disproportionately wide.
- The maximum stress does not exceed the proportional limit.
- Applied loads are not impact loads.

Other Allowances for Calculations

In addition to assumptions about material properties and the beam, we must accept and make other allowances for the material and for the lock feature. They include:

- For practical reasons, we generally ignore differences in the length of the deflecting portion of the beam vs. the location of the applied force, see Fig. 13.2 and the note following the figure. At longer beam lengths, the effect on the calculations is minimized.
  - We ignore the change in the moment arm in locks where the point at which separation or assembly force is applied moves across the assembly and retention faces.
  - We ignore the difference between the deflecting beam length used in the calculations and the actual length of the moment arm.
- Consistency of part material properties due to process variation is unknown and can change over time.
- Parent material behavior, (deflection) at the beam's base is not part-specific, but estimated from limited empirical data.
- The stress-strain curve is linear in the range of our calculations.
- Accuracy and consistency of coefficient-of-friction values.
- The true or effective beam length vs. the end of the beam at the starting point of the retaining member.
- Ignoring beam curvature in the bending calculations.
- Assuming normal ambient temperatures in most calculations.
- Assuming material property data collected in laboratory testing matches real life conditions.
- The material's time-rate of strain or visco-elastic behavior is unknown. We assume rapid deflection for assembly and disassembly, but this may not match the test conditions that generated the data. An extended deflection time, (possible during disassembly) can create higher strain than rapid disassembly deflection.
- Effects of material creep, aging and property degradation.

We sometimes use simplified equations for beam behavior.

Keep all this in mind when setting design targets and safety factors for your applications. The more assumptions violated, the less representative of actual feature behavior the calculations *may* become.

Given all the assumptions and allowances for unknowns, one might wonder why we bother with feature analysis at all. Despite everything, feature analysis is worthwhile:

- Some of the unknowns will offset each other.
- Some feature behaviors during deflection will tend to make the maximum strain calculations more conservative.
- The understanding gained through analysis of the feature's behavior and sensitivity to changes is valuable.
- Differential analysis of design options is quite useful for both original design and for diagnosing and fixing a snap-fit problem.
- Maximum allowable strain values themselves are not necessarily exact numbers and actual values may be higher or lower.
- One can use calculations to optimize a design solution based on a conservative estimate of maximum allowable strain.

Software exists and is also easy to write to perform these calculations. However, before one turns to software, a true understanding of lock behavior can only be gained by first mastering the hand calculations.

Without calculations, the developer is working in the dark. The author is very uncomfortable not using calculations and, in all cases, treats feature calculations as *reasonable indications* of actual feature behavior as well as sensitivities to design changes and behavioral trends.

The nature of the polymer being analyzed can also influence judgments about calculation accuracy. For example, there are engineering and commodity polymers, ductile, tough, and rigid polymers. Some will be less sensitive to the unknowns and variables and more predictable than others.

When more confidence in an analysis of feature performance is required, finite analysis should be considered. But, even FEA will require making assumptions.

Again and always, end-use testing is recommended.

13.2 The Deflecting Member: Cantilever Beam

Recall from Chapter 12 that locks consist of a deflecting member and a retaining member and that most locks use a cantilever beam as the deflecting member.

This section presents equations for calculating beam bending behavior. Traditional equations for rectangular-section cantilever beams fixed at one end are described and special adjustments to those equations are introduced. We will then step through an example calculation for beam strain and deflection force. Once the maximum beam strain is determined to be acceptable, the deflection force calculated for the beam will be used in the retaining member calculations for assembly and separation force.

We then move to calculations for the retaining member, using a catch (a locator) as the retaining member on a beam or as a stand-alone locator in a lock pair.

The examples here are limited to a catch-beam configuration in the familiar form of a cantilever hook, but the fundamental principles and the equations are easily adapted to other beam/catch configurations like traps and loops.

As we saw in Chapter 11, the stress, strain, and modulus for a material are represented in a stress-strain curve. The modulus (E) may be developed through bending, tensile or compression testing. The material data should indicate how the modulus value was determined. Of course, for calculating beam bending performance, a modulus generated by bending tests is preferred. All material property data should be based on standardized tests and the standard should be cited with the data.

The calculations presented here will use the modulus (E) and strain values (ε). The general equation for the modulus is:

$$E = \frac{\text{stress}}{\text{strain}} = \frac{\sigma}{\varepsilon} \tag{13.1}$$

The calculations are presented in metric units. In Eq. 13.1, stress is expressed as Newton per mm^2 (MPa). Strain is dimensionless so the modulus is also in units of MPa.

There are several values for a modulus available when we work with polymers, depending on the shape of the stress-strain curve and our needs. For these calculations, we'll use the secant modulus, E_S, Eq. 13.2, the slope of a line from the origin through the design point as explained in Chapter 11.

$$E_S = \frac{\sigma_{\text{design-point}}}{\varepsilon_{\text{design-point}}} \tag{13.2}$$

In some cases, a very linear stress-strain curve for example, the modulus chosen for the calculations may be the slope of the stress strain curve itself, or the modulus may be established at a predetermined strain offset from the origin.

13.2.1 General Equations for Rectangular Sections

Locks based on rectangular section beams are, by far, the most common lock configuration and are the only lock style discussed here in detail. Formulae for calculating properties of other common sections can be found in many structural engineering references.

For any beam with a rectangular section, Fig. 13.1, the section properties are given below. They are the basis for the beam behavior equations we will be using and are included here for reference.

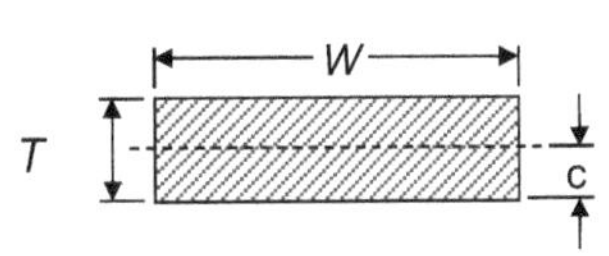

W = beam width, (section width)

T = beam thickness, (section height)

$c = ½\,T$

Figure 13.1 Beam section

Where W is the beam width, T is the beam thickness, and c is the distance of the outer surface from the neutral axis. In a rectangular section, c is one-half the beam thickness, (T), or one-half the section height. (Common practice in section equations is to identify what we will be calling *beam thickness* or T as *height* or *h,* thus the discrepancies between Fig. 13.1 and Eqs. 13.3–13.5.)

$$c = \frac{height}{2} = \frac{T}{2} \tag{13.3}$$

The outer surface is where the highest tensile and compressive stresses occur. We care about the tensile stress side of the section because that is where strains causing lock damage and failure occur.

I is the section moment of inertia (mm^4).

$$I = \frac{\text{base} \times \text{height}^3}{12} \tag{13.4}$$

Z is the section modulus (mm^3).

$$Z = \frac{\text{base} \times \text{height}^2}{6} \tag{13.5}$$

13.2.2 Constant Section Beam Bending

Figure 13.2 shows the variables used in the calculations. Dimensions for both the deflecting member (beam) and the retaining member (catch) are shown, but here we are only interested in the beam.

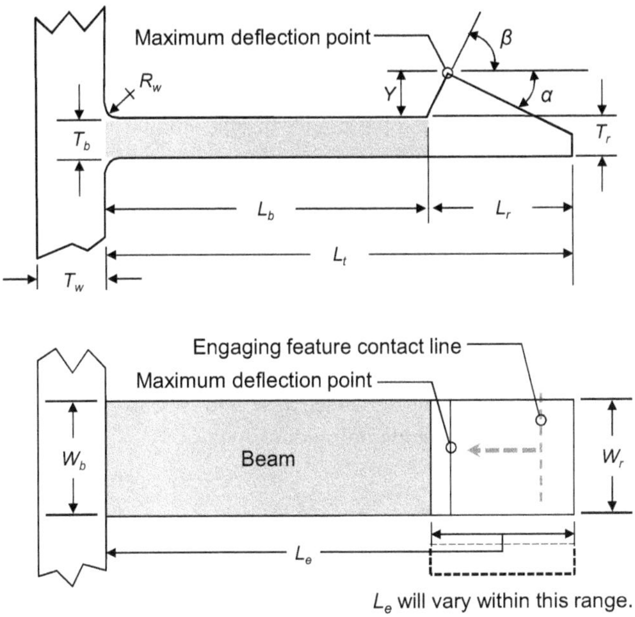

L_b Beam length.
L_r Retention feature length.
L_e Effective beam length: The distance from the beam's base to the mating feature's point of contact. It includes deflecting and non-deflecting areas of the lock and is the true moment arm of the deflection force.
L_t Total lock feature length.
T_w Wall thickness at the beam.
T_b Beam thickness at the wall.
T_r Beam thickness at the retention feature.
R_w Radius at the beam to wall intersection.
W_b Beam width at the wall.
W_r Beam width at the retention feature.
Y Undercut depth.
δ Assembly deflection, often equal to Y.
α Insertion face angle, as designed with the lock in its free state.
β Retention face angle, as designed with the lock in its free state.

Figure 13.2 Beam-based lock feature dimensions

A note about beam length, L_b vs. L_e

Refer to Fig. 13.2.

During assembly or separation movement, the engaging edge of the mating feature moves across the insertion and retention face surfaces in a direction perpendicular to the beam's base. Thus the actual length of the deflecting force's moment arm is constantly changing. This variable length is the beam's *effective length* or L_e. In any calculations involving beam length, the effective length *could* be used.

We have also defined the distance from the beam's base to the retention feature as L_b, which is the beam's flexible length only, L_e includes the nonflexible length, and L_r is the retaining feature's length. If we use L_e we will be misrepresenting the beam's flexible length.

Both situations introduce some inaccuracy in the calculations. The effects are greater for shorter beams and lesser for longer beams. It is up to the developer to decide which length to use.

Sometimes, L_e may be useful when estimating a force-deflection signature or calculating effective angles. However, for the sake of simplicity in the equations and example calculations presented here, we will only be using L_b.

This is not an issue with side-action locks or when a loop engages a catch because the beam length does not change.

For reference, some fundamental beam bending equations are included here.

Moment, *M*, Eq. 13.6, is a moment in the beam due to a concentrated force applied to the beam at some point. For our purposes, the force is typically applied at the beam's end.

$$M = \left(F_p L_b\right) \tag{13.6}$$

Aside from calculating the secant modulus, stress, σ, Eq. 13.7, is not used in the calculations in this chapter. It is included here to show its relationship to the section properties, the modulus, and to strain.

$$\sigma = \frac{Mc}{I} = \frac{F_p L_b}{Z} \tag{13.7}$$

Deflection, δ, Eq. 13.8, is the beam's movement orthogonal to its undeflected position.

$$\delta = \frac{F_p L_b^3}{3EI} \tag{13.8}$$

Angle, θ, Eq. 13.9, is the angle between the beam's long axis in its undeflected state and a tangent to the beam surface when deflected by a concentrated force at the end. In our calculations, we normally care about the maximum angle at the end of the beam. Angle is used in the calculations only in those circumstances where we must take beam end rotation into account.

$$\theta = \frac{F_p L_b^3}{2EI} \tag{13.9}$$

Where θ is expressed in radians:

- radian = arc length/arc radius
- 2π radians = 360°
- 1 radian = 57.3°

All of the above equations can be applied at *any point* along the beam's length and the results of those calculations will represent behavior at that point.

Be sure that you understand what L_b represents in the equations and calculations.

The convention shown in Fig. 13.2 is that L_b is the distance from the beam's base to its end: the *maximum* beam length. It does not include the retaining member.

The subscript conventions used here are often the author's own and are necessary to identify the various strain, deflection and force values as we step through the calculations. The reader is, of course, free to use any labeling convention they wish.

13.2.3 Adjusting the Design Strain for Stress Concentration

The stress concentration adjustment is described in the literature and is commonly applied in traditional snap-fit feature calculations for beams. It is *not* one of the unique or special adjustments to the snap-fit calculations we will be discussing.

Chapter 11, Sections 11.3 and 11.4, explained how to start the process by determining a value for the *design point* or *design strain*, (ε_{design}), using material property data. This design point is the starting point for our calculations. Once it is adjusted for stress concentration effects, it will become our *maximum allowable strain* or *target strain*.

Beam deflection will create either *dynamic* or *static* strain. Dynamic strain occurs when a deflection force is quickly applied and removed, as would be the case with lock features when parts are assembled or releasing locks deflect during part separation. Static strain occurs when the applied deflection force is sustained for a period of time. Sometimes for disassembly, a non-releasing lock must be held in its deflected position while other non-releasing locks are released. Both strain values are nice to have if they are available and it is good to know if the strain value from the material property information is static or dynamic. For most plastics, the allowable static strain value will be lower than the dynamic strain and a lock material that survives rapid assembly may not survive prolonged deflection or slow disassembly.

A stress concentration occurs where a sudden section change in a loaded area, in this case a beam-wall intersection, causes higher local stress. Stress concentrations will make the actual strain greater than the strain value calculated from beam theory.

For beams in general, we are concerned with the area under tensile stress where the beam meets a wall or any other change in section thickness. A radius at that location will reduce the stress concentration effect, but it cannot be totally eliminated.

Figure 13.3 shows the stress concentration curve commonly found in the plastic part design literature. Although K is called the *stress concentration factor*, it can also be applied to strain in calculations where we are not working with stress.

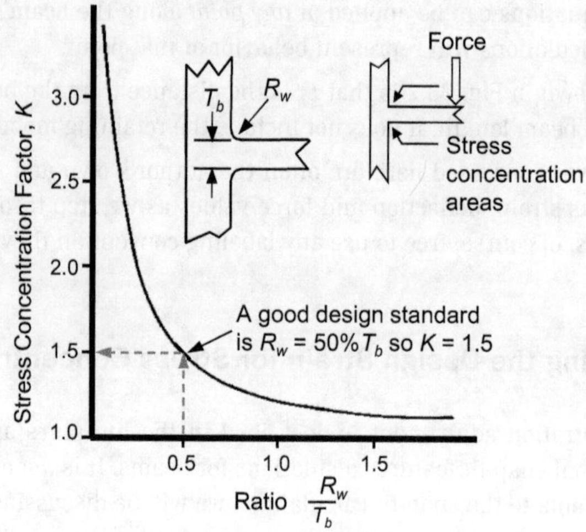

Figure 13.3 Stress concentration factor curve. Adapted from an illustration in *Designing With Plastic – the Fundamentals* from Ticona LLC

As shown, a value of $K = 1.5$ is reasonable with a radius at 50% of beam thickness. A K value of 1.0, although ideal, is impractical because of the very large radius it would require. Rules for process-friendly design also limit the radius to about 50% of beam thickness.

There also seems to be consensus in the literature that:

- A minimum radius of 0.5 mm (0.020 in.) is permitted for stressed areas.
- A minimum radius of 0.13 mm (0.005 in.) is permitted for unstressed areas.

Also consider whether the material is relatively ductile, tough or rigid when choosing a radius.

The stress concentration factor can be applied in two ways:

- A calculated strain is multiplied by (K) to get an actual strain value, which will be higher than the calculated value.
- The original design strain is divided by (K) to get a lower maximum allowed strain or target strain value.

Because we will be doing strain calculations based on beam dimensions and comparing those results to the maximum allowable strain, we'll use the second approach as shown in Eq. 13.10.

If a beam extends in-plane from a wall of the same thickness, then $K = 1$.

$$\varepsilon_{\text{max}} = \frac{\varepsilon_{\text{design-point}}}{K_{\text{stress-concentration}}} \tag{13.10}$$

We will refer to ε_{max} as the *target strain.*

As we step through these calculations, the goal is to ensure that the final calculated beam strain, is less than or equal to the target strain or ε_{max}.

Equation 13.11 is the acceptability criteria for allowable maximum beam strain.

$$\varepsilon_{\text{calc}} \leq \varepsilon_{\text{max}} \tag{13.11}$$

13.2.4 Calculating the Initial Beam Strain

We care about strain at the beam's base which will occur at maximum beam deflection. To avoid confusion as we step through a series of strain calculations, the subscript in the ε_{calc} term will have a hyphenated addition as shown in Eq. 13.12.

In our first beam strain calculation, a strain value, ($\varepsilon_{calc\text{-}initial}$) is calculated for the beam dimensions we have chosen as a starting point, Eq. 13.12.

$$\varepsilon_{calc\text{-}initial} = 1.5 \frac{T_b \delta_{design}}{L_b^2} \tag{13.12}$$

Now, compare this result with the target strain (ε_{max}) using Eq. 13.11.

This is our first look at the maximum strain associated with our lock design and it gives us an idea of how close the design is to the target strain. However, there are adjustments yet to be made that will reduce the strain, so a calculated strain value higher than ε_{max} is not a big concern unless the difference is quite high. How high is *quite high*? Familiarity with the effects of the upcoming adjustments will help you develop a feeling for acceptable values of $\varepsilon_{calc\text{-}initial}$ at this point.

A calculated strain lower than or equal to the target strain means there is an opportunity to make the beam and the lock stronger, but it is best to wait until all calculations, including assembly force and separation strength, have been completed before making that decision. Again, the upcoming adjustments will further reduce the calculated strain.

13.2.5 Adjusting for Deflection at the Beam's Base

When a beam protruding from a part deflects, a moment occurs at the beam's base. That moment can result in local part deflection. In some cases, this local deflection will have a significant effect on the beam strain, see Fig. 13.4, and should be taken into account. We'll refer to this deflection as *wall deflection*.

This adjustment is unique to snap-fit calculations.

Empirical testing quantified these effects for some general beam-part configurations and the *Q-factor* was introduced as an adjustment to the strain calculation. For a given applied force and the associated strain, wall deflection increases actual beam deflection beyond that predicted by beam theory. Therefore, Q is called the *deflection magnification factor* [1].

However, rather than deflection magnification, we are more interested in the strain and deflection force *reductions* in the beam as a result of wall deflection. We will use Q to reduce the strain in our calculations for a given deflection. This will, in turn, reduce the calculated beam deflection force.

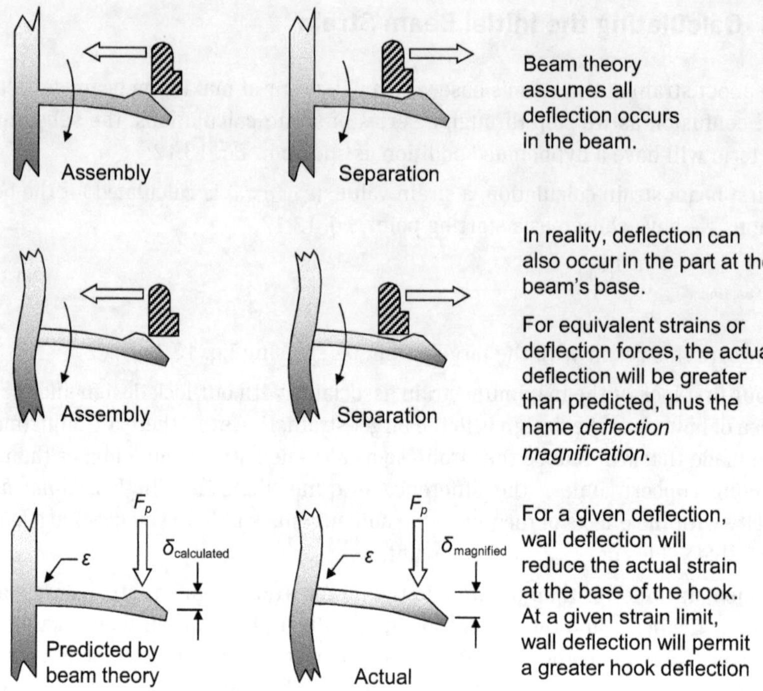

Figure 13.4 Effect of beam deflection

A beam's length/thickness (*L/T*) ratio or *aspect ratio* determines a value for *Q*. Figure 13.5 illustrates the significant differences in the contributions of bending, wall, shear, and plate behavior as the beam's *L/T* ratio varies from 1 to 10.

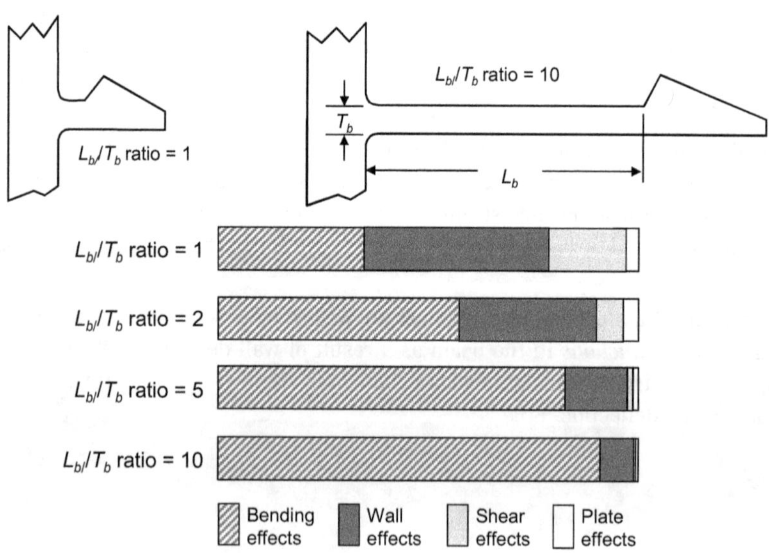

Figure 13.5 Effects of beam *L/T* ratio on behavior at the beam's base

Recall that beam L/T ratios below 5 are not recommended, primarily due to strain issues, but the significant shear and wall effects shown here are another reason - the equations for beam bending behavior become meaningless when that ratio is below 5. At L/T ratios above 5, the shear and plate effects are minimized and accounting for bending and wall effects in the calculations provides reasonable accuracy.

Figure 13.6 shows the five beam to wall configurations for which Q values are available. Remember, these values were developed through empirical testing and, while not specific to every possible combination of beam thickness and wall thickness, they will help bring the beam calculations closer to reality.

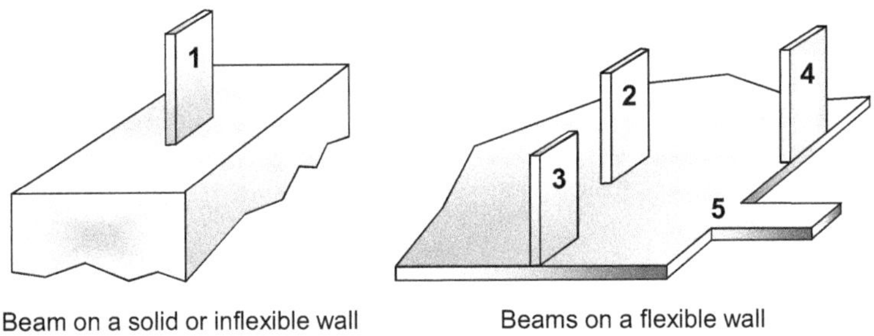

Figure 13.6 Beam to part configurations for determining the Q-factor

Table 13.1 provides Q values for use in the strain calculation, Eq. 13.12. Study the ranges of these values for the various configurations and L/T ratios for an idea of the magnitude of strain reduction possible.

The preferred range of 5 to 10 for the L/T ratio is outlined with a bold border.

If the Q-factor is not to be used in the calculations or values are not available, either leave the Q term out of the equations or set Q = 1.0. If representative parts are available, it is possible to perform some simple tests to determine a Q-factor value(s) specific to your application.

Table 13.1 Q-factor Values for Straight Beams

	Beam to wall configuration (See Fig. 13.6)				
	1	2	3	4	5
Beam Aspect Ratio L_b/T_b	Beam ⊥ to a solid wall	Beam ⊥ and in interior area of wall	Beam ⊥ to wall and ⊥ to edge	Beam ⊥ to wall and parallel at edge	Beam in-plane with wall at edge
1.5	1.60	2.12	2.40	6.50	8.00
2.0	1.35	1.70	1.90	4.60	5.50
2.5	1.22	1.45	1.65	3.50	4.00
3.0	1.17	1.35	1.45	2.82	3.15
3.5	1.15	1.28	1.38	2.4	2.65

Table 13.1 *Q*-factor Values for Straight Beams *(Continuation)*

	Beam to wall configuration (See Fig. 13.6)				
	1	2	3	4	5
Beam Aspect Ratio L_b/T_b	Beam ⊥ to a solid wall	Beam ⊥ and in interior area of wall	Beam ⊥ to wall and ⊥ to edge	Beam ⊥ to wall and parallel at edge	Beam in-plane with wall at edge
4.0	1.14	1.25	1.36	2.25	2.40
4.5	1.13	1.23	1.33	2.10	2.20
5.0	1.12	1.21	1.28	1.95	2.10
5.5	1.11	1.19	1.27	1.85	1.95
6.0	1.10	1.17	1.25	1.75	1.85
6.5	1.09	1.15	1.24	1.70	1.80
7.0	1.08	1.13	1.22	1.65	1.75
7.5	1.07	1.11	1.2	1.60	1.70
8.0	1.06	1.10	1.19	1.55	1.65
8.5	1.05	1.09	1.18	1.5	1.60
9.0	1.04	1.08	1.17	1.45	1.57
9.5	1.03	1.07	1.16	1.4	1.55
10.0	1.02	1.06	1.16	1.38	1.52
10.5	1.01	1.05	1.15	1.36	1.50
11.0	1.00	1.04	1.15	1.35	1.47

Values interpreted from *Q*-factor graphs in the *Modulus Snap-Fit Design Manual*, Allied Signal Plastics, 1997

Equation 13.13 is the same as Eq. 13.12 but it has the *Q*-factor included in the denominator. If *Q* is included in the calculations, you may skip Eq. 13.12 and use Eq. 13.13 to immediately calculate a strain value, *but this is not the suggested approach.*

$$\varepsilon_{\text{calc-}Q} = 1.5 \frac{T_b \delta_{\text{design}}}{L^2 Q} \tag{13.13}$$

The *preferred* approach is to divide the result of Eq. 13.12 by *Q* as shown in Eq. 13.14. This yields the same result as Eq. 13.13, but this approach is recommended because the *Q*-factor's effect will be visible in your series of strain calculations. This method is used in the example calculations.

The strain is now called $\varepsilon_{\text{calc-}Q}$ to indicate the *Q*-factor adjustment has been made.

Preferred equation.

$$\varepsilon_{\text{calc-}Q} = \frac{\varepsilon_{\text{calc-initial}}}{Q} \tag{13.14}$$

Again, compare the new strain value $\varepsilon_{\text{calc-}Q}$ to the target strain using Eq. 13.11.

A point to make again is that the beam's width does not appear in the strain equations. This means that the beam's width (thus the lock's strength) can be changed without affecting the strain.

We've done all we can at this point with the strain calculations. Now we will turn our attention to the beam's deflection force.

13.2.6 Calculating the Initial Beam Deflection Force

Knowing the adjusted strain value and the initial beam dimensions we can now calculate a deflection force using Eq. 13.15. We are interested in the maximum force so we will use maximum beam length and the strain we have just calculated (also using the maximum beam length).

The secant modulus (E_S) was explained in Section 13.2 and calculated using Eq. 13.2.

Because we have *already* used the Q-factor to adjust the calculated strain value, we *cannot* use it again. IMPORTANT!

$$F_{p\text{-}Q} = \frac{W_b T_b^{\,2} E_S \varepsilon_{\text{calc-}Q}}{6L_b} \tag{13.15}$$

We'll identify the deflection force as $F_{P\text{-}Q}$ because it includes the Q-factor adjustment.

To avoid confusion and to ensure the correct strain and deflection values are used in the remaining calculations, it is best to use Q to adjust the strain as we have shown above and then use Eq. 13.15.

The author has seen instances where the Q-factor was mistakenly used in both the strain and the deflection force calculations. *Be aware that this error may occur if others are performing these calculations for you and watch for it.*

The Q-factor could be used in the denominator of the deflection force equation as shown in Eq. 13.16, and there may be times when it may be appropriate to do so. However, if you have already used Q to adjust the strain, *do not* use it again here.

$$F_{p-Q} = \frac{W_b T_b^{\,2} E_S \varepsilon_{\text{calc-initial}}}{6L_b Q} \tag{13.16}$$

Equation 13.16 is *not* the preferred method.

13.2.7 Adjusting for Mating Feature/Part Deflection

We normally assume the lock is engaging an inflexible locator as the other member of the lock pair, but this is not always the case.

This adjustment is also unique to snap-fit calculations.

The Q-factor discussion in Section 13.2.5 dealt with deflection of the part or wall on which the beam is mounted. Likewise, if the *mating feature* in the lock pair is flexible or is mounted on a flexible area of the other part, then some deflection will occur at that feature. If the mating feature/part deflection is thought to be significant, we can include its effect on the beam's deflection force and strain.

The effect of mating feature/part deflection on the beam's deflection force and strain is the same as described in the *wall deflection* discussion in Section 13.2.5. The actual beam deflection is reduced so strain and deflection force are reduced.

We have already calculated values for the beam's maximum deflection force and an associated maximum strain. We'll assume linear behavior and construct a force-deflection signature as shown in the first graph in Fig, 13.7. (However, calculating a mid-point value for the beam signature may reveal some *nonlinear* behavior.)

The maximum deflection is shown in these graphs as equivalent to *Y*. Because *Y* is the retention feature undercut, it is a reasonable starting point for a deflection value. However, the deflection scale ultimately depends on the lock's expected deflection as designed.

Next, through mechanical testing or analysis, we can develop a force-deflection signature for the mating feature/part as shown in the second graph. This force-deflection signature uses the design deflection point (in this case *Y*) as its origin and is plotted with a negative slope since it is in opposition to the beam's deflection.

In practice, the signatures are plotted together as shown in the third graph where their intersection is the resulting maximum force and deflection for the beam. This result will not be perfect, but as with the wall deflection adjustment, it will give us more accurate results than we would obtain otherwise.

If necessary, with more complex calculations, we can determine a few intermediate points for the beam's force-deflection signature and also add more points to construct a more accurate signature for the mating part.

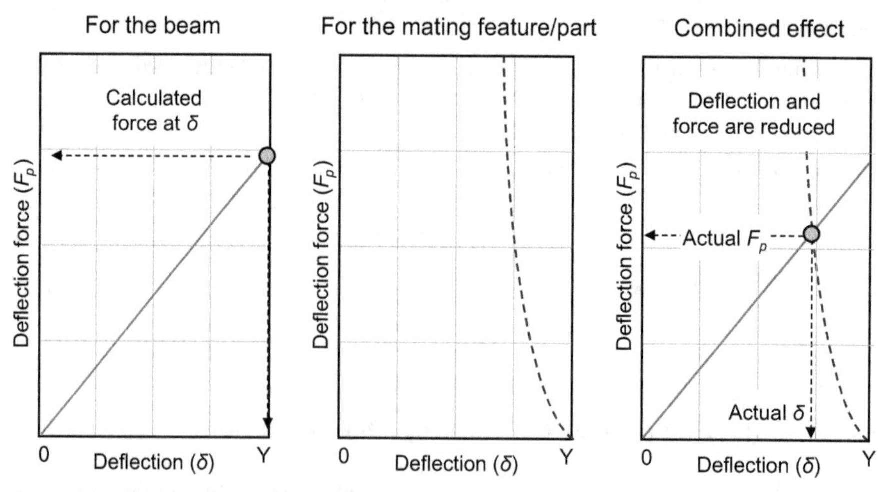

Figure 13.7 Mating feature/part deflection effects on beam behavior

The adjusted value for maximum deflection, (δ_{actual}) obtained from this step will be used to calculate a final value for maximum beam strain (Eq. 13.17). The adjusted deflection force, ($F_{P\text{-final}}$) will be used in the assembly and separation force calculations that follow.

$$\varepsilon_{\text{final}} = \frac{\delta_{\text{actual}}}{\delta_{\text{design}}}\left(\varepsilon_{\text{calc-}Q}\right) \tag{13.17}$$

Compare $\varepsilon_{\text{final}}$ to the target strain using Eq. 13.11. Changes to the design may be indicated at this point if the final calculated strain is *higher* than the target strain.

In Summary:

Two adjustments, one for wall deflection at the beam's base and one for mating feature/part deflection have been made. Both reduced the original strain and force calculated using the traditional bending equations. Including these adjustments in the calculations results in:

- Lower strain and beam deflection force at a given deflection than predicted by traditional beam calculations.
- Higher allowable beam deflection for a given strain than predicted by beam theory.
- Lower assembly force and separation force than predicted by traditional beam calculations.

Conversely, ignoring these adjustments results in:

- Higher strain and beam deflection force at a given deflection than predicted by traditional beam calculations.
- Lower allowable beam deflection for a given strain than predicted by beam theory.
- Higher assembly force and separation force than predicted by traditional beam calculations.

13.2.8 Example Beam Strain and Deflection Calculations

This example applies to only the beam portion of a cantilever lock. This beam could be part of a cantilever hook, loop or trap style lock. Figure 13.31 in Section 13.9 shows how to construct a spreadsheet to perform these calculations.

As we will do here, it is common practice to begin calculations with a given deflection and solve for strain. To minimize work, particularly if the calculations are manual, it is worthwhile to begin with initial lock dimensions that are as close as possible to the final desired dimensions. Unless lock dimensions are already established by other requirements, the rules-of-thumb presented in the previous chapter can be used to establish a starting point for the lock design. This is less important when analysis software is available and many design alternatives can be evaluated quickly, but it is still a good idea.

However, when using software for beam analysis, be aware that most of the available beam analysis software *does not* comprehend the adjustments discussed here. The user must apply these adjustments manually to the results of software-based calculations.

Because we are interested in the highest possible strain in the beam, select dimensions for the calculation that represent maximum material conditions for beam thickness, undercut, and mating feature interference.

Draft angle warning:

We generally ignore the effects of draft angle in these calculations. However, depending on part and feature orientation in the mold, draft angles can affect any dimension of the

An example application is described in Section 13.3.1.

beam, length, thickness or width. Any draft angle that causes smaller cross-section or beam length dimensions at the beam's base than at its end will increase the stress and strain at the beam's base and must be avoided. If this situation is unavoidable, the draft angle must be minimized and the lock feature designed with an extra margin of safety. The effects of excessive draft angles are discussed in Section 13.3.1.

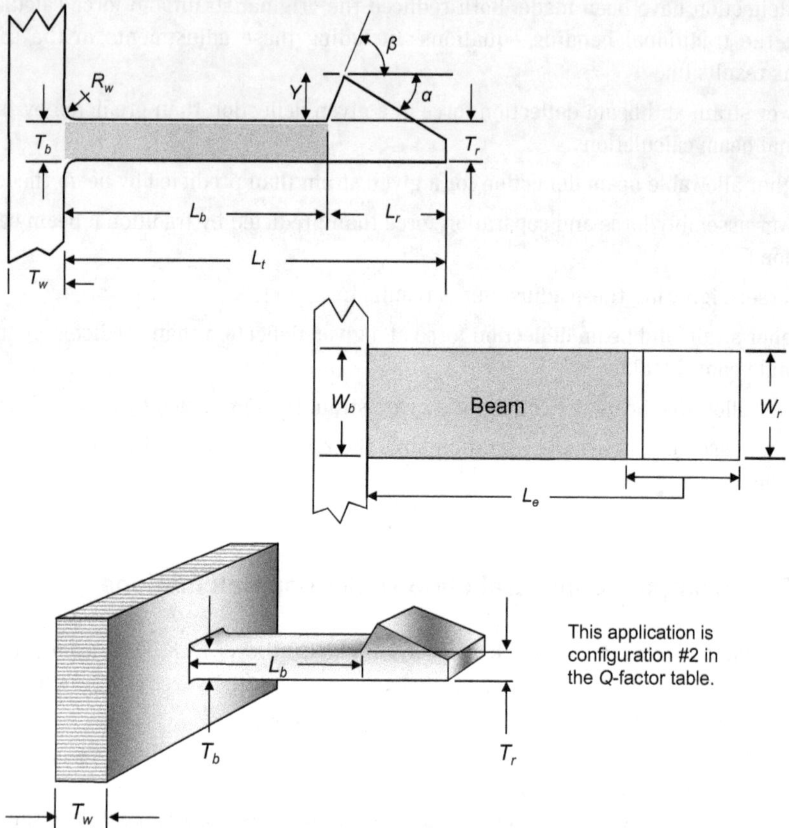

Beam Dimensions

Beam length (L_b)	15 mm
Wall thickness at beam (T_w)	4 mm
Beam thickness at wall (T_b)	2 mm
Beam thickness at retention feature (T_r)	2 mm
Radius at beam/wall intersection (R_w)	1 mm
Beam width at wall (W_b)	3 mm
Beam width at retention feature (W_r)	3 mm
Undercut depth = deflection ($Y = \delta_{design}$)	2 mm

Material information

Secant modulus (E_s)	2000 MPa
Design point (ε_{design})	3%
Friction coefficient (μ)	~ .3

Given the variability and uncertainty of friction coefficient information, we will not distinguish between static and dynamic friction in this example.

These initial dimensions are based on the rules-of-thumb introduced in Chapter 12.

Figure 13.8 Beam dimensions for the example calculation

The example calculations will proceed in this order. Chapter sections are referenced.

1. Get the design point as described in Section 11.4.
2. Find the stress concentration factor (K), Section 13.2.3.
3. Calculate the target strain, (ε_{max}).
4. Initial calculation of beam strain ($\varepsilon_{calc\text{-}initial}$), Section 13.2.4.
5. Compare ($\varepsilon_{calc\text{-}initial}$) to the target strain ($\varepsilon_{max}$).
6. Adjust the calculated strain for wall deflection, Section 13.2.5.
7. Compare ($\varepsilon_{calc\text{-}Q}$) to the target strain ($\varepsilon_{max}$).
8. Calculate the initial maximum deflection force ($F_{p\text{-}Q}$), Section 13.2.6.
9. Construct force-deflection curves, (signatures).
10. Adjust values for mating feature/part deflection, Section 13.2.7.
11. Compare (ε_{final}) to the target strain (ε_{max}).

The dimensions in Fig. 13.8 are the starting point for the calculation examples that follow.

In applications with some pre-existing requirements or limits on the beam dimensions, it is still a good idea to apply the rules-of thumb to the initial design to the extent possible.

The calculation examples shown here were carried out by hand on a calculator. The calculation spreadsheets described in Section 13.9 may give slightly different results because of round-off to different decimal places in the two methods. Given the assumptions we are making and general plastic property variation, differences in accuracy between the tenth or hundredth decimal place are meaningless.

Find the stress concentration factor (K)

Given:

Beam thickness at wall (T_b) = 2.0 mm

Radius at beam/wall intersection (R_w) = 1.0 mm

Using the stress concentration factor curve, Fig. 13.3:

$$\frac{R_w}{T_b} = \frac{1.0}{2.0} = 0.5$$

So K = 1.5

Calculate the target strain (ε_{max})

Given:

Design point (ε_{design}) = 3% = 0.03

K = 1.5

Using Eq. 13.10:

$$\varepsilon_{max} = \frac{\varepsilon_{design\text{-}point}}{K_{stress\text{-}concentration}} = \frac{0.03}{1.5} = 0.02$$

So the target strain = ε_{max} = 2%

Caution

As you step through this process, you will be calculating a series of adjusted values for strain, deflection force and deflection. Be careful to always carry the new values forward into the later calculations. It is easy to get lost and forget to use the most recent value(s).

Calculate the initial beam strain ($\varepsilon_{calc\text{-}initial}$)

Given:

Beam thickness at wall (T_b) = 2.0 mm

Beam length (L_b) = 15.0 mm

Deflection (δ_{design} = Y) = 2.0 mm

Using Eq. 13.12:

$$\varepsilon_{calc\text{-}initial} = 1.5\frac{T_b{}'\ _{design}}{L_b{}^2} = 1.5\frac{2.0 \times 2.0}{15.0} = 0.0267$$

So $\varepsilon_{calc\text{-}initial} = 2.67\%$

Compare ($\varepsilon_{calc\text{-}initial}$) to the target strain ($\varepsilon_{max}$)

Given:

Target strain = ε_{max} = 2 %

Using Eq. 13.11:

$$\varepsilon_{calc\text{-}initial} \leq \varepsilon_{max}$$

Is:

2.67 % ≤ 2 %

No, it is not. The initial calculation shows the beam strain is higher than the target strain.

This *may* be a problem, but there are adjustments yet to be made, so we will continue with the calculations.

Adjust the calculated strain for wall deflection

Given:

Beam length (L_b) = 15.0 mm

Beam thickness at wall (T_b) = 2.0 mm

Beam-wall configuration is #2 in Table 13.1

$\varepsilon_{calc\text{-}initial}$ = 2.67 %

Using Table 13.1, the beam aspect ratio is:

$$\frac{L_b}{T_b} = \frac{15.0}{2.0} = 7.5$$

So Q = 1.11

Using Eq. 13.14 to apply the Q-factor:

$$\varepsilon_{calc\text{-}Q} = \frac{\varepsilon_{calc\text{-}initial}}{Q} = \frac{0.0267}{1.11} = 0.024$$

So $\varepsilon_{calc\text{-}Q} = 2.4\%$

Compare (ε_{calc}-Q) to the target strain (ε_{max})

Using Eq. 13.11:

$$\varepsilon_{calc-Q} \leq \varepsilon_{max}$$

Is:

2.4% ≤ 2%

No, it is not. The beam strain is further reduced, but it is still higher than the target strain. This *may* be a problem, but more adjustments are possible. We will continue with the calculations.

Calculate the maximum deflection force, ($F_{p\text{-}Q}$)

Given:

Beam width at wall (W_b) = 3.0 mm

Beam thickness at wall (T_b) = 2.0 mm

Secant modulus (E_s) = 2000 MPa

$$\varepsilon_{\text{calc-}Q} = 2.4\%$$

Beam length (L_b) = 15.0 mm

Using Eq. 13.15:

$$F_{p-Q} = \frac{W_b T_b^2 E_S \varepsilon_{\text{calc}-Q}}{6L_b} = \frac{3.0 \times 2.0^2 \times 2000 \times 0.024}{6 \times 15.0} = 6.4 \text{ N}$$

So $F_{p-Q} = 6.4$ N

Construct force-deflection signatures

Refer to Fig. 13.9.

For the beam's signature, use the origin and:

Deflection (δ_{design}) = 2.0 mm

Deflection force ($F_{p\text{-}Q}$) = 6.4 N

For the mating feature/part's signature, construct a curve of opposite slope using analysis or test data. The origin will be at the current design deflection. In this case, (δ_{design}) = 2.0 mm.

The intersection will be the actual beam deflection and deflection force:

δ_{actual} = 1.48 mm

F_{final} = 4.8 N

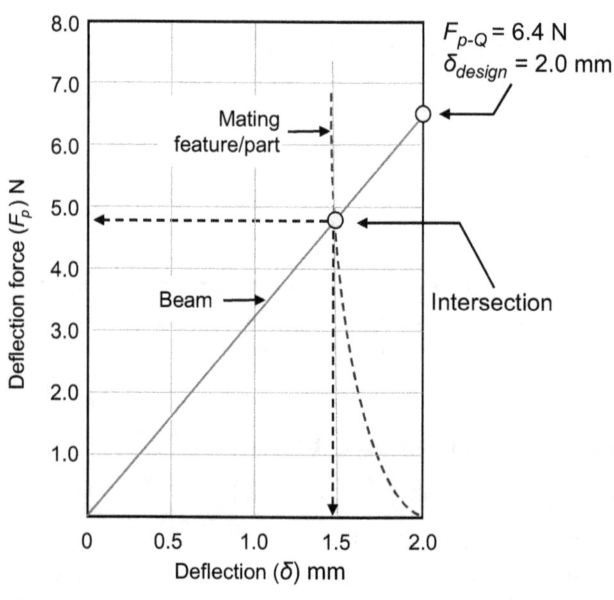

Figure 13.9 Evaluating the effects of mating feature/part deflection

Adjust the strain for mating feature/part deflection

Given:

δ_{actual} = 1.48 mm *From the mating feature/part deflection analysis.*

δ_{design} = 2.0 mm

$\varepsilon_{calc\text{-}Q}$ = 2.4%

Using Eq. 13.17:

$$\varepsilon_{final} = \frac{\delta_{actual}}{\delta_{design}}\left(\varepsilon_{calc-Q}\right)$$

Calculate a final strain value, (ε_{final}):

$$\varepsilon_{final} = \frac{\delta_{actual}}{\delta_{design}}\left(\varepsilon_{calc-Q}\right) = \frac{1.48}{2.0} \times 0.024 = 0.0178$$

So $\varepsilon_{final} = 0.0178 \approx 1.8\%$

Compare (ε_{final}) to the target strain (ε_{max})

Using Eq. 13.11:

$$\varepsilon_{final} \leq \varepsilon_{max}$$

Is:

1.8% ≤ 2%

Yes, the *final* calculated strain is less than the target strain.

Summary of straight cantilever beam example

We started the calculations with:

Deflection, δ_{design} = Y = 2.0 mm

First calculation of strain $\varepsilon_{calc\text{-}initial}$ = 2.67%

First calculation of deflection force $_{p\text{-}Q}$ = 6.4 N

We finished the calculations, after adjustments, with:

Actual deflection, δ_{actual} = 1.48 mm

Final strain, ε_{final} = 1.8%

Final deflection force, $F_{p\text{-}final}$ = 4.8 N *From the mating feature/part deflection analysis.*

It will not always be possible to make some of these adjustments, so use the equations appropriately. The adjustments for wall deflection and mating feature/part deflection reduced both the actual beam strain and the actual deflection force. Strain reduction is always a good thing. A reduced deflection force is good for assembly, but it will also reduce the separation strength.

The values for the actual deflection and the final deflection force will be used in the assembly and separation force calculations described in Section 13.5.

If you think you will need the extra strength and because the final strain is less than the target strain, you may wish to go back and increase the beam's thickness and undercut by a few tenths of a millimeter and recalculate the final strain. However, it is suggested that you finish the assembly and separation force calculations before doing any recalculations. Don't forget, you can always get more lock strength by increasing the beam's width with no increase in strain.

13.2.9 Deflection Graphs for a Straight Beam

The following figures show how beam deflection in a straight constant-section beam is related to the beam's length/thickness ratio, (L/T) and the beam material's maximum allowable strain. The graph in Fig. 13.10 shows a range of L/T values from 2 to 30, which are well beyond normal beam L/T ranges. One can see in this graph how the recommended range of $5 \leq L/T \leq 10$ optimizes trade-offs between the beam's length and thickness and the maximum allowable stress.

Note that the deflection factor on the graphs is identified as F_δ. Do not confuse this with the use of F in the beam equations where it represents force.

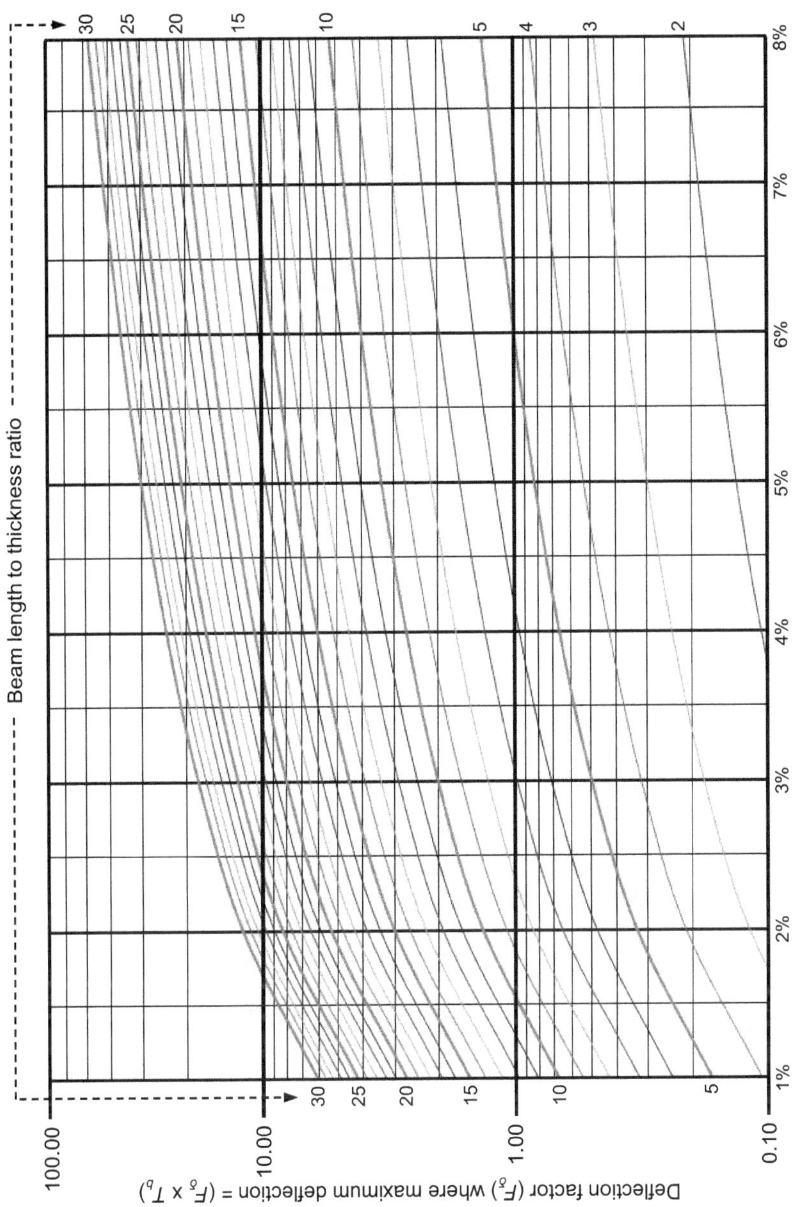

Figure 13.10 Maximum allowable deflection graph for straight beams

Figure 13.11 shows an expanded area of Fig. 13.10 enlarging the recommended L/T range. Both graphs can help the developer understand beam behavior and sensitivity to the L/T ratio, particularly if beam designs that are outside of the recommended L/T range are necessary.

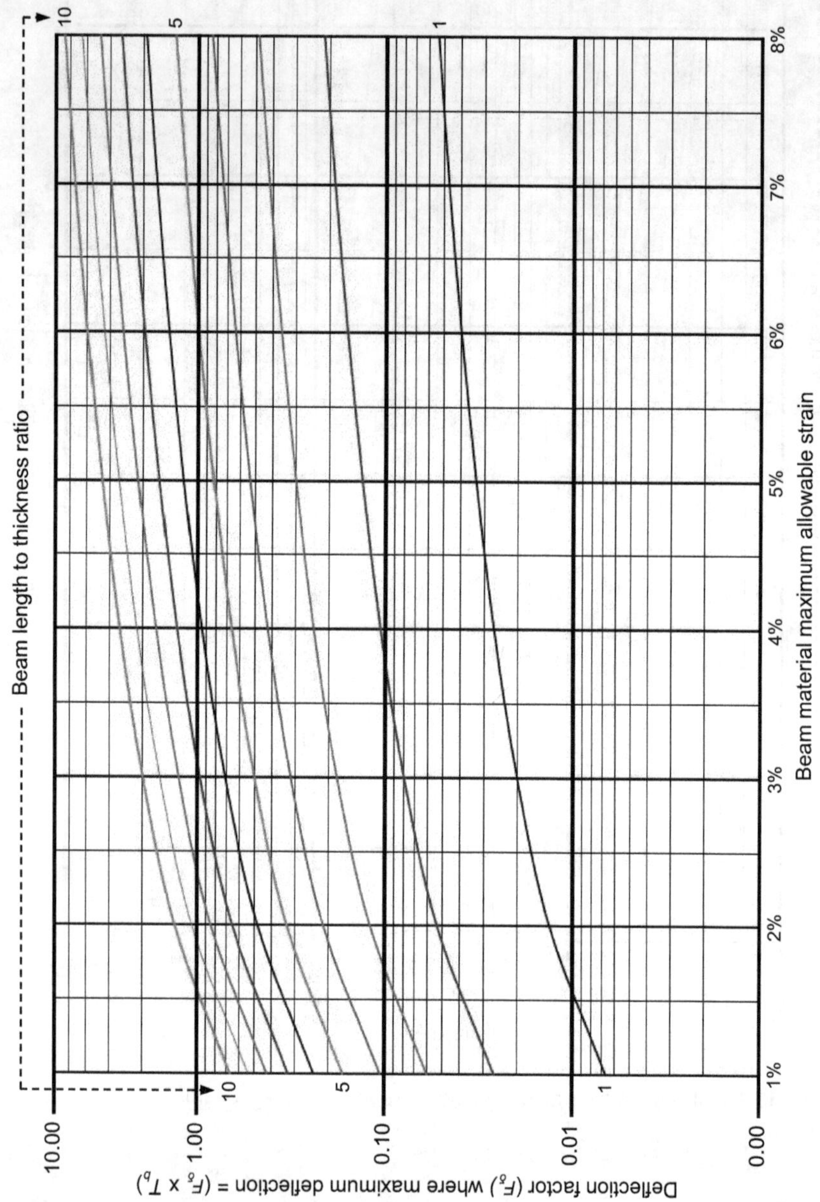

Figure 13.11 Expanded maximum allowable deflection graph for straight beams

These graphs only apply to straight beams, not to thickness or width-tapered beams.

The beam calculations we have discussed use predetermined beam dimensions and a predetermined *beam deflection* to arrive at a *final strain* value close to the target strain.

These graphs provide a different starting point. They allow the user to begin with a *target strain* and a predetermined beam length and thickness to find a *maximum allowable beam deflection*. However, the graphs do not include any of the adjustments for wall deflection and mating feature/part deflection that follow the initial calculations for strain. If you use these graphs, you could then enter the calculations at that point.

This approach is also possible by manipulating the beam equations for different known and unknown variables, but it is nice to be able to visualize the relationships on the graph.

Note:

- The deflection factor value of 1.0 is of interest because a common rule-of-thumb is to design so that deflection = beam thickness.
- Another common rule-of-thumb is that when deflection = beam thickness, the L/T ratio should be no less than 5 with values closer to 10 preferred.
- The intersection of the $F_b = 1$ line with the $L/T = 5$ and the $L/T = 10$ curves represents the limits of a common design space for cantilever beam-based locks with respect to the allowable strain which ranges from 1.5 % to 6 %.

For the example shown in Fig. 13.12, we will use values from the straight beam example calculations we have just completed.

Using the maximum allowable beam deflection graph, see Fig. 13.12

Given

From the straight beam calculation example:

Target strain = 2.67 % *(This is the initial calculated strain from the example calculations above. Using it here as the target strain permits a direct comparison to the calculations.)*

The beam's length, L_b = 15.0 mm

The beam's thickness at the base, T_b = 2.0 mm

The beam L/T ratio = 7.5

(A) Find the maximum allowable strain on the *x*-axis (2.67 %)

(B) Find the curve for an L/T ratio of 7.5 on the graph

(C) At the intersection of the maximum allowable strain and the L/T curve find the deflection factor F_δ on the *y*-axis log scale.

(D) In this example, F_δ on the *y*-axis log scale = 1.0

(E) Using Eq. 13.18, calculate the maximum allowable deflection:

$$\delta_{\max} = F_\delta \times T_b = 1.0 \times 2.0 = 2.0 \text{ mm}$$

So, for the given beam geometry and a strain of 2.67 %, the maximum allowable deflection is 2.0 mm. This agrees with the original allowable deflection in the calculation example.

$$\delta_{\max} = F_\delta \times T_b \tag{13.18}$$

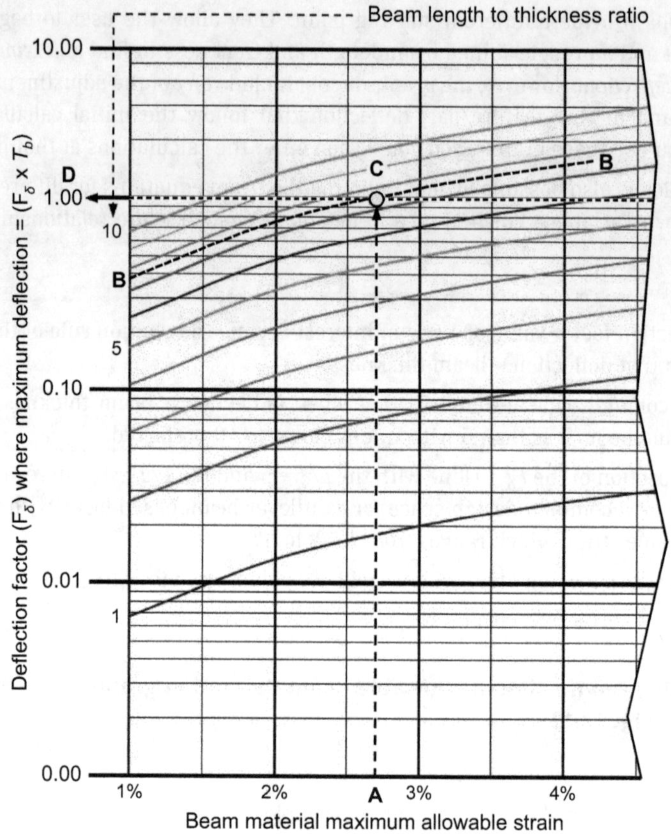

Figure 13.12 Using the deflection graph

We are not yet finished with cantilever beams.

Next, we will present the equations and example calculations for cantilever beams tapered in thickness and in width. Just as with the straight cantilever beams, we will calculate values for actual strain, deflection and deflection force in these beams.

Again, we'll verify that the actual strain does not exceed the target strain.

13.3 Deflecting Member: Tapered Beams

Rectangular section beams may also be tapered in thickness, width, or both. We'll discuss thickness and width-tapered beams. Beams tapered in both dimensions, while possible, are rare and the calculations are complex. Refer to [2] for information about those calculations. Some of the analysis tools and design guidelines listed in the appendix also provide more information about tapered beam calculations.

13.3.1 Taper Error Example

Beams can be incorrectly tapered in thickness and width, what we'll call a *negative* taper. Beams with a negative thickness taper ($T_b < T_r$) were mentioned in Section 12.1.3, and shown in Fig. 12.6.

Examples of beams improperly tapered in other dimensions are shown in Fig. 13.13. In both examples, it *appears* that improper use of draft angle is responsible. Both applications are backs for exterior reflectors and use trap style locks. The reflector itself is included in the first but not the second application.

Reflector base applications

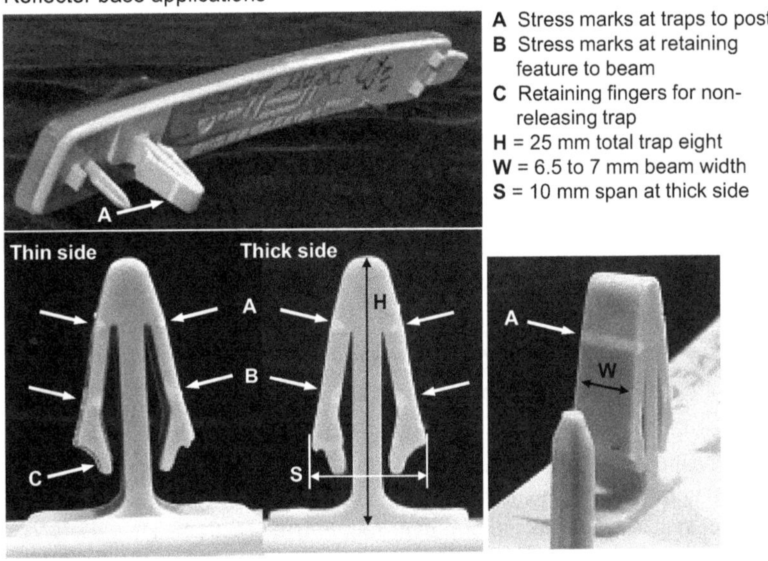

In both designs, the traps are width-tapered beams with a negative taper. This also increases stress and strain at the beam's base.

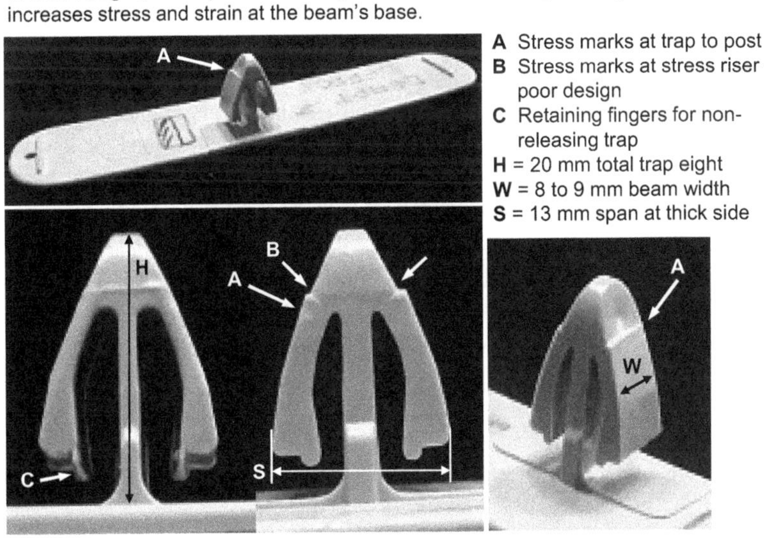

Figure 13.13 Beams with incorrect tapers

As with a thickness-tapered beam, a width-tapered beam must have its greatest width at the beam's base with the width decreasing toward the beam's end. A negative taper with a smaller width at the base than at the end will concentrate more stress and strain at the beam's base and increase the likelihood of beam damage. In both of these applications, there is a slight negative taper in the D dimension or beam width. In beams with other design issues, this only makes the high strain problem worse.

The beams in these applications are also tapered in thickness but at a 90° angle to the beam's long axis. This creates a cross-section similar to a trapezoid where the beam is much thicker at one side than the other with the difference extending the length of the beam. Beam strain calculations for a rectangular section beam are no longer possible. The trapezoidal shape of the section creates higher strain at the thicker side of the beam's base. The effect of the higher strain can occur at the beam's base as well as near the retention feature at the end of the beam. Note the white stress marks in the illustrations.

The second application also has a *designed-in* stress riser where the beam meets the center post, providing another location for high strain in this application. This is a sure sign of a lack of understanding of beam behavior.

Note the extensions at the ends of the beams that act as retaining fingers for these nonreleasing traps. This is a good attribute in these trap applications.

In summary:

- Negative tapers in beam width or thickness are not allowed.
- A trapezoid-like beam section will concentrate stress and strain at one side of the beam and should be avoided.

  If such a beam design is necessary, use strain calculations for the appropriate beam section. In addition to the expected stress concentration at the beam's base, consider an additional stress concentration due to the unusual section shape, design conservatively and test.

These applications are from two different manufacturers and two different product engineering sources. How these mistakes have come about?

As traps, these are very easy locks to design properly.

One possibility is error in the original design of these traps. However as we have learned, traps are a very robust lock style and can have very high efficiency ratios. There is no reason these traps should have been improperly designed for low-demand applications like these reflectors. This is a case where effective traps could have been designed by simply following the beam design rules-of-thumb in Chapter 12.

A more likely scenario is *developer failure to communicate with the mold-maker*. For the sake of this discussion, let us assume the second scenario. Mold design decisions are made by the mold-maker and previous discussions have stressed the importance of communication between the two parties.

A scenario for the design errors in Figure 13.13

Let us assume the developers provided nominal dimensions for these traps in the application's detailed designs and then turned them over to the mold-maker.

The mold-maker, not understanding the effects of beam taper on lock feature performance, made rational decisions about part orientation in the mold, die-action and draft angle based on standard mold design practices. The applications in Fig. 13.13 are the result.

Preventative measures the developers could have taken include:

- Noting the allowable directions for draft angle on the drawings and explaining the reasons why.
- Specifying allowable die-action directions that would eliminate draft angle concerns.
- Specifying minimal draft angles for critical dimensions. Draft angles as low as 1.5° to 2° are possible and, in these applications, would have minimized the negative effects. However, low draft angles cause more rapid die wear and may make part extraction difficult; these are issues to be discussed with the mold-maker.
- Being ultimately responsible for the product, the developers should have had direct communication with the mold-maker to explain the reasons for the special requirements.
- Recognizing these glaring design errors when prototype parts became available and insisting on mold changes although, by this time, mold changes would have been more difficult and costly.

13.3.2 Beams Tapered in Thickness

Tapered beams do a better job than nontapered beams of distributing stress and strain over the length of the beam. This reduces strain at the beam's base during assembly and separation deflection, which is good. It reduces deflection force, which is good for assembly but it also reduces separation force which can be bad if retention strength is important. Beams can be tapered in either thickness or width or both. Beams with both a thickness and width taper are rare and the calculations are beyond the scope of this book.

We'll begin with a discussion of thickness-tapered beams, Fig. 13.14. These beams typically have T_b:T_r ratios ranging from 1.25:1 up to 2:1, with higher ratios being more effective in reducing strain. The shorter the beam relative to its length, the greater the effect of taper on strain and deflection force reduction.

Use caution, a taper may cause a beam to become too weak at the retaining member end, resulting in excessive rotation and damage at that junction. A strain calculation at this area may be necessary.

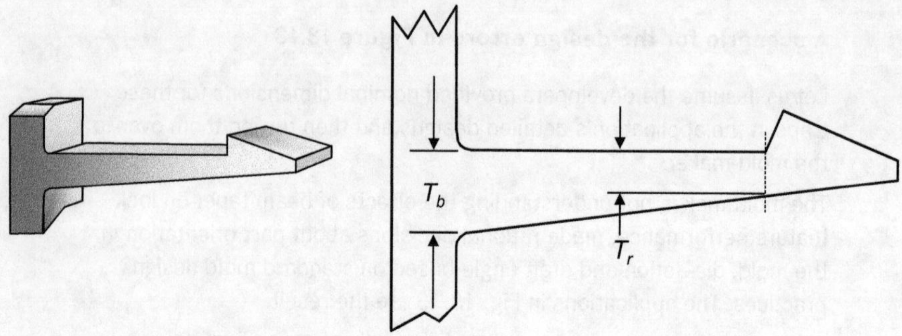

Figure 13.14 Thickness-tapered beam

The procedures and calculations for determining the design strain, stress concentration effect, target strain, and deflection force for thickness-tapered beams are the same as for constant-section beams and will not be repeated here. The procedures for adjusting the calculations for wall and mating feature/part deflection are also the same.

However, the *strain equation*, Eq. 13.19, is different:

$$\varepsilon_{\text{calc-initial}} = 1.5 \frac{T_b \delta_{\text{design}}}{L_b^{\,2} K_{\text{taper}}} \tag{13.19}$$

Where K_{taper} is found in the chart shown in Fig. 13.15.

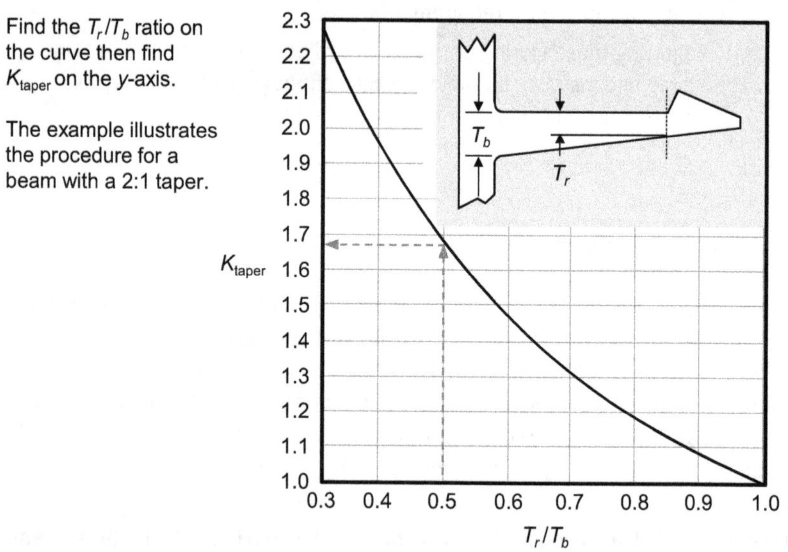

Figure 13.15 The proportionality constant (K_{taper}) for thickness tapered beams, adapted from Ticona LLC, *Designing With Plastic – the Fundamentals*

Strains for beams of any T_r/T_b ratio can be calculated. However, *Q*-factor information for wall deflection is available only for beams with a 2:1 thickness taper ratio, see Fig. 13.16 and Table 13.2.

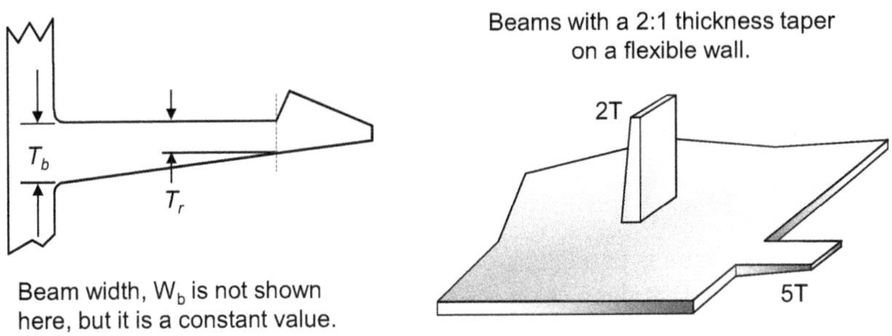

The identifying labels correspond to similar beam-wall configurations on the straight beam Q-factor chart, Table 13.1. The *T* indicates a thickness-tapered beam

Figure 13.16 Thickness-tapered beam configurations for the *Q*-factor

Table 13.2 *Q*-factor Values for Beams with a 2 : 1 Thickness Taper

	Beam - wall configuration (See Fig. 13.16)			Beam - wall configuration (See Fig. 13.16)	
	2T	5T		2T	5T
Beam Aspect Ratio L_b/T_b	Beam ⊥ and in interior area of wall	Beam in-plane with wall at edge	Beam Aspect Ratio L_b/T_b	Beam ⊥ and in interior area of wall	Beam in-plane with wall at edge
2.0	1.60	3.5	7.0	1.14	1.52
2.5	1.50	3.0	7.5	1.13	1.47
3.0	1.40	2.5	8.0	1.13	1.43
3.5	1.33	2.25	8.5	1.12	1.40
4.0	1.25	2.05	9.0	1.12	1.38
4.5	1.22	1.90	9.5	1.11	1.35
5.0	1.20	1.80	10.0	1.11	1.32
5.5	1.17	1.70	10.5	1.10	1.30
6.0	1.15	1.65	11.0	1.10	1.28
6.5	1.14	1.58	-	-	-

Values interpreted from *Q*-factor graphs in the Modulus Snap-Fit Design Manual, Allied Signal Plastics, 1997

The *Q*-factor could be applied as shown in Eq. 13.20, but as with the straight beam calculations, we prefer to make the *Q*-factor visible in our chain of calculations, so we'll use Eq. 13.21 instead.

$$\varepsilon_{\text{calc}-Q} = 1.5\frac{T_b\delta_{\text{design}}}{L_b^2 Q_{\text{taper}} K_{\text{taper}}} \tag{13.20}$$

$$\varepsilon_{\text{calc}-Q} = \frac{\varepsilon_{\text{calc-initial}}}{Q_{\text{taper}}} \tag{13.21}$$

Preferred equation.

The strain calculation has already accounted for the effect of tapering the beam, so the deflection force equation, Eq. 13.22, for thickness-tapered beams is the same as Eq. 13.15 for constant section beams.

$$F_{p-Q} = \frac{W_b T_b^2 E_S \varepsilon_{\text{calc}-Q}}{6L_b} \quad (13.22)$$

Although any taper ratio is possible, a 2:1 taper is common. If a 2:1 thickness taper is applied to the original straight beam example and all other dimensions remain the same, the new calculations are as shown here. We will create the taper by reducing the beam thickness at its end.

Example calculations for a thickness-tapered beam

Given:

Design point (ε_{design}) = 3% = 0.03

$K_{stress\text{-}concentration}$ = 1.5

Target strain = ε_{max} = 2%

Thickness taper ratio = 2:1 *New design requirement.*

Beam thickness at wall (T_b) = 2.0 mm

Beam thickness at the retaining member (T_r) = 1.0 mm *New value.*

Beam length (L_b) = 15.0 mm

Beam width at wall (W_b) = 3.0 mm

Deflection (δ_{design} = Y) = 2.0 mm

K = 1.67 from Fig 13.15

Q = 1.13 from Table 13.2

Calculate strain and deflection force for a thickness-tapered beam

Using Eq. 13.19:

$$\varepsilon_{\text{calc-initial}} = 1.5\frac{T_b \delta_{\text{design}}}{L_b^2 K_{\text{taper}}} = 1.5\frac{2.0 \times 2.0}{15.0^2 \times 1.67} = 0.016 = 1.6\%$$

Using Eq. 13.21:

$$\varepsilon_{\text{calc-}Q} = \frac{\varepsilon_{\text{calc-initial}}}{Q_{\text{taper}}} = \frac{0.016}{1.13} = .014$$

So $\varepsilon_{\text{calc}-Q} = 1.4\%$

Compare using Eq. 13.11:

$$\varepsilon_{\text{calc-}Q} \le \varepsilon_{\max}$$

Is:

1.4% ≤ 2.0%

Yes it is.

Using Eq. 13.22:

$$F_{p-Q} = \frac{W_b T_b^2 E_S \varepsilon_{calc-Q}}{6L_b} = \frac{3.0 \times 2.0^2 \times 2000 \times 0.014}{6 \times 15.0} = 3.7\text{ N}$$

As we did in the straight beam example, see Section 13.2.8, we can now use the calculated deflection force and design deflection values to construct a graph for evaluating mating feature/part deflection and determine the final values for deflection force, strain, and deflection. The mating feature/part has not changed from the straight beam example, so its force deflection signature is the same.

Force-deflection curves for the thickness-tapered beam example:

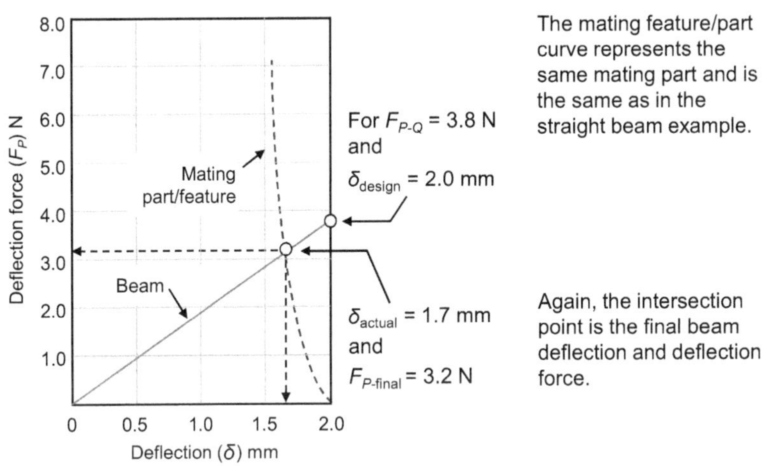

Figure 13.17 Plotting the effect of mating feature/part deflection

Adjust the strain for mating feature/part deflection

Given:

δ_{actual} = 1.7 mm *From the mating feature/part deflection analysis.*

δ_{design} = 2.0 mm

$\varepsilon_{calc\text{-}Q}$ = 1.4%

Using Eq. 13.17:

$$\varepsilon_{final} = \frac{\delta_{actual}}{\delta_{design}}(\varepsilon_{calc-Q})$$

Calculate a final strain value, (ε_{final}):

$$\varepsilon_{final} = \frac{\delta_{actual}}{\delta_{design}}\left(\varepsilon_{calc-Q}\right) = \frac{1.7}{2.0} \times 0.014 = 0.012 = 1.2\%$$

So ε_{final} = 1.2 %

Comparing straight vs. thickness-tapered beam results

The only change to the original straight beam was tapering the thickness, (2:1) by making the beam thinner at the end. Note the significant reductions in strain and force at this stage of the calculations:

For the straight beam:

$\delta_{actual} = 1.48$ mm

$\varepsilon_{final} = 1.8\%$

$F_{p-final} = 4.8$ N

For the 2:1 thickness-taper beam:

$\delta_{actual} = 1.7$ mm

$\varepsilon_{final} = 1.2\%$

$F_{p\text{-}final} = 3.2$ N *From Fig. 13.17.*

Note how the actual deflection is higher but the strain and deflection force are lower for the tapered beam.

Judgments about the final strain vs. the target strain would now dictate if any dimensional changes are required. If so, new calculations would be performed.

13.3.3 Beams Tapered in Width

When tapering beam thickness is not possible, as when a beam is an in-plane extension of a wall, tapering the beam's width is an option, see Fig. 13.18. Again, the shorter the beam relative to its length, the greater the strain reduction.

Although any taper ratio is possible, a 4:1 taper is common and a strain calculation for this taper ratio (Eq. 13.23) can be found in the literature.

Also, for retaining feature and beam stability during deflection, the beam width at the end should be at least equal to or preferably greater than the beam thickness.

If we begin with identical beam base dimensions for thickness and width, we find that tapering the beam in width is less effective than tapering a beam in thickness. For example, a 2:1 thickness taper will be more effective in reducing strain and deflection force than a 2:1 width taper.

Unlike for thickness-tapered beams, the author has been unable to find a taper factor (K_{taper}) graph for width-tapered beams. We are limited to the strain calculation for a 4:1 width-tapered beam as given in Eq. 13.23.

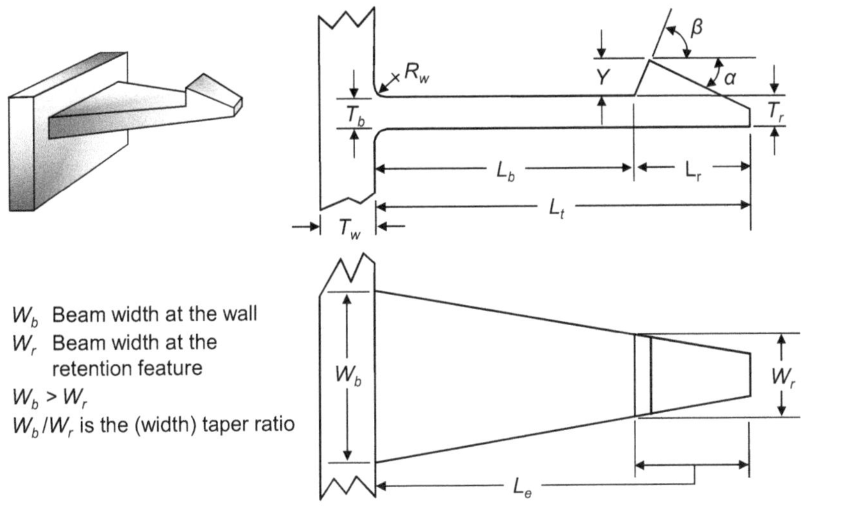

Figure 13.18 Beam tapered in width

The procedures and calculations for determining the design strain, stress concentration effect, target strain, and deflection force for width-tapered beams are the same as for thickness-tapered and constant-section beams and will not be repeated here.

There are currently no values of Q available for width-tapered beams, but an *estimate* for a 4:1 width-tapered beam could be made based on a conservative interpretation (use lower Q values) of the information in Fig 13.16 and Table 13.2 for a 2:1 thickness-taper beam. If such an estimate is made, use it to adjust $\varepsilon_{\text{calc-initial}}$ as shown above in Eq. 13.21 for thickness-tapered beams.

Procedures for adjusting for mating feature/part deflection are also the same as above for thickness-tapered and constant-section beams and will not be repeated here.

The *strain equation* for a width-tapered beam, Eq. 13.23 is

$$\varepsilon_{\text{calc-initial}} = 1.17 \frac{T_b \delta_{\text{design}}}{L^2} \tag{13.23}$$

As with thickness-tapered beams, the strain calculation has already accounted for the effects of tapering the beam. The deflection force equation, Eq. 13.23, for width-tapered beams is the same as that for thickness-tapered and constant section beams.

$$F_p = \frac{W_b T_b^{\,2} E_S \varepsilon_{\text{calc-initial}}}{6 L_b} \tag{13.24}$$

When a 4:1 width taper is applied to the original straight beam example and all other dimensions remain the same, the new calculations are as shown here.

The 4:1 taper ratio is created by increasing beam width at the base to 12 mm from its original 3 mm.

Example calculations for a width-tapered beam

Given:

Design point (ε_{design}) = 3% = .03

$K_{stress\text{-}concentration}$ = 1.5

Target strain = ε_{max} = 2%

Beam thickness (T_b) = 2.0 mm

Width taper ratio = 4:1 *New design requirement.*

Beam width at retaining member (W_r) = 3.0 mm

Beam width at base (W_b) = 12.0 mm *From 4 x W_r = 12.0.*

Beam length (L_b) = 15.0 mm

Deflection (δ_{design} = Y) = 2.0 mm

Q = not available

Calculate strain and deflection force for a width-tapered beam

Using Eq. 13.23:

$$\varepsilon_{\text{calc-initial}} = 1.17\frac{T_b\delta_{\text{design}}}{L^2} = 1.17\frac{2.0\times 2.0}{15.0^2} = 0.021 = 2.1\%$$

A wall deflection, (Q-factor) adjustment is not made here.

Using Eq. 13.11:

$$\varepsilon_{\text{calc-initial}} \leq \varepsilon_{\max}$$

Is:

2.1% ≤ 2.0%

No, but it is very close and knowing that wall deflection (which we did not consider) would lower the calculated strain value, we can consider the strains equivalent. Adjustments for mating part/feature deflection (also not included here) would also lower the strain and the resulting deflection force.

Using Eq. 13.24:

$$F_p = \frac{W_b T_b^2 E_S \varepsilon_{\text{calc-initial}}}{6L_b} = \frac{12.0\times 2.0^2\times 2000\times 0.021}{6\times 15.0} = 22.4\,N$$

Because we have no Q-factor information for width-tapered beams and because the adjustments for wall deflection have already been shown twice in previous examples, this example will stop here. We can compare the width-tapered beam results to the straight beam results at the same point in the calculations - after the initial strain and a corresponding deflection force are calculated:

Comparing straight vs. width-tapered beam results

Comparison:

For the straight beam:

$\varepsilon_{\text{calc-initial}} = 2.67\%$ *No adjustments made to the calculated strain.*

$F_p = 7.1\ \text{N}$ *Calculated without any strain adjustments.*

For the 4:1 width-tapered beam:

$\varepsilon_{\text{calc-initial}} = 2.1\%$

$F_p = 22.4\ \text{N}$ *No adjustments were made for wall deflection or mating feature/part deflection.*

The only change in the original straight beam was changing the beam width at the base to get a 4:1 taper. Note the reduction in strain but also the significant increase in deflection force.

As was shown in the straight beam example in Section 13.2.8, we can now use the calculated deflection force and design deflection values to construct a graph for evaluating mating feature/part deflection and determine the final values for deflection force, strain and deflection.

Judgments about the final strain vs. the target strain would now dictate if any dimensional changes are required to the tapered beam. If so, new calculations would be performed.

13.4 Beam Calculation Summary

Certain assumptions and accommodations were made here to permit the use of classic beam equations and simplify the calculations so they can be done by hand to enable understanding of beam behavior.

The effects of beam curvature and end rotation on the insertion and retention face angles, for example, have been ignored. It is possible to develop more complex equations which will take that behavior into account. Finite element analysis is also possible and is discussed in a later section.

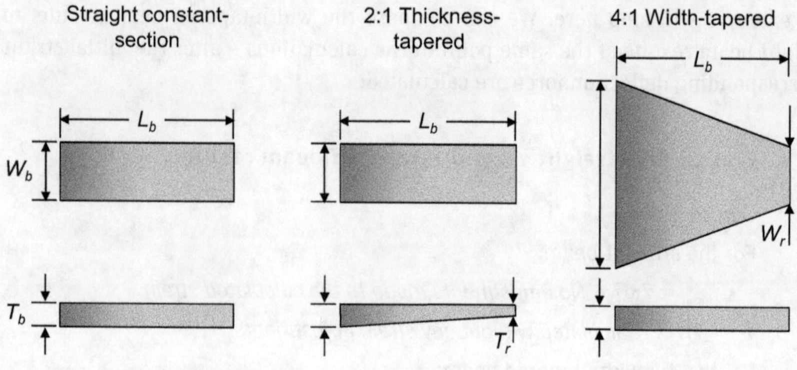

Figure 13.19 Beam configurations used in the examples

Table 13.3 Beam Performance Summary

From beam calculations with all adjustments made.				
	At beam base/end	Deflection (mm)	Final strain (%)	Deflection force (N)
Straight beam	Thickness = 3.0/3.0 mm	1.48	1.8	4.8
2:1 thickness taper	Thickness = 3.0/1.5 mm	1.7	1.2	3.2
From initial beam calculations with no wall deflection or mating feature/part deflection adjustments made.				
	At beam base/end	Design deflection (mm)	Initial calculated strain %)	Deflection force (N)
Straight beam	Width = 3.0/3.0 mm	2.0	2.8	7.1
4:1 width taper	Width = 12.0/3.0 mm	2.0	2.1	22.4

13.5 Other Deflecting Member Styles

13.5.1 Other Beam-Based Styles: Loops and Traps

Both loops and traps were described in Sections 6.2.2 and 6.2.3. Most of the analysis procedures described previously can be applied to these lock styles. Figures 13.20 and 13.21 describe the methods for doing so.

For beams in loops and traps, variables for the maximum strain and deflection force calculations are shown here. Use them in the same manner as discussed above for a straight or tapered beam. Once an acceptable strain is identified, the associated deflection force can be applied to assembly and separation calculations as discussed in Section 13.6 if a catch is used as the retaining member.

For a beam (the deflecting member) in loop style locks:

Loops with full length beams

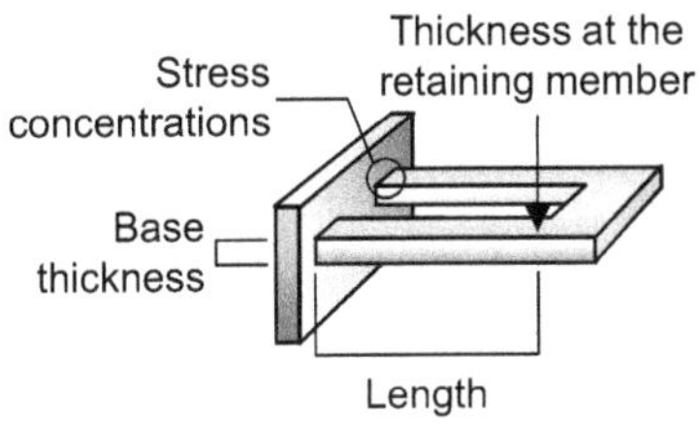

Calculate the strain and deflection force for one of the beams (we'll assume they are identical). They may also be tapered in width or thickness.

Once the strain requirement is satisfied, add the beam deflection forces together for a total deflection force to be used in the assembly and separation calculations.

Loops with a complex beam:

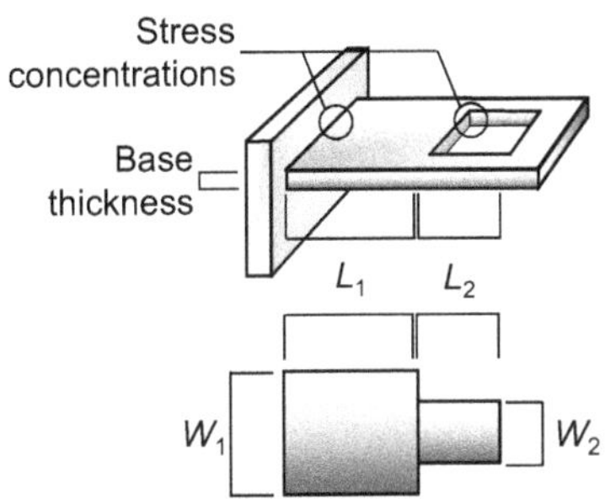

Calculate behavior of the single beam with length L_1. Treat the two beams as a second beam with length L_2. The total deflection will be shared by the two beams and will require iterative hand calculations. Once the deflection of the large beam is known, the strain at its base can be calculated.

Finite element analysis might be indicated here. Especially if wall and mating feature/part deflections are to be considered.

Loop deflection:

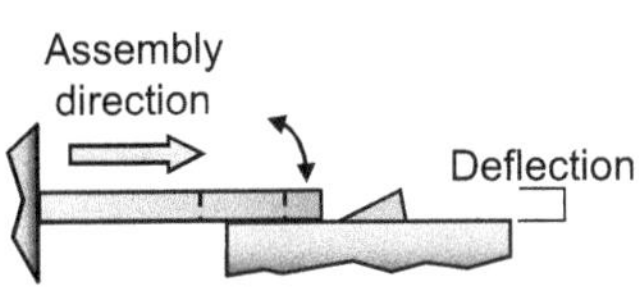

Loop deflection depends on the mating feature's height. Usually the mating feature will be a catch locator on a surface.

Figure 13.20 Beam analysis for loop style locks

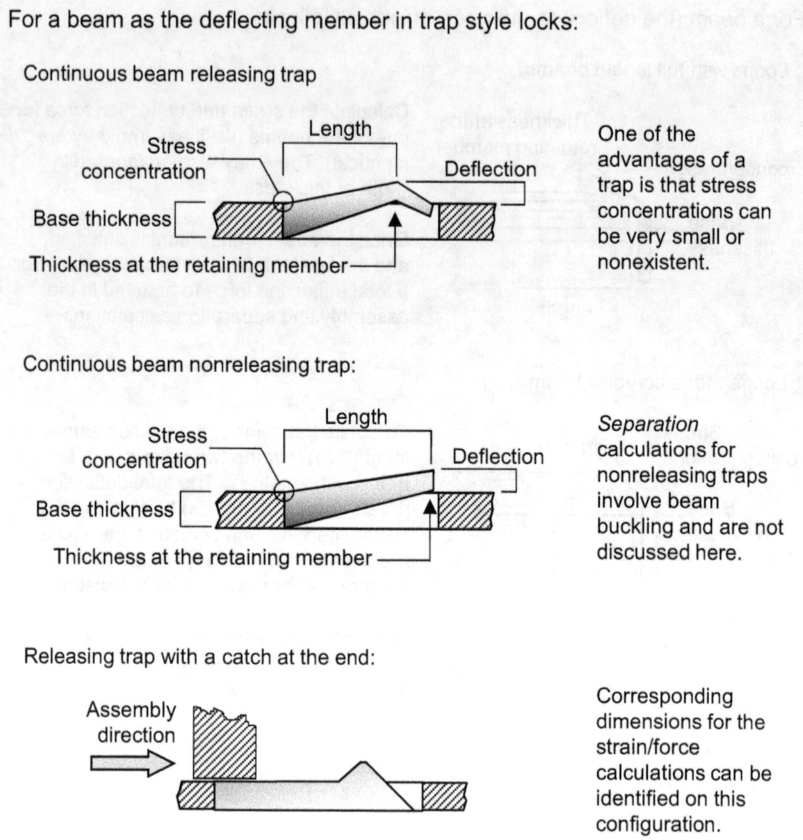

Figure 13.21 Beam analysis for trap style locks

13.5.2 Other Styles: Torsional, Annular, and Planar Deflection

See the appendix to find analysis resources for other styles.

Other lock styles were introduced in Chapter 6, but the locks based on straight cantilever beam deflection are, by far, the most common and received detailed attention here. We also did not discuss analysis methods for the more complex beam-based locks having 90° or 180° profiles or combinations of those angles.

Equations and analysis procedures for some of these other lock styles are available in the literature and online.

An online search will turn up numerous sites offering snap-fit equations as well as some analysis tools. Beware that few, if any, of these calculations and tools include the adjustments that were described here for calculating deflecting member behavior.

Adjustments to the assembly and separation force calculations are also required and are described in the next section. These adjustments are also missing from the published calculations and the on-line tools.

13.6 The Retaining Member: Catch

Catches are the most common retaining feature used with cantilever beam-based locks and are frequently the mating locator in many lock pairs. The catch as a retaining member also appears in torsional and planar lock styles. Regardless of the lock style, once a deflection force has been calculated for its deflecting member, that force will be used in the equations presented here for a catch used as either the retaining member of the lock itself or as a mating locator feature.

While we call it the *retaining* member, the catch affects both assembly and separation behavior. A catch, shown in Fig. 13.22, has both assembly and retention faces. We will discuss assembly behavior first.

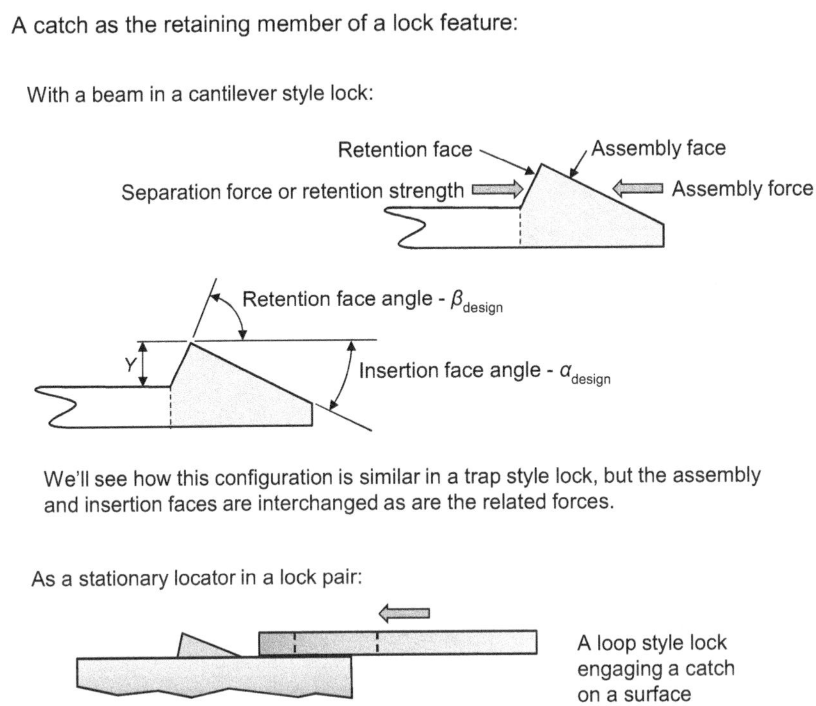

Figure 13.22 Features of the common catch

First we will consider lock assembly and separation with a catch as the retaining member on the familiar and common cantilever hook. We will then explain how the same calculations, with minor changes, apply to trap locks and to stationary catches engaging a loop.

13.6.1 Lock Assembly Force

We calculate maximum assembly force to ensure a lock can be assembled without violating ergonomic rules for forces applied by fingers, thumbs, or hands. Even automatic or robotic operations require consideration of assembly force. Issues could include the size/capacity of the assembly machine or possible part damage if assembly forces are too high.

For assembly, the insertion face is a ramp on which the mating feature slides and Eq. 13.25 is the basic equation for assembly force.

$$F_{\text{assembly}} = F_p \frac{\mu + \text{Tan}\alpha_{\text{design}}}{1-(\mu \,\text{Tan}\,\alpha_{\text{design}})} \tag{13.25}$$

An important adjustment to this equation is required because the insertion face *design* angle is commonly *and improperly* used in assembly force calculations.

13.6.1.1 Adjusting for the Insertion Face Effective Angle

Effective insertion face angle is one of the required calculation adjustments.

The author is not aware of any published or online calculations that consider the effect of beam deflection on the *insertion* and *retention* face angles. When sample calculations are shown, they typically use angle values for the lock in its free, (or *as-designed*) state as shown in Fig. 13.22 and Eq. 13.25. In reality, these angles can change significantly as the beam on which they are mounted deflects and those changes will affect the force calculations. The design angles must be adjusted to reflect the insertion and retention face *effective angles*. If these changes are ignored, then the calculated assembly force will be lower and the calculated separation force will be higher than the actual values.

Insertion and retention face angles were also discussed in Section 9.1.6 and Fig. 9.10, and in Section 12.2.

The adjustment described in Fig. 13.23 assumes no retaining member rotation and no beam curvature during deflection. When a beam is long relative to its thickness or when a beam is tapered, rotation and curvature may be significant. However, this simplified calculation will bring the calculated assembly force much closer to reality than ignoring the angle changes altogether. A more complex calculation that takes beam curvature and end rotation is possible but normally not necessary.

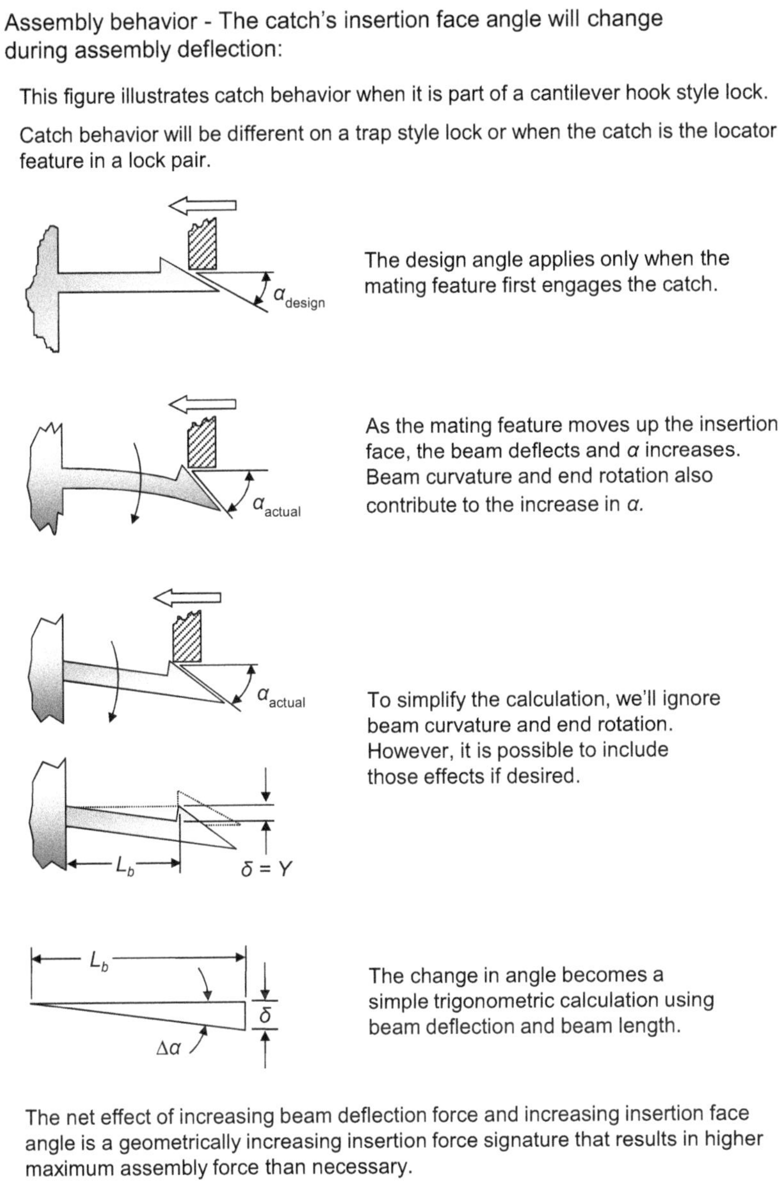

Figure 13.23 Catch insertion face behavior and the effective angle

To calculate the maximum assembly force, we must know the effective angle at that point. First calculate the change in insertion face angle at maximum deflection using Eq. 13.26. Use the maximum deflection from the deflecting member calculations, or for a more conservative result, use the design retention face height (Y) in the calculation.

$$\Delta\alpha = \mathrm{Tan}^{-1}\left(\frac{\delta}{L_b}\right) \qquad (13.26)$$

For simplicity in the discussion, we will always show L_b in the equations and use it in the calculations. There may be times when using L_e in these calculations would be appropriate for the additional precision it could provide in the profile calculations. See the discussion about beam length in Section 13.2.2 and Fig. 13.2.

Add the change in angle to the original design angle to find the effective insertion face angle, Eq. 13.27.

$$\alpha_{\text{effective}} = \alpha_{\text{design}} + \Delta\alpha \tag{13.27}$$

Because both beam deflection force and insertion face angle increase as the lock deflects for assembly, their effects are additive and maximum assembly force always occurs at maximum deflection. In the calculations, we'll use:

- Maximum deflection force ($F_{p\text{-final}}$) calculated for the beam deflecting member.
- Effective insertion face angle ($\alpha_{\text{effective}}$) calculated using Eq. 13.27.
- A friction coefficient (μ) based on test data, tabulated values or our own experience.

Reread the discussion about friction coefficient uncertainty in Section 11.5.

Find a friction coefficient in Chapter 11 in Table 11.3 or from supplier data. However, friction coefficient data for plastics can be highly variable and truly accurate values are difficult to find. If friction coefficient data is not available, make a judgment from the available data depending on the lubricity of the material(s), surface roughness, and a bias toward a high or low estimate of force depending on the application.

Note the friction coefficient in Eq. 13.28 is not labeled as *static* or *dynamic*. Because the surfaces are sliding across each other during assembly, we should be using a dynamic friction coefficient value; if one is available, use it.

In the author's opinion, given the nature of friction data and the other assumptions and variables associated with these calculations, distinguishing between static and dynamic friction coefficients is generally unnecessary.

Maximum assembly force is found using Eq. 13.28, which is identical to Eq. 13.25 but uses the *effective* insertion face angle rather than the design angle.

$$F_{\text{assembly-max}} = F_p \frac{\mu + \text{Tan}\alpha_{\text{effective}}}{1 - (\mu \text{Tan}\alpha_{\text{effective}})} \tag{13.28}$$

This is the maximum assembly force for one lock feature. When multiple locks engage simultaneously, multiply the result by the number of locks.

13.6.1.2 Example Assembly Force Calculations

Figure 13.32 in Section 13.9 shows how to construct a spreadsheet to perform these calculations.

Example assembly force calculation

Given:

Insertion face angle (α_{design}) = 25° *From Chapter 12, Rules-of-Thumb.*

From Fig. 13.8, application data:

Friction coefficient, (μ) = 0.3

Beam length, (L_b) = 15.0 mm

Results from straight beam example calculation:

Deflection, δ_{actual} = 1.48 mm

Deflection force, $F_{p\text{-actual}}$ = 4.8 N

Using Eq. 13.26:

$$\Delta\alpha = \text{Tan}^{-1}\left(\frac{\delta}{L_b}\right) = \text{Tan}^{-1}\left(\frac{1.48}{15.0}\right) = \text{Tan}^{-1}(0.0987) = 5.6^\circ$$

Using Eq. 13.27:

$$\alpha_{\text{effective}} = \alpha_{\text{design}} + \Delta\alpha = 25^\circ + 5.6^\circ = 30.6^\circ \approx 31^\circ$$

So $\alpha_{\text{effective}} = 31^\text{o}$

For use in Eq. 13.28:

$$\text{Tan}\alpha_{\text{effective}} = \text{Tan}(31^\circ) = 0.6009$$

Using Eq. 13.28:

$$F_{\text{assembly-max}} = F_p \frac{\mu + \text{Tan}\alpha_{\text{effective}}}{1 - (\mu \text{Tan}\alpha_{\text{effective}})} = 4.8 \times \frac{0.3 + 0.5922}{1 - (0.3 \times 0.5922)} = 5.14 \text{ N}$$

So $F_{\text{assembly-max}} = 5.14 \text{ N}$ per lock

13.6.1.3 Modifying the Insertion Face Profile

The above example shows how to use the change in insertion face angle at maximum deflection to find a more accurate and higher value for maximum assembly force. But the insertion face is flat and will have an assembly force signature with an increasing rate of change. See Fig. 13.24 and the discussion of assembly feedback in, Section 9.1.6.

We can also use the concept of effective angle to design a profile for the insertion face to offset the deflection effect and reduce assembly force without changing beam deflection or affecting the separation force.

Section 9.1.6, discussed the assembly force signature and its effect on assembly feedback. With our knowledge about the effective angle we can use it, if we wish, to modify the insertion face profile to improve the insertion force-deflection signature.

The profile is designed by calculating an effective local angle, α_{local} at selected deflection points ($\delta_{\%}$) as the mating feature moves up the insertion face, (Eq. 13.29). Maximum deflection would be $\delta_{100\%}$. Using these effective angle values, we could generate a curved profile tangent to those points as shown in Fig. 13.24, the calculation example that follows, and Fig. 13.25. This process will give a profile with a constant rate force-deflection signature.

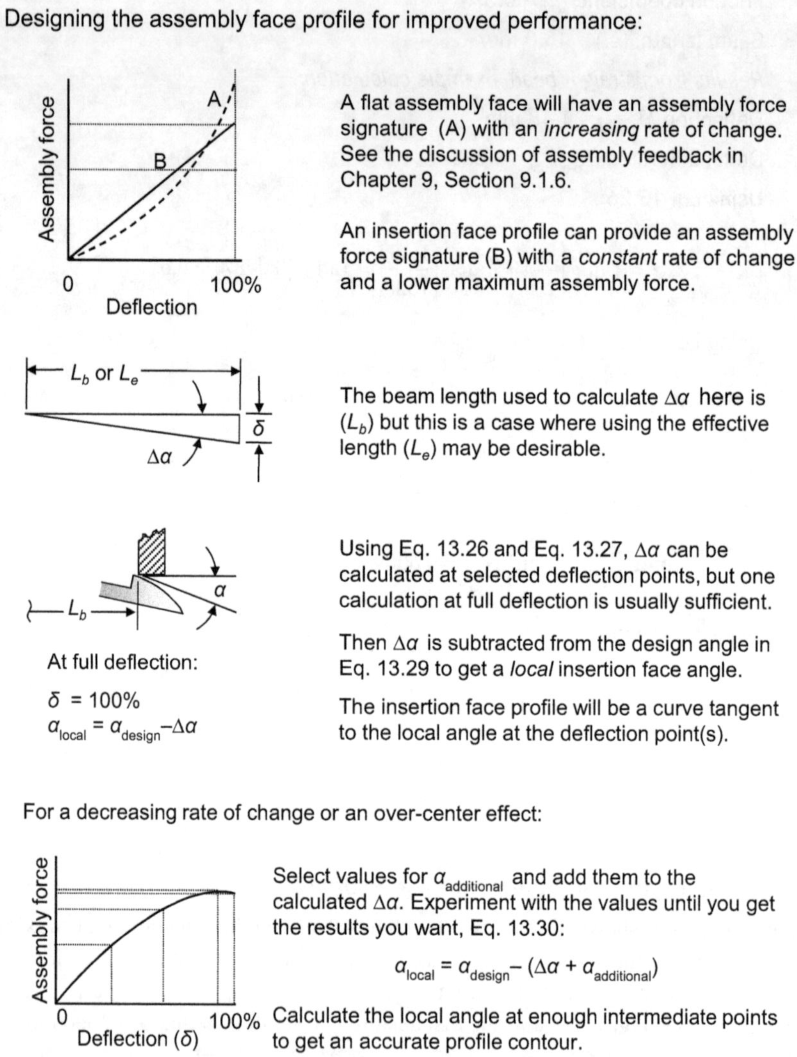

Figure 13.24 Adding a profile to the assembly face

$$\alpha_{local} = \alpha_{design} - \Delta\alpha_{\delta\%} \tag{13.29}$$

For a force-deflection signature having a decreasing rate of change, use Eq. 13.30 where additional angles are added to the calculation. The resulting signature is shown at the bottom of Fig. 13.24.

$$\alpha_{\text{local}} = \alpha_{\text{design}} - \left(\Delta\alpha_{\delta\%} + \alpha_{\text{additional}}\right) \tag{13.30}$$

The net effect of adding a profile is the same as designing the catch with a very low insertion face angle to compensate for beam deflection. Such an angle would be very low, something on the order of 5 to 10°. This is impractical because of the excessive length such a low angle would add to the retaining feature.

Adding a profile solves this problem. A profile may add some mass and length to the retaining member and this must be taken into account in the design. Extra mass in the retaining member may require coring out the area.

After all modifications are made to the assembly face, we can calculate the maximum assembly force and the assembly force at multiple deflection points, Eq. 13.31. This will give us a force-deflection signature for the lock. The value of F_p used in those calculations must reflect the actual deflection at those points

$$F_{\text{assembly-max-profile}} = F_p \frac{\mu + \text{Tan}\,\alpha_{\text{local}}}{1 - \left(\mu\,\text{Tan}\,\alpha_{\text{local}}\right)} \tag{13.31}$$

The sometimes complex subscripts are used here to differentiate between the many variables in these explanations. In practice, the user can simplify these subscripts to suit themselves.

Here we will calculate the angle at maximum deflection only and use it as shown in Fig. 13.25. If angles are calculated at additional deflection points, the process still *begins* at maximum deflection and works down the insertion face toward its end.

Developing an insertion face profile to offset deflection effects

Given:

$\alpha_{\text{design}} = 25°$

$L_b = 15.0$ mm

$\delta = 1.48$ mm at full (100%) beam deflection

$\Delta\alpha = 5.6°$ *From Eq. 13.26 in above calculation.*

Using Eq. 13.29 at maximum deflection:

$$\alpha_{\text{local}} = \alpha_{\text{design}} - \Delta\alpha_{\delta\%} = 25° - 5.6° = 19.4°$$

We'll use this value to create the profile in Fig. 13.25.

The profile shown in Fig. 13.25 will have a *reasonably* constant rate of change since it was developed using only one deflection point. (For more profile accuracy, use additional intermediate deflection points in the calculations.) This profile is based on the initial design angle of 25° adjusted by the effective angle. The resulting maximum assembly force will reflect a true, (effective) insertion angle equal to the initial design angle of 25°.

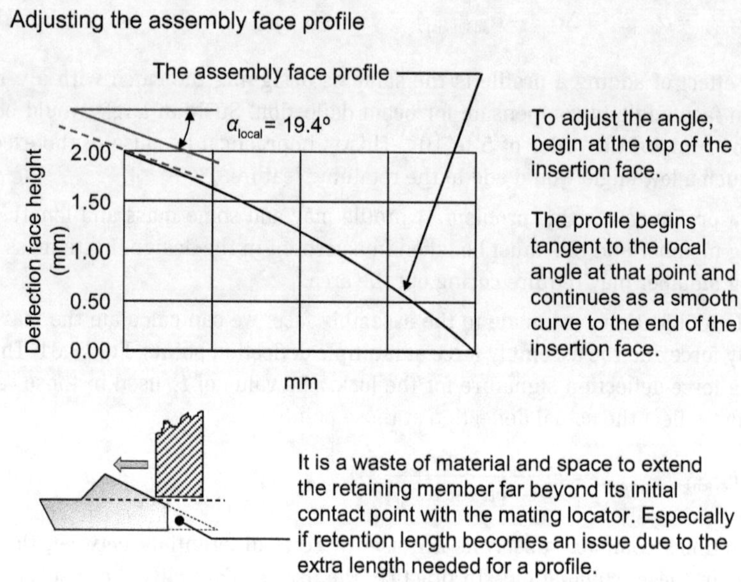

Figure 13.25 Creating an insertion face profile for the example

This is important!

The effective angle is *calculated* at the actual deflection as determined in the deflecting member calculations. But for designing the profile, the effective angle is *applied* at the top of the retention face which, in this example, is 2.0 mm deep.

The original design angle of 25° is now somewhere between the beginning and end of the insertion face profile. This results in a slightly steeper angle toward the end but the deflection force is also much lower at that point. The profile determines the final insertion face length. But as shown above, the insertion face does not need to extend very far beyond its initial contact point with the mating feature.

Recalculating maximum assembly force for the new profile

Refer to the example in Section 13.6.1.2 for details.

From the above profile calculations:

$$\alpha_{local} = 19.4^{o}$$

$$\Delta\alpha = 5.6^{o}$$

Using α_{local} as α_{design} in Eq. 13.27:

$$\alpha_{effective} = \alpha_{local} + \Delta\alpha = 25^{o}$$

So $\alpha_{effective} = 25^{o}$

For use in Eq. 13.28:

$$\mathrm{Tan}\alpha_{effective} = \mathrm{Tan}(25^{\circ}) = 0.4663$$

Using Eq. 13.28:

$$F_{\text{assembly-max}} = F_p \frac{\mu + \mathrm{Tan}\alpha_{\text{effective}}}{1-(\mu \mathrm{Tan}\alpha_{\text{effective}})} = 4.8 \times \frac{0.3 + 0.4663}{1-(0.3 \times 0.4663)} = 4.28\ \mathrm{N}$$

So $F_{\text{assembly-max}} = 4.28\ \mathrm{N}$ with the profile

Compared to the calculation without the profile where:

$F_{\text{assembly-max}} = 5.14\ \mathrm{N}$ per lock

Using this profile simply brought the *effective* insertion face angle at maximum deflection back to the original design angle of 25°. The continuous curvature of the insertion face maintains an effective angle that is close to the original design angle. Using a larger value for the adjustment $\Delta\alpha$ as shown in Eq. 13.30 would have further reduced the assembly force.

The difference in the assembly forces in this example may not seem like much, but it could be significant when multiple locks are engaged simultaneously.

The geometry of this example is also fairly robust with a long beam and an L/T ratio within recommended limits, as well as a design deflection, ($\delta = Y$) and insertion face angle of 25°, based on the rules-of-thumb. With shorter beams and higher insertion face angles, the difference between the maximum assembly forces with and without the profile would be higher.

13.6.2 Catch Separation Force

Lock release behavior has many names including *separation force, retention strength,* and *release force.* We will use *separation force* in this discussion.

The calculations apply to a single lock. For an attachment with multiple lock features releasing simultaneously, the total separation force is the sum of the individual forces. The basic separation force is calculated using Eq. 13.32.

Eq. 13.32 uses the retention face angle, β, but is otherwise identical to Eq. 13.25 for the insertion face.

$$F_{\text{separation}} = F_p \frac{\mu + \mathrm{Tan}\beta_{\text{design}}}{1-(\mu\,\mathrm{Tan}\,\beta_{\text{design}})} \tag{13.32}$$

13.6.2.1 Adjusting for the Retention Face Effective Angle

Beam deflection during separation will cause the retention face angle to change. The effect is to *decrease* the retention face angle. This results in *lower* separation force and retention strength than we would calculate using the retention face angle as-designed.

Separation force calculations for catches with high retention face angles are possible and the results may be meaningful, but this is strictly a discussion of releasing locks for several reasons:

- Unless we want a retention face angle greater than 90° (a whole different retention concept), the best we can do is a 90° retention face and we cannot adjust it to a greater angle.
- As retaining features approach a nonreleasing geometry, (a high retention face angle) shear forces and distortion begin to dominate the retaining and deflecting member behavior over the effects of friction at shallower angles and the calculation results become less accurate and ultimately meaningless.

For calculation purposes, realistic retention face angles will vary with friction. Perform these calculations at higher angles with the understanding that their accuracy will deteriorate at some point. The author suspects this deterioration begins at angles somewhere above 45°.

We will use a retention face angle of 45° in the following example calculations. Even releasing locks must often resist applied forces and an adjustment to the retention face angle in these cases makes sense. The process is similar to that for the insertion face but the objective will be to increase rather than decrease the local retention face angle as deflection occurs.

Unlike assembly where deflection force and insertion face angle both increase, as separation movement occurs, the continuously *decreasing* retention face angle, see Fig. 13.26, and the *increasing* beam deflection force interact in ways that cannot be known without calculations. It is possible that, in some cases, we may not know where the maximum separation force occurs unless calculations are performed at multiple points on the retention face. Here, however, we will do a calculation at only one point - the maximum catch height.

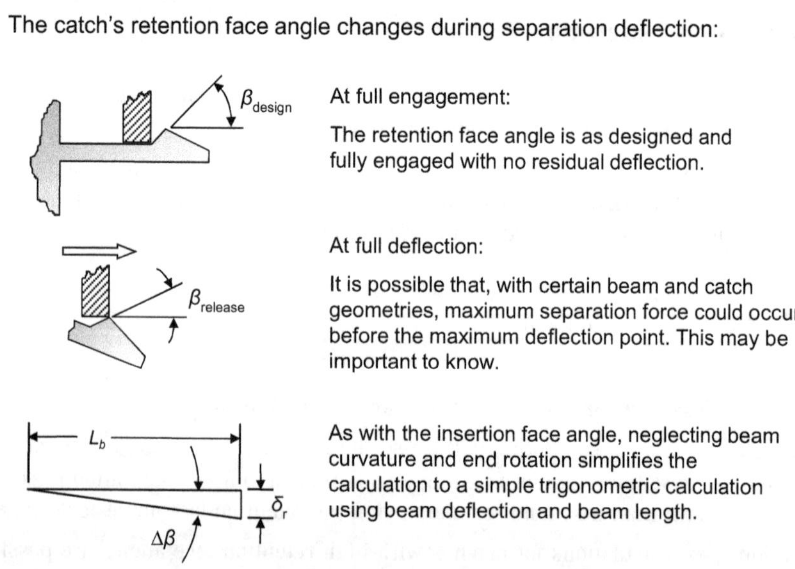

Figure 13.26 Catch separation behavior and the effective angle

Figure 13.26 illustrates catch separation behavior when it is on a cantilever hook style lock. Catch behavior will be different on a trap lock or when the catch is stationary as the locator feature in a lock pair.

As with assembly, beam deflection will cause the design retention face angle to change. This change should be included in the separation calculations.

Unlike with assembly where the insertion face angle increases, during separation the retention face angle will decrease. This causes a *lower* separation force than predicted when the original design angle is used in the calculations.

Calculate the change in the retention face angle at maximum deflection using Eq. 13.33. The only difference from Eq. 13.26 (for the change in insertion face angle) is that we are now calculating $\Delta\beta$. At equivalent deflections, unless we choose to use L_e rather than L_b in these two equations, we will find that $\Delta\beta$ is equal to $\Delta\alpha$.

Effective retention face angle is one of the required calculation adjustments.

$$\Delta\beta = \mathrm{Tan}^{-1}\left(\frac{\delta}{L_b}\right) \tag{13.33}$$

Calculate the effective retention face angle at maximum deflection, see Eq. 13.34. This equation is similar to Eq.13.37 for the insertion face but with two differences: we are using β instead of α, and we are *subtracting* the change from the design angle rather than adding it.

$$\beta_{\text{effective}} = \beta_{\text{design}} - \Delta\beta \tag{13.34}$$

Calculate the maximum separation force using Eq. 13.35, which is the basic equation for an object sliding on a ramp, but we are now using $\beta_{\text{effective}}$ calculated above.

$$F_{\text{separation-max}} = F_p \frac{\mu + \mathrm{Tan}\beta_{\text{effective}}}{1-\left(\mu\,\mathrm{Tan}\,\beta_{\text{effective}}\right)} \tag{13.35}$$

13.6.2.2 Example Assembly Force Calculations

Figure 13.33 in Section 13.9 shows how to construct a spreadsheet to perform these calculations.

Separation force calculation using the effective angle

Given:

Separation face angle (β_{design}) = 45°
From Chapter 12, Rules-of-Thumb.

From Fig. 13.8. application data:

Friction coefficient, (μ) = 0.3

Beam length (L_b) = 15.0 mm

Results from straight beam example calculation:

Deflection, δ_{actual} = 1.48 mm

Deflection force, $F_{p\text{-actual}}$ = 4.8 N

Using Eq. 13.33:

$$\Delta\beta = \mathrm{Tan}^{-1}\left(\frac{\delta}{L_b}\right) = \mathrm{Tan}^{-1}\left(\frac{1.48}{15.0}\right) = \mathrm{Tan}^{-1}(0.0987) = 5.6°$$

Using Eq. 13.34:

$$\beta_{\text{effective}} = \beta_{\text{design}} - \Delta\beta = 45° - 6.6° = 39.4°$$

So $\beta_{\text{effective}} = 39.4°$

For use in Eq. 13.35:

$$\text{Tan}\beta_{\text{effective}} = \text{Tan}(39.4°) = 0.8214$$

Using Eq. 13.35:

$$F_{\text{separation-max}} = F_p \frac{\mu + \text{Tan}\beta_{\text{effective}}}{1 - (\mu \,\text{Tan}\,\beta_{\text{effective}})} = 4.8 \times \frac{0.3 + 0.8213}{1 - (0.3 \times 0.8213)} = 7.1\ \text{N}$$

So $F_{\text{separation-max}} = 7.1$ N per lock

To illustrate the difference in separation force values when the effective retention face angle is not accounted for, the calculation below does not include that adjustment.

Separation force calculation using the design angle

Given:

Separation face angle (β_{design}) = 45°

From Chapter 12, Rules-of-Thumb.

From Fig. 13.8, application data:

Friction coefficient, (μ) = 0.3

Beam length, (L_b) = 15.0 mm

Results from straight beam example calculation:

Deflection, δ_{actual} = 1.48 mm

Deflection force, $F_{p\text{-actual}}$ = 4.8 N

For use in Eq. 13.35:

$$\text{Tan}\beta_{\text{design}} = \text{Tan}(45°) = 1.0$$

Using Eq. 13.35:

$$F_{\text{separation-max}} = F_p \frac{\mu + \text{Tan}\,\beta_{\text{design}}}{1 - (\mu \,\text{Tan}\,\beta_{\text{design}})} = 4.8 \times \frac{0.3 + 1.0}{1 - (0.3 \times 1.0)} = 8.9\ \text{N}$$

So $F_{\text{separation-max}} = 8.9$ N per lock

Compared to the more accurate, (and lower)separation force as calculated using the effective angle:

$F_{\text{separation-max}} = 7.1$ N per lock

The separation force calculated without accounting for the effective angle is 25% higher than the more realistic force calculated using effective angle. This difference of 1.8 N per lock could have a significant effect on separation performance, especially if multiple locks are involved in the attachment.

As with the insertion face, the retention face can be modified to offset this difference by adding a profile.

13.6.2.3 Modifying the Retention Face Profile

The idea of a more desirable retention face profile for improved retention performance was mentioned in Section 12.2.2. In a manner similar to that for the insertion face profile, the retention face profile is determined by calculating a local angle(s) and then constructing the profile as a curve tangent to the angle(s).

The process is similar to that for designing a profile for the insertion face, but we need to offset the reduction in retention face angle so we will *add* an adjustment angle at the top of the retention face to the design angle. Begin by using Eq. 13.33 to calculate $\Delta\beta$.

Calculate a local angle at the maximum deflection point on the catch or at several additional deflection points along the retention face:

$$\beta_{\text{local}} = \beta_{\text{design}} + \Delta\beta_{\delta\%} \tag{13.36}$$

Or, calculate a slightly higher local angle by adding some additional angle:

$$\beta_{\text{local}} = \beta_{\text{design}} + \left(\Delta\beta_{\delta\%} + \beta_{\text{additional}}\right) \tag{13.37}$$

Calculate the maximum separation force:

$$F_{\text{separation-max-profile}} = F_p \frac{\mu + \text{Tan}\beta_{\text{local}}}{1 - \left(\mu \,\text{Tan}\,\beta_{\text{local}}\right)} \tag{13.38}$$

We will show two examples of a profile calculation. The first is a continuation of the above example.

Developing a retention face profile to offset deflection effects

Given:

$\beta_{design} = 45°$

$L_b = 15.0$ mm

$\delta = 1.48$ mm at full (100%) beam deflection

$\Delta\beta = 5.6°$ *From Eq. 13.33 in above calculation*

Using Eq. 13.36:

$$\beta_{\text{local}} = \beta_{\text{design}} + \Delta\beta_{\delta\%} = 45° + 5.6° = 49.6 \approx 50°$$

So a local angle of about 50° at the top of the retention face can compensate for the deflection.

Given the lock feature geometry we are using in this example, the difference in retention face angle is relatively small. We could simply change the design angle from 45° to 50° to achieve the desired separation force and, in this case, that may be a reasonable solution.

However, with shorter beam lengths and different design angles than we are using here, a much larger angle adjustment may be necessary and simply making the design angle steeper may cause the lock to lose its releasing capability.

The following example, Fig. 13.27, uses a *shorter* beam length of 7 mm. This will increase the deflection effect on the angle and make the profile adjustment more visible in the illustration. The profile developed here is still based on the initial design angle of 45°and all other dimensions except beam length remain the same.

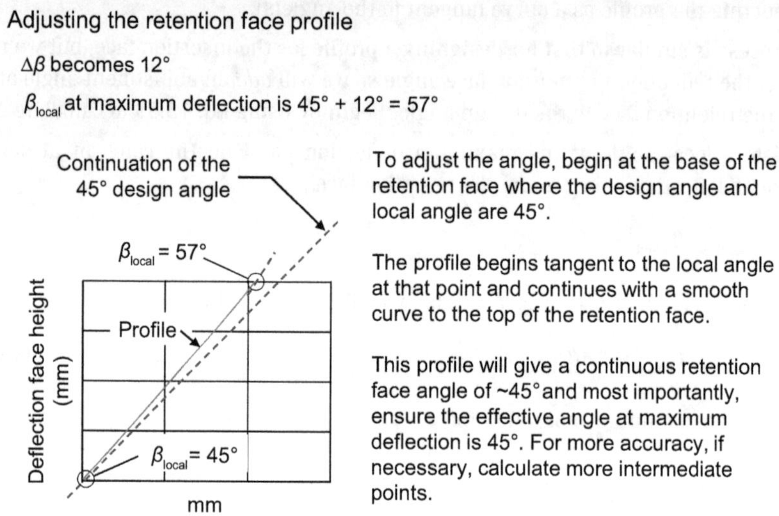

Figure 13.27 Designing a retention face profile

This is important!

The effective angle is *calculated* at the actual deflection as determined in the deflecting member calculations. But, for the profile design, the effective angle is *applied* at the top of the retention face which, in this example, is 2 mm deep.

The resulting profile is not very dramatic but, with all other variables being equal, if we calculate the profile's effect on the separation force for this 7 mm long beam, we will see a significant effect:

Comparing separation force results for a *shorter* beam

Without considering the change in effective angle, the separation force for a 7 mm long beam is 8.9 N.

When considering the change in effective angle, the separation force calculated for a 7 mm long beam is 5.7 N.

Comparing the results: 8.9 N >> 5.7 N

We can see the additional sensitivity to retention face angle reduction during deflection in this shorter beam. This is a significant difference in separation force. Ignoring the effective angle will again result in an unrealistically high value for the separation force which could lead to an ineffective attachment. In the shorter beam, adding a profile to compensate for this difference is highly recommended.

Comparing these results to those for the 15 mm long beam in the above example calculations also illustrates the increasing sensitivity as the beam gets shorter.

13.7 Stationary Catches and Traps as Retaining Members

A catch is sometimes mounted on a surface and remains stationary during assembly and separation. The engaging feature is often a loop style lock, Fig. 13.29.

When a catch is stationary, the deflection effects we were concerned about with beams do not apply. Use the calculations for cantilever beam bending to evaluate the loop for strain, deflection, and deflection force. Then use the basic equations, (repeated here) for a feature sliding up a ramp, to calculate assembly, and separation forces. They are identical in form, the only difference being the insertion and assembly face angles. For stationary catch calculations, use the design angle because there will be no retaining feature deflection.

$$F_{\text{assembly}} = F_p \frac{\mu + \text{Tan}\alpha_{\text{design}}}{1 - (\mu \,\text{Tan}\,\alpha_{\text{design}})} \tag{13.39}$$

$$F_{\text{separation}} = F_p \frac{\mu + \text{Tan}\beta_{\text{design}}}{1 - (\mu \,\text{Tan}\,\beta_{\text{design}})} \tag{13.40}$$

Profiles can be designed into catches for special applications. Use the equations and methods related to the effective angle to design the profile and calculate force-deflection signatures. Some applications that could benefit from profiles may be:

- User-activated applications where a user-friendly assembly and/or separation signature can improve perceived quality.
- Applications that must absorb a very short-term separation force, as in an impact or drop test, without separating. A properly designed retention face profile may allow some part movement and energy-absorption and then return the parts to their original assembled condition.
- Applications that require special attention to assembly feedback where an enhanced force-deflection signature would assist the assembler.

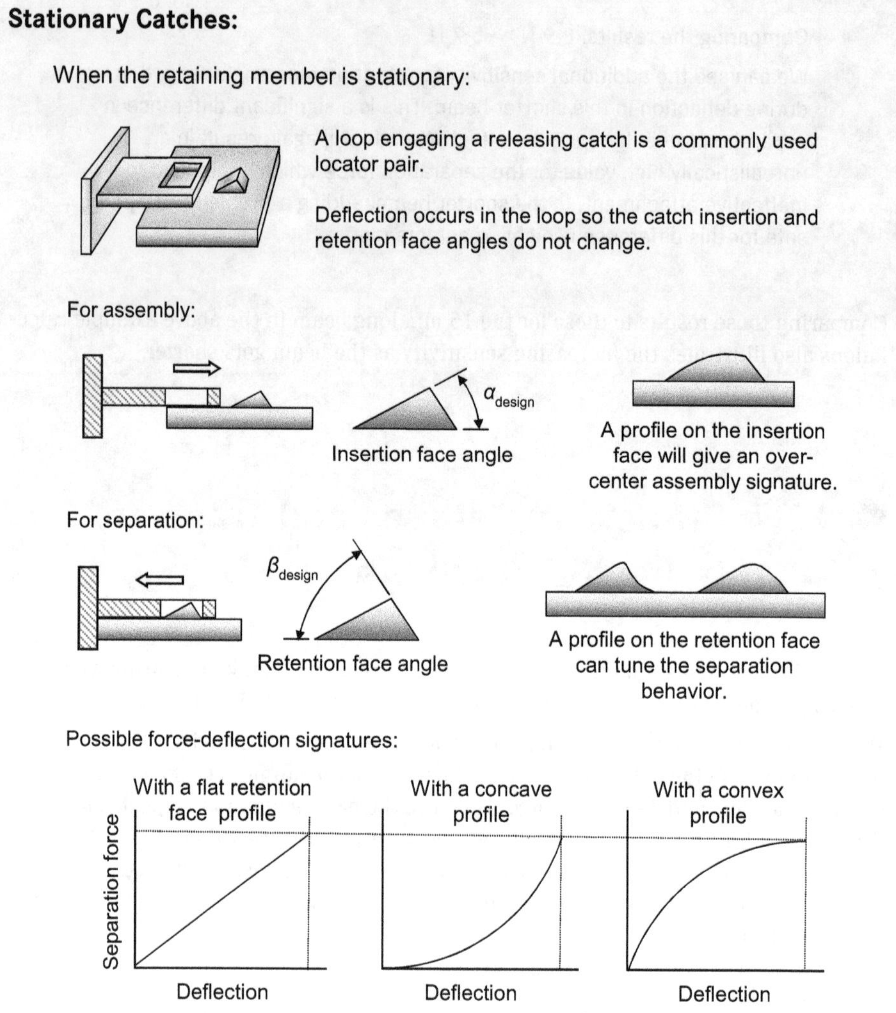

Figure 13.28 A stationary catch as a locator engaging a loop

Traps:

Trap style locks involve cantilever beam deflection so the calculations will be similar to those described previously for catches on cantilever beams. However, there will be some important differences:

- Unlike other cantilever beam-based locks, in traps the assembly direction is *away* from the beam's base and the separation direction is *toward* the base. This means that:
- With traps, the insertion face angle will decrease during assembly and the retention face angle will increase. This is the opposite of catch behavior on a cantilever hook and changes how we use the separation force equations.
- With some traps, the maximum assembly force may occur at the beginning of engagement then rapidly drop off.

- Because of behavior differences, for calculation purposes it helps to break traps down into three configurations.
- A trap with a *nonreleasing* shape could engage a catch designed to allow that trap to *release*.

The good news is that there are no new equations involved.

Figure 13.29 shows trap configurations and identifies the insertion and retention faces. See Chapter 6, Section 6.3.3, for more examples of traps and discussion of trap behavior.

Trap configuration 1

The trap beam is both the deflecting member and the retaining member:

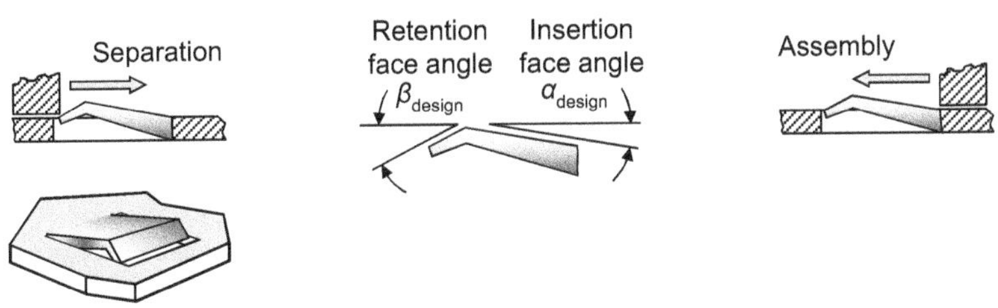

Trap configuration 2

The catch retaining member sits at the end of the deflecting member:

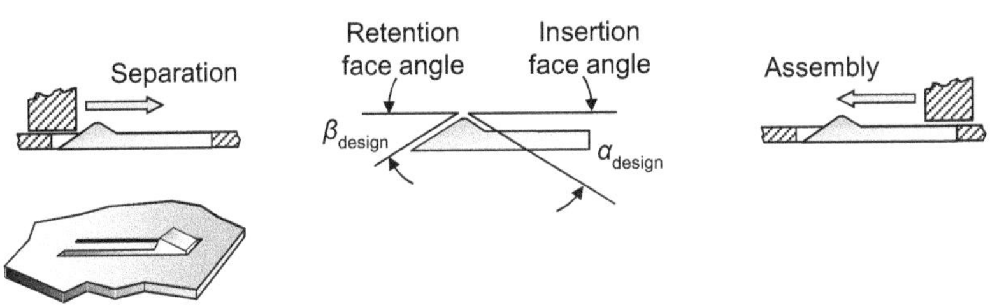

Trap configuration 3

A trap beam engages a catch or other feature on a mating locator:

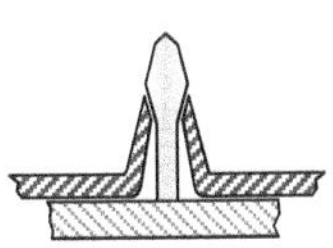

In this prong and trap lock pair, the trap beams are the deflecting members for assembly.

The retaining member(s) are catches on the prong. The catch retention face angles will determine if this trap is releasing or nonreleasing.

Figure 13.29 Catches and traps

Beam deflection equations and the assembly/separation equations just described will apply to all trap configurations for assembly calculations and will also apply to releasing traps for separation calculations. The separation behavior calculations were presented here using catches for the examples but the equations also apply to a feature sliding on a ramp as is the case with all traps, see Fig. 13.30.

The force-deflection signature for trap engagement:

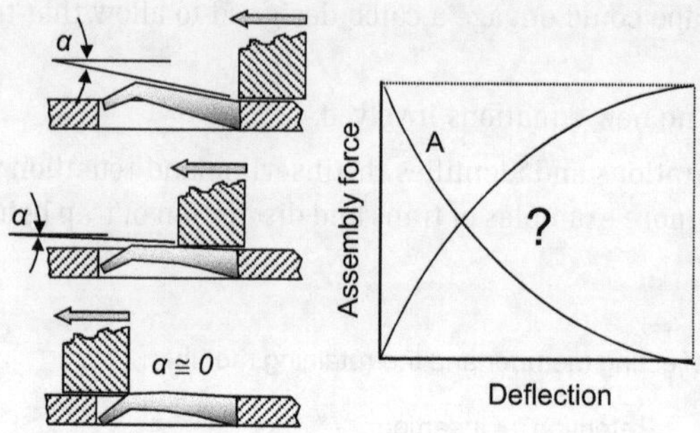

Force-deflection signatures for trap engagement are not easy to predict. In this and most situations, high deflection forces occur at initial engagement so the signature will look something like (A) but there can be variation depending on the trap configuration.

Calculating the force-deflection signature at several points may be necessary.

The force-deflection signature for trap separation:

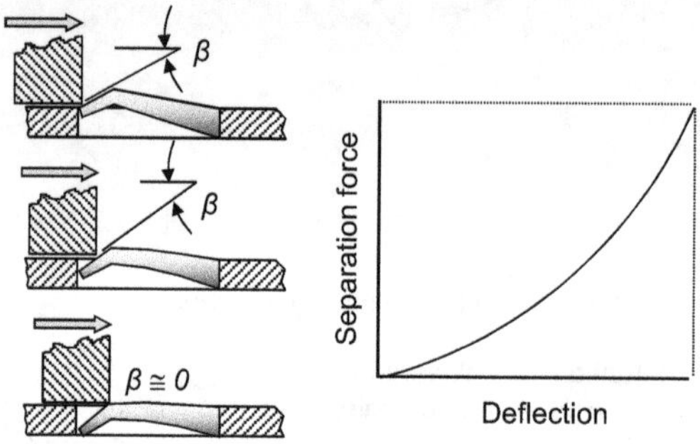

This signature is typical of all releasing traps during *separation* because the retention face angle increases as the beam deflection force increases.

(Similar to cantilever hook behavior during *engagement*.)

Insertion and retention face profiles can also be used to tune the force-deflection signatures.

Figure 13.30 Typical trap engagement and separation behavior

13.7.1 Other Separation Considerations

For releasing locks, disassembly involves applying a force in the separation direction to one of the parts so the locking features release. For non-releasing locks, separation strength calculations may become more difficult and require consideration of several possible separation modes. For cantilever beam-based and for most locks, these are:

- *Bending*, where the mating feature slides over the retention face, the beam bends and releases. For a releasing lock, this is the common separation behavior and it is the behavior discussed in this section.

  For cantilever hooks with a 90° angle that are *intended* to be non-releasing, retaining member and beam end distortion followed by bending and release is possible when the applied force exceeds the expected maximum.

- *Shear*, where some portion of the constraint pair fails in shear. Shear calculations are simply based on the applicable cross-sectional area and the shear strength of the material. Because of their simplicity, they are not discussed here.

- *Tension*, where some portion of the constraint pair fails in tension. These calculations are based on the applicable cross-sectional area and an appropriate tensile strength limit (yield, maximum, or ultimate) of the material. Like shear calculations, because of their simplicity, they are not discussed here.
- *Combination*, which is a complex set of effects where a combination of bending, shear, tension, and retaining mechanism rotation cause distortion and release. Calculations of this behavior are beyond the scope of this chapter and are normally beyond the capability of simple hand calculations. They may require finite element analysis.

For permanent and nonreleasing locks, shear or a combination of behaviors as discussed above are more likely and hand calculations become difficult.

If release involves the same deflection and behaviors as assembly (bending along the same beam axis, for example), then the maximum allowable strain is already known. If release is quick, as for a releasing lock, then the comparison to the dynamic strain limit made for assembly strain will still apply.

Nonreleasing locks require manual deflection for separation and several factors may affect the maximum strain calculation and must be considered.

Manual deflection may result in greater deflection than assembly simply because lock deflection is not based on a limiting physical attribute (Y). Use the maximum possible manual deflection to calculate strain or use *guard* enhancements to limit manual deflection to a safe level.

Slower manual deflection can hold the lock feature at its maximum deflection for a longer period of time than (rapid) assembly. Dynamic strain limits used for assembly calculations may not be acceptable for evaluating disassembly strain. Calculations may require use of a static or a low-rate strain limit.

13.8 Using Finite Element Analysis

Finite element analysis of feature behavior should be considered when:

- Too many materials or analytical assumptions are violated
- The application and/or its snap-fit features are complex
- The application is high on the demand-complexity matrix
- Complex stress/strain conditions exist
- Deflections are large
- Plate-like deflections occur (the beam is wide relative to its length)
- Planar deflecting members are involved

Find a discussion of finite element analysis for snap-fit features in [2]. Also, see the appendix for information about available finite element tools for snap-fits.

Remember that proper constraint in the attachment is always a requirement. While finite element analysis is capable of analyzing improperly constrained attachments, the attachment itself is fundamentally incorrect and likely to have problems or require costlier design features.

■ 13.9 Calculation Spreadsheets

We must make assumptions to do these calculations and the property values we use typically reflect testing under specific conditions, not the actual working environment.

Despite their shortcomings, we regard calculation results as *reasonable* estimates of actual behavior design. Performing calculations may also expose enough uncertainty or feature sensitivity that the use of finite element analysis is indicated.

The author believes that a significant roadblock to good feature design is the time required to perform the multiple hand calculations required to design and then evaluate multiple lock feature options.

There is high value in having a calculation tool for quick evaluation and comparison of lock dimensions during lock feature design. With such a tool, a developer can quickly look at the effects of multiple design options on assembly strain and assembly/separation forces and select a preferred design. The developer may even find that the chosen lock feature style cannot be made to function reliably under any circumstances.

This chapter has been about feature design, but a calculation tool is also very valuable for diagnosing existing lock feature problems. These are usually high strain and feature damage/failure or application assembly/separation force issues. (Chapter 14 discusses the snap-fit diagnostic process in detail.)

If feature level issues are the root cause of a problem and calculations are indicated, verify the feature's material, (it may have changed from the original specification) and get the most accurate material property data possible. For a high impact application (recall the demand-complexity matrix), special testing for property data may be required. Then use the feature's actual dimensions and deflection as measured on the application to calculate strain and force values. Those dimensions and calculated values provide a baseline for evaluating design changes.

It is worthwhile to compare the calculated strain value to published and recommended strain limits – this may provide valuable information. However, the one thing known for certain is that regardless of any assumptions or uncertainties, *that particular feature design did indeed fail.* Design changes and their effects relative to the baseline values can be evaluated and an optimal change(s) identified.

These spreadsheets include the adjustments missing from other published or on-line calculations.

Figures 13.31–13.33 show how to construct spreadsheets in Microsoft Excel© to perform the basic cantilever beam strain and bending calculations as well as the assembly and separation force calculations for a catch at the end of a beam. A reader with a little Excel© experience will be able to create these spreadsheets fairly easily and if they choose, adapt them to perform calculations for tapered beams as well as for traps and loops. The basic equations are the same as those presented for the straight beam and the catch.

The rows at the bottom of each figure summarize how the input and result values from some of the examples presented above would appear in each spreadsheet. *The rounding off of some values in the example calculations above (done by hand) may result in some minor disagreement with the spreadsheet results.*

Create the top calculation row in the spreadsheet as shown and then drag the contents down to create a final spreadsheet with as many rows as desired. Different design options can then be entered into each row for easy comparison of results.

In calculating cantilever beam behavior (Fig. 13.31), the user may not have access to some information (Q-factor or mating feature/part deflection) needed to carry the calculation all the way to the final force and strain values. If Q-factor information is not available, simply enter 1.00 in the spreadsheet for the Q-factor value. The resulting force and strain values will be your final results unless information is available from curves generated for mating feature/part deflection, in which case, that adjustment is the final step in the calculations.

The resulting strain value can then be compared to the target strain.

Bending force and strain for a straight cantilever beam:

- Begin the spreadsheet at cell B5.
- All dimension and strain cells – 2 decimal places.
- Modulus input cell – no decimal places.
- Q-Factor – 2 decimal places.
- Format strain cells as %.

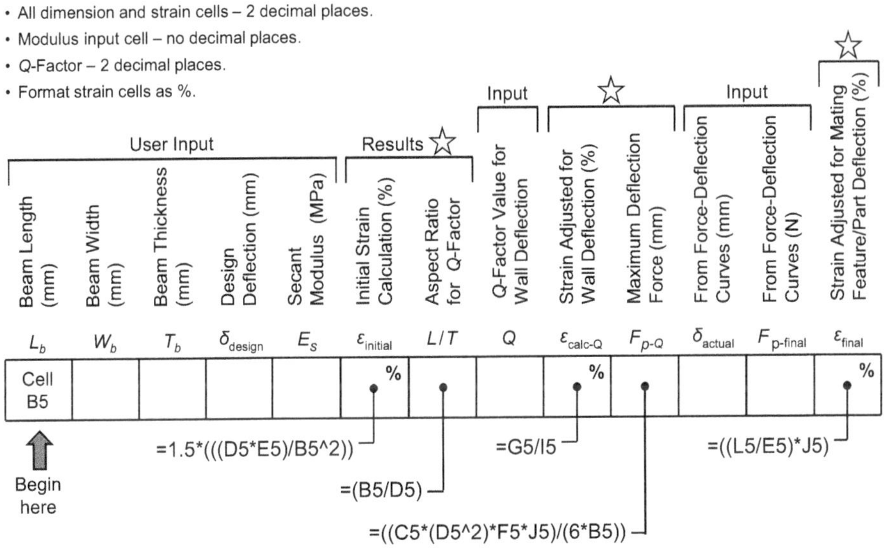

Beam calculations from the example in Section 13.2.8. Input values are in bold font.

L_b	W_b	T_b	δ_{design}	E_S	$\varepsilon_{initial}$	L/T	Q	$\varepsilon_{calc\text{-}Q}$	$F_{p\text{-}Q}$	δ_{actual}	$F_{p\text{-}final}$	ε_{final}
15.00	**3.00**	**2.00**	**2.00**	**2000**	2.67%	7.50	**1.11**	2.40%	6.41	**1.48**	**4.80**	1.78%

Figure 13.31 Spreadsheet for straight cantilever beam behavior

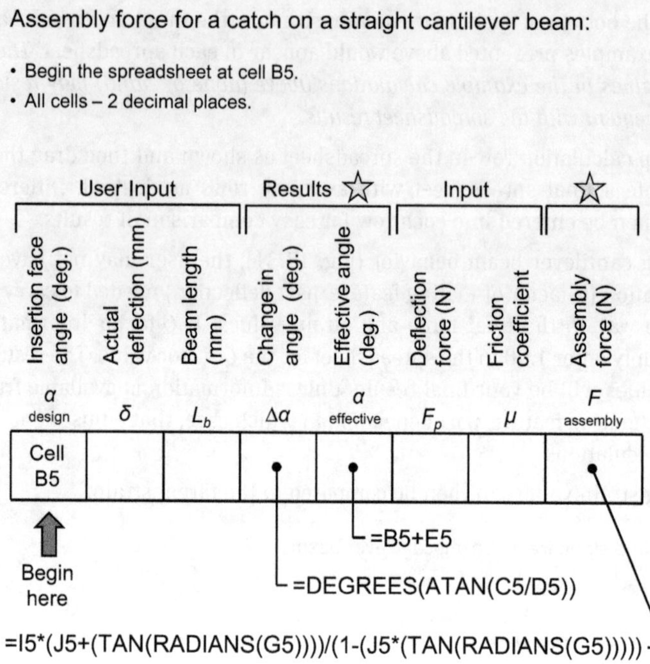

Assembly force calculations from the example in Section 13.6.1.2.
Input values are in bold font.

25.00	**1.48**	**15.00**	5.63	30.63	**4.80**	**.30**	5.21

Figure 13.32 Spreadsheet for catch assembly force

Separation force for a catch on a straight cantilever beam:

- Begin the spreadsheet at cell B5.
- All cells – 2 decimal places.
- The only difference from the assembly force calculation is that E5 is *subtracted* from B5 rather than added in the effective angle calculation.

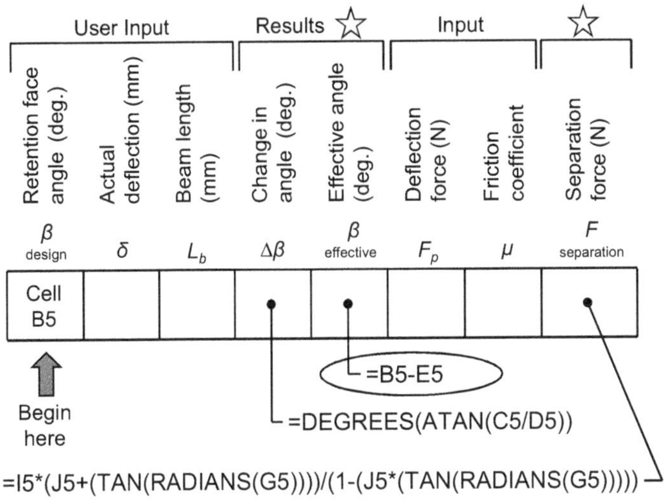

Separation force calculations from the example in Section 13.6.2.2.
Input values are in bold font.

45.00	**1.48**	**15.00**	5.63	39.37	**4.80**	**.30**	7.13

Figure 13.33 Spreadsheet for catch separation force

13.10 Summary

This chapter describes calculations for the majority of snap-fit lock styles used in product applications. Equations for other, less common lock styles such as torsional, annular, and planar are beyond the scope of this book. However, they are available online and can be found in some of the resource material listed in the appendix.

One of the most important points in this chapter is that most of those snap-fit analysis resources *do not* include certain important adjustments in the calculations they present. The exception is the well-known adjustment for stress concentration. The adjustments introduced in this chapter are summarized in Table 13.4.

These adjustments are also included in the spreadsheets described in Section 13.9.

Table 13.4 Summary of Snap-Fit Calculation Adjustments

	Effect on actual strain	Effect on assembly force	Effect on separation force
Stress concentration (*k*)	increase	–	–
Deflection magnification (*Q*)	reduce	reduce	reduce
Mating feature/part deflection	reduce	reduce	reduce
Effective insertion face angle ($\alpha_{effective}$)	–	increase	–
Effective retention face angle ($\alpha_{effective}$)	–	–	reduce

A couple of additional comments about calculations and beam behavior:

- The wall deflection, (or *Q*-factor) effect is simply the result of the moment at the base of a deflected beam. Extending the deflection factor information should not be difficult using either finite element analysis or beam calculations based on the moment created at the wall by known beam configurations and deflections. FEA would provide the opportunity for analysis of the effects on wall deflection of the relationship between beam thickness and wall thickness.
- In a manner similar to the effect of beam bending on the insertion and retention face angles, tilting of the catch at the end of a deflecting beam will reduce the effective *height* of the insertion and retention faces. This change in deflection, (which can be calculated) is small but by reducing strain, it provides another small margin of safety against over-strain in lock feature design.
- A better understanding of the friction/retention face angle relationship and its effect on shear, tensile, and bending interactions with respect to the reliability of separation force calculations would be useful.

Other important points in Chapter 13 include:

- Use the cantilever beam and catch rules-of-thumb to establish a reasonable lock design as a starting point for the analysis process. The rules-of-thumb are also useful for initial diagnosis of some feature problems in existing applications.
- Many of the rules of thumb and the calculations given here for cantilever beam style locks are applicable to other lock styles.
- Use stress-strain data that represent actual application conditions whenever possible.
- Tapering the beam can significantly reduce strain at the beam's base.
- Strain at the base of the beam is independent of beam width. Separation strength can be increased by increasing the width with no increase in strain. This change will increase the assembly and separation forces.
- Calculations are useful for both initial design and for problem diagnosis.
- Calculation results may not be absolute, consider them to be estimates of performance. Actual feature assembly and separation performance is a function of many variables, many of which are beyond the scope of hand calculations.
- Regardless of the confidence level in the calculation results, end-use testing is the only way to verify feature performance.
- Consider the demand-complexity matrix when making decisions about end-use testing.

References

[1] Allied Signal Plastics, *Modulus Snap-Fit Design Manual*, Morristown, NJ (1997)

[2] Tres, P., *Designing Plastic Parts for Assembly*, Hanser, Munich, Germany (2000)

Bibliography

The reader will note the age of most of these publications. In some cases, new editions of the books are available and, the internet is a limitless source of current and old information on the subject. Some of the technical reports published by plastic suppliers are still available although sometimes suppliers have merged and/or changed names and the reports may be hard to find. Some are available online. Sources for calculations and analysis are given in the appendix. Any links provided in the appendix were verified active as of April 2016.

In any case, while new materials are developed, material testing techniques can change, and new or improved property data can become available, the classic equations for mechanical behavior have not changed.

Publications that the author found to be particularly useful are preceded by an asterisk (*). These tend to be books or reports targeted to product developers, which provide practical information for everyone, not necessarily polymer experts.

* Allied Signal Plastics, *New Snap-Fit Design Guide*, Society of Plastics Engineers ANTEC (1987)

* Hoechst Technical Polymers, *Designing With Plastic - The Fundamentals*, Design Manual TDM-1, Ticona LLC (Formerly Hoechst Celanese Corporation, now a division of Celanese AG), Summit, NJ (1996)

* Malloy, Robert A., *Plastic Part Design for Injection Molding*, Hanser Publications, Munich (1994)

* Molders Division of The Society of the Plastics Industry Inc., *Standards and Practices of Plastics Molders*, 1998 ed., Washington, D.C (1998)

* *Snap-Fit Joints for Plastics - a Design Guide*, Polymers Division of the Bayer Corporation, Pittsburgh, PA (1998)

Borg-Warner Chemicals, *Designing Cantilever Snap-Fit Latches for Functionality*, Tech. Publ. #SR-402

Chow, W.W., *Snap-Fit Design*, University of Illinois, Department of Mechanical Engineering, Urbana, IL, July (1977)

Kar, P., Renaud, J., *Parametric Investigations of Integrated Plastic Snap Fastener Design*, Univ. of Notre Dame, Proceedings of S.M. Wu Symposium on Manufacturing Science at Northwestern University (1994)

Lee, C.S., Duban, A., Jones, E.D., *Improving Snap-Fit Design*, Plast. Des. Forum, Sept./Oct. (1987)

Noller, R., *Understanding Tight-Tolerance Design*, Plast. Des. Forum, March/April (1990) pp. 61–72

Rackowitz, D.R., *Beyond the Data Sheet - Designer's guide to the interpretation of data sheet properties*, BASF Plastic Materials, Wyandotte, MI

Roy, Dhirendra C., *Snap-Finger Design Analytics and Its Element Stiffness Matrices*, United Technologies Automotive, SAE Tech. Pap. Ser. (SP-1012), International Congress and Exposition (1994)

Smith, Z., *It's a SNAP!*, Hoechst Celanese Corporation, Assembly Mag., Summit, NJ, Oct. (1994)

Smith, Z., Fletcher, M., Sopka, D., *The Give and Take of Plastic Springs*, Mach. Des., Nov. (1997) pp. 69–72

Standard Test Method for Kinetic Coefficient of Friction of Plastic Solids, ASTM Standard D 3028, ASTM Committee D-20 on Plastics

Throne, James L., *Plastics Process Engineering*, Marcel Dekker Inc., New York (1979)

Trantina, G. G., Minnicbelli, M. D., *Automated Program for Designing Snap-Fits*, GE Plastics, Plast. Eng., Pittsfield, MA, Aug. (1987)

14 Diagnosing Snap-Fit Problems

Understanding failure modes and their relationship to the most likely root causes can help with a quick and accurate diagnosis. This minimizes the cost and time impact but it also helps ensure that the proposed changes will indeed fix the problem. Nothing is worse than making changes to a product and finding that the problem still exists or has even gotten worse. (Yes, this does happen.) Accurate diagnosis is particularly valuable during product development when prototype testing may indicate the need for improvements, yet time and cost constraints limit the available options.

The root causes of many snap-fit problems are at the *attachment* level. Yet, because the attachment level is not understood, attempt(s) to fix the problems are at the feature level, thus doomed to failure, or the fix will cost more than it should.

Snap-fit problems are frequently misdiagnosed.

When evaluating any snap-fit problem, *even feature failure or damage*, first verify that all attachment level requirements have been satisfied. If not, address them before attempting a feature level fix [1].

A second common cause of problems is failure to select the proper lock feature style. This mistake is often compounded by unsuccessful attempts to force that feature to work anyway. This happens much too often. Look back at the example in Chapter 6, Fig. 6.9.

The term *problem* means more than simply feature failure or breakage. A snap-fit problem is identified by one or a combination of these symptoms:

- Difficult assembly
- Short-term feature failure or damage
- Long-term feature failure or damage
- Part distortion or damage
- Part loosening and/or squeaks or rattles
- Unintended part release
- Service difficulty
- Customer complaints about usage

These are *symptoms* of problems; they are not the root cause. Simply treating the symptoms may not fix the real problem or it may create other problems in the attachment. Many of these symptoms can have both attachment and feature level root causes. Sometimes, the root cause is a combination of multiple shortcomings.

Figures 14.1 and 14.2 reflect the author's personal experience trouble-shooting snap-fit problems in products that were already in production. Note the high incidence of causes related to attachment level issues and the high frequency of multiple root causes for problems.

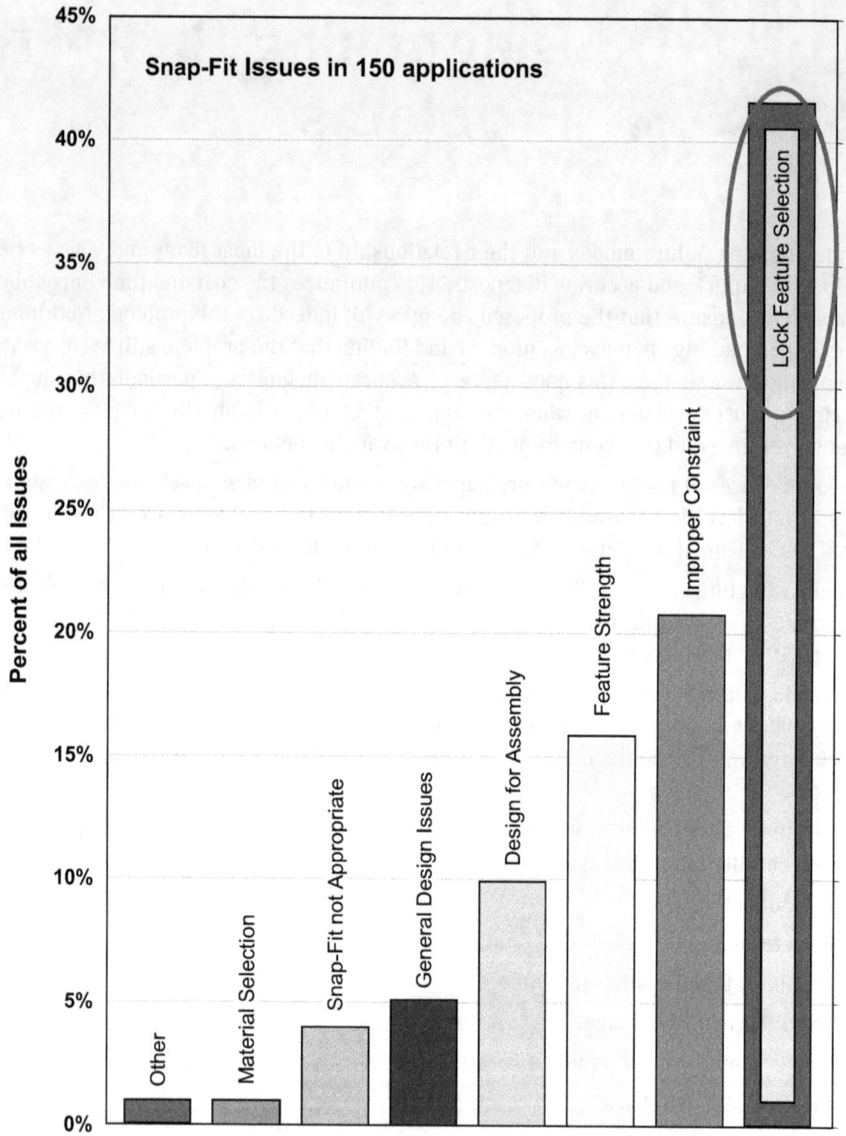

Figure 14.1 Frequency of various snap-fit problems

The information in Fig. 14.1 has evolved over the years. The author's early impression of snap-fit issues was that *improper constraint* was the single biggest cause of snap-fit problems. Over time, increasing awareness of the cantilever hook lock's short-comings and the importance of decoupling to lock performance lead to the creation of a new category called *lock feature selection*. In the author's opinion, this new category represents the single major cause of snap-fit problems, with improper constraint now being the second.

While they are related, lock feature (style) selection and lock feature strength (the third highest cause of problems) are not the same thing. Regardless of style, the lock must still be designed for strength.

Figure 14.2 illustrates the author's observations on root causes from a different perspective. Both figures are intended give the reader a starting point for prioritizing their own snap-fit diagnosis process.

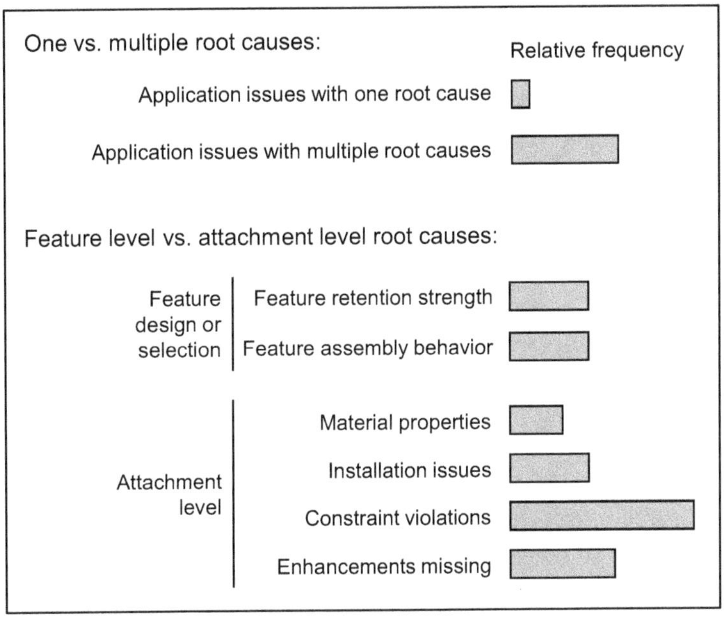

Figure 14.2 Distribution of snap-fit problem root causes

14.1 Common Snap-Fit Mistakes

Most problems, whether they are attachment or feature level, are the result of mistakes made during the development process. Some of the more common mistakes are:

- Over-constraint where features fighting each other can cause immediate feature damage or breakage or longer-term damage due to thermal expansion. Over-stress due to residual assembly forces can cause long-term feature failure.
- Under-constraint where features (usually locks) are improperly used to resist applied forces or part movement.
- Weak or compliant parts are expected to provide a rigid base for lock or locator features.

- Failure to fully consider material properties, including:
  - Incomplete material property data
  - Plastic creep and thermal effects
  - Material degradation over time
- Failure to anticipate assembly variables like compatibilities involving engage direction, assembly motion, feature positioning, and feature style.
- Difficult or frustrating assembly.
- Failure to anticipate all possible end-use conditions, including:
  - Improper disassembly and service
  - Unexpected but possible events (such as dropping or striking a product)
  - Improper customer usage

14.2 Attachment Level Diagnosis

Attachment level problems are often independent of the lock. In other words, they will occur regardless of the locking features style used. Understanding the key requirements of constraint, compatibility, and robustness can help one recognize and resolve many attachment level problems.

Certain enhancements are required for every application. Always verify they are present. If any are missing, problems are likely. These enhancements include:

- Guides
- Clearance
- Operator feedback
- Process-friendly design

The four most common symptoms related to attachment level problems include:

- Difficult assembly
- Parts distorted
- Loose parts
- Feature damage – remember, *feature damage does not necessarily indicate a feature problem*. This is one of the most common errors in diagnosis. Feature damage may be a symptom, not a root cause.

For each of these four attachment level symptoms, the most likely root causes are shown in Fig. 14.3.

Remember, attachment level problems must be fixed first and testing may be required. If the problem remains, fixing the attachment level problems was still a necessary step before proceeding with any feature level fixes.

Difficult Assembly

- Over-constraint
- Assembly motion and constraint feature incompatibility
- Basic shape and assembly motion incompatibility
- Access and basic shape incompatibility
- Access and assembly motion incompatibility
- Parts warped
- Simultaneous engagement of several features
- No guide or clearance enhancements
- No operator feedback and/or feedback interference
- Mating part is hard to hold or handle

Distorted Parts

- Parts warped when production
- Parts distorted in assembly
- Feature tolerances and position robustness
- Over-constraint
- Compliant (flexible) parts, often panels are not constrained at enough points

Loose Parts

- Feature damage (see below)
- Weak feature mounting area(s) on mating and base parts
- Difficult assembly (see left)
- Under-constraint
- Compliant parts are not a strong base for the constraint features

Feature Damage

- Over-constraint
- Under-constraint
- Incompatibility between features and assembly motion
- Long-term creep or yield
- Damaged during assembly (see above)
- Material degradation
- Damaged in shipping and handling
- Improper processing
- Not process-friendly
- Abuse in usage
- Abuse or damage during service/removal
- Missing guide or clearance enhancements

Figure 14.3 Likely root causes of attachment level problems

14.3 Feature Level Diagnosis

Consider feature level root causes only after attachment level root causes are either fixed or ruled out.

Obviously, it is preferable to identify and make feature level changes before new parts are made. If the problem is indeed a feature problem, simple changes to some feature dimensions may be possible. If they do not solve the problem, other changes to the lock feature or to the entire lock pair are needed.

The most common feature level problems include:

- High assembly force, see Table 14.1 for recommended fixes.
- High feature strain or feature damage during assembly or disassembly, see Table 14.2.
- Low retention strength and lock damage under loads, see Table 14.3.
- High separation force, see Table 14.4.

In these tables, possible feature changes are listed from top to bottom, beginning with those that are easier to implement. Changes become more difficult, costly, or unlikely as we move down the table lists. Ease of implementation is based on this reasoning:

- Changes to the lock retention mechanism are generally the easiest.
- Changes to the lock deflection mechanism are difficult.
- Changes to the lock style are more difficult.
- Changes to the attachment system are very difficult.
- Material changes or major part changes are very unlikely.

Make a good lock style choice to start with. It's easier.

Fixing one problem may create another. For example, making a lock stronger to solve a low retention strength problem may increase the assembly force and the assembly strain. If the assembly force becomes too high or strain is excessive, a new set of problems will be created.

In addition to ranking changes by ease of implementation, each table shows how that change will affect the other three common feature level problems. A change may have a positive, neutral or negative effect on feature performance in the other areas.

The recommended changes and the predicted interactions in these tables are written primarily with beam-based locks in mind. However, many apply to all lock styles. The thought process itself certainly applies to all styles.

Within each *difficulty* group noted in the tables, the suggested changes are ranked by the number of additional positive or negative interactions they can have on the attachment. Usually a negative effect is simply an incremental shift in a particular characteristic. A negative effect does not guarantee a new problem, just a movement toward a condition that will increase the likelihood of a problem.

Table 14.1 Feature Level Solutions for High Assembly Force

Level of difficulty	Make change to:	Recommended changes for high assembly force:	Reducing a high assembly force may also have these effects:		
			Feature strain or damage during assembly or disassembly	Retention strength or lock damage under loads	Separation force
1	Retaining mem.	Reduce insertion face angle	x	x	x
1	Retaining mem.	Add contour to insertion face	x	x	x
1	Retaining mem.	Modify retention feature engagement to reduce the additive effects of simultaneous engagement	x	x	x
1	Retaining mem.	Make retention face shallower (decrease deflection)	reduce	worse	reduce
2	Deflecting mem.	Make beam longer	reduce	worse	reduce
2	Deflecting mem.	Reduce beam thickness overall	reduce	worse	reduce
2	Deflecting mem.	Reduce beam thickness at end by tapering	reduce	worse	reduce
2	Deflecting mem.	Reduce beam width overall	x	worse	reduce
2	Deflecting mem.	Reduce beam width at end by tapering	x	worse	reduce
3	Locking system	Decouple insertion and retention behaviors	reduce	improved	reduce
3	Locking system	Design for sequential lock engagement	x	x	x
3	Locking system	Redesign for a tip assembly motion	x	x	x
3	Locking system	Decrease mating feature stiffness (increase deflection)	reduce	worse	reduce
3	Locking system	Make base area more flexible (Q-factor)	reduce	worse	reduce
3	Locking system	Change lock style	x	x	x
4	Parts	Change part material or design	x	x	x

An '*x*' in the effects columns indicates either no effect or an effect cannot be predicted.

Table 14.2 Feature Level Solutions for High Feature Strain or Damage during Assembly or Disassembly

Level of difficulty	Make change to:	Recommended changes for high feature strain or damage during assembly or disassembly:	Reducing high feature strain or damage may also have these effects:		
			Assembly force	Retention strength or lock damage under loads	Separation force
1	Process	Verify part manufacturing process is correct	x	x	x
1	Retaining mem.	Make retention face shallower (decrease deflection)	reduce	worse	reduce
2	Deflecting mem.	Make beam longer	reduce	worse	reduce
2	Deflecting mem.	Reduce beam thickness overall	reduce	worse	reduce
2	Deflecting mem.	Reduce beam thickness at end by tapering	reduce	worse	reduce
2	Deflecting mem.	Increase beam thickness at base by tapering	increase	improved	increase
3	Locking system	Verify part design is process-friendly	x	x	x
3	Locking system	Decouple insertion and retention behaviors	reduce	improved	reduce
3	Locking system	Add guidance enhancement feature	x	x	x
3	Locking system	Add visual enhancement feature	x	x	x
3	Locking system	Decrease mating feature stiffness	reduce	worse	reduce
3	Locking system	Make base area more flexible (Q-factor)	reduce	worse	reduce
3	Locking system	Add guard enhancement feature	increase	x	increase
3	Locking system	Change lock style	x	x	x
4	Parts	Change part material or design	x	x	x

An '*x*' in the effects columns indicates either no effect or an effect cannot be predicted.

Table 14.3 Feature Level Solutions for Low Retention Strength or Lock Damage under Load

Level of difficulty	Make change to:	Recommended changes for low retention strength or lock damage under load:	Fixing low retention strength or lock damage may also have these effects: Feature strain or damage during assembly or disassembly	Assembly force	Separation force
1	Retaining mem.	Load beam closer to neutral axis	x	x	x
1	Retaining mem.	Increase retention face angle	x	x	increase
1	Retaining mem.	Add contour to the retention face	x	x	increase
1	Retaining mem.	Make retention face deeper (increase deflection)	increase	increase	increase
2	Deflecting mem.	Increase beam thickness at base by tapering	reduce	increase	increase
2	Deflecting mem.	Increase beam width at base by tapering		increase	increase
2	Deflecting mem.	Make beam shorter	increase	increase	increase
2	Deflecting mem.	Increase beam thickness overall	increase	increase	increase
2	Deflecting mem.	Increase beam width overall	increase	increase	increase
3	Locking system	Decouple insertion and retention behavior	reduce	reduce	reduce
3	Locking system	Reorient lock to carry less load	x	x	x
3	Locking system	Add more lock features	x	increase	increase
3	Locking system	Add retainer enhancement feature	x	increase	increase
3	Locking system	Increase mating feature stiffness	increase	increase	increase
3	Locking system	Make base area less flexible (Q-factor)	increase	increase	increase
3	Locking system	Change lock style	x	x	x
4	Parts	Change part material or design	x	x	x

An '*x*' in the effects columns indicates either no effect or an effect cannot be predicted.

Table 14.4 Feature Level Solutions for High Separation Force

Level of difficulty	Make change to:	Recommended changes for high separation force:	Fixing high separation force may also have these effects:		
			Feature strain or damage during assembly or disassembly	Assembly force	Retention strength or lock damage under loads
1	Retaining mem.	Make retention face shallower (decrease deflection)	reduce	reduce	worse
1	Retaining mem.	Reduce retention face angle	x	x	worse
2	Deflecting mem.	Make beam longer	reduce	reduce	worse
2	Deflecting mem.	Reduce beam thickness overall	reduce	reduce	worse
2	Deflecting mem.	Reduce beam thickness at end by tapering	reduce	reduce	worse
2	Deflecting mem.	Reduce beam width overall	x	reduce	worse
2	Deflecting mem.	Reduce beam width at end by tapering	x	reduce	worse
3	Locking system	Decouple insertion and retention	reduce	reduce	improved
3	Locking system	Add assist enhancement feature	x	x	x
3	Locking system	Decrease mating feature stiffness (increase deflection)	reduce	reduce	worse
3	Locking system	Make base area more flexible (Q-factor)	reduce	reduce	worse
3	Locking system	Change lock style	x	x	x
4	Parts	Change part material or design	x	x	x

An '*x*' in the effects columns indicates either no effect or an effect cannot be predicted.

14.4 Summary

This chapter described an attachment level approach to diagnosing and fixing the most common snap-fit problems. Problems were first defined as a broad range of situations including, but not limited to, the more obvious ones involving snap-fit feature damage and failure. Most importantly, an approach of addressing systemic (attachment level) causes before attempting feature level fixes is emphasized. The diagnostic process is summarized in Fig. 14.4.

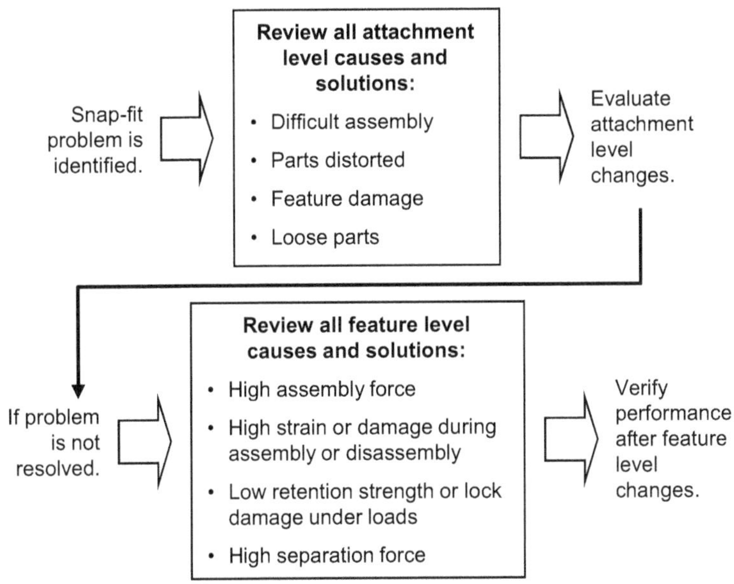

Figure 14.4 The diagnostic process for snap-fits

Important points in Chapter 14:

- Do not mistake a symptom for a root cause. This is important in any problem-solving effort.
- The root cause of feature failure or damage may not be the feature itself. Do not assume a feature failure has a feature level root cause.
- The root cause of many if not most snap-fit problems is at the attachment level, not the feature level. Some problems will be a combination of both feature and attachment level issues.
- Many attachment level problems result from improper lock feature selection and/or improper constraint.
- Resolve all attachment level causes of a problem before attempting any feature level fixes.
- Some problems will be combination of both feature and attachment level causes.

- Always try the easiest fixes first.
- Most feature level changes in a snap-fit will have multiple effects. A change to fix one problem is very likely to change other behaviors and may create new problems.

Reference

[1] Bonenberger, P.R., *Solving Common Problems in Snap-Fit Designs*, Western Plastics Expo, Long Beach, CA, Jan. (1999)

15 Gaining a Competitive Advantage in Snap-Fit Technology

Engineering managers and executives should read this chapter because of the business advantages that are possible when an organization goes beyond simply teaching *individuals* about good snap-fit practices and becomes *snap-fit capable* [1].

If you read nothing else in this chapter, read Section 15.5.3 and the introductions for Sections 15.6 and 15.7.

This chapter addresses two fundamental questions: *How do we start?* and *Where do we go?* Most business leaders and engineering managers will not have the time or inclination to read this entire book, much less pick out the information needed to create detailed short and long-term implementation strategies.

The ideas and recommendations in this chapter reflect the author's experience in a very large engineering organization with access to extensive resources. This will not be the case in many smaller organizations. However, the core principles are available for adaptation to any product engineering organization, regardless of size. Some initiatives will be more important than others and engineering managers and organizations can pick and choose which are applicable to their business and how they wish to execute them. A simplified or basic plan is shown in Table 15.1, Section 15.5.3.

The recommendations in this chapter reflect 37 years of product assembly and fastening experience, including 15 years of snap-fit experience and 20/20 hindsight. Critical technical points are integrated with management and organizational development strategies as a starting point for bringing snap-fit capability into an organization and then leveraging that capability for engineering and business advantage.

For an organization with a product design culture based on loose fasteners, venturing into snap-fit technology carries the fear of wasting resources on an uncertain outcome: *What if we try it and it doesn't work?* Even worse: *What if we try it and discover a problem after delivering thousands of these products?* On the other hand, the company may be missing out on significant savings or business opportunities by not using snap-fits. After all, snap-fits are now a proven technology and there are many very good and successful snap-fit product applications.

Product engineering organizations are familiar with the use of loose-threaded fasteners and other mechanical joining methods (clips, rivets, push-in fasteners, etc.) as a means of joining one part to another. Most of the principles and strategies discussed here for moving into snap-fits are appropriate regardless of which attachment technologies are already in use.

Because it is the most common scenario, the emphasis here is on replacing threaded fasteners with snap-fit attachments. This is a significant technical shift from a *clamp-load* based attaching method to one which, most of the time, *should not* be expected to provide or to rely on clamp load.

Table 3.5 in Chapter 3 lists advantages and disadvantages of snap-fits vs. threaded fasteners.

Some applications are relatively easy to adapt to a snap-fit attachment. For a novice (individual or organization), these applications are a good starting point for gaining experience and confidence. At the other end of the spectrum are applications involving a complex interface and critical performance/reliability requirements. These applications are not for *beginners*. Figure 15.1 shows some applications representing different levels of difficulty.

Figure 15.1 Applications with different levels of difficulty

Chapter 3, Section 3.4, introduced the concepts of interface complexity and application demand in a *demand-complexity matrix* as an aid to understanding different levels of snap-fit application difficulty. The subject is discussed in detail in Section 15.4.

Individuals can become better at snap-fits by applying the principles in this book: Why not a company?

With years of experience in both threaded fastener and snap-fit technologies, the author does not favor one fastening method over the other. Attachment related quality and reliability problems are the result of selecting the wrong fastening method, poor product design and/or poor execution of the fastening process, not an inherent inferiority in a particular fastening technology.

Questions addressed in Chapter 15 include:

- What can an organization do to support its people as they learn to use snap-fits?
- What can management expect during the early learning phase?
- What are potential pitfalls and what are enablers for success?
- How can management ensure continuity in *organizational snap-fit capability* for a long-term and sustainable business advantage?

For a more historical and technical background, a review of these chapters and sections is suggested:

- Prefaces to the first and second editions (included in this edition). They explain the background of the snap-fit knowledge in this book. The author was the initiator and project leader as an engineering organization worked to *bootstrap* itself to a higher level of snap-fit capability.
- Chapter 1 - The key to successful implementation of snap-fit technology is understanding the *systems* aspects of the technology. Chapter 1 introduces the idea of a systematic way of thinking about snap-fits. This is a short but important chapter.
- Chapter 3 - An introduction to a logical development process leading to fundamentally sound snap-fit concepts and reliable attachment designs.

15.1 Terminology

Some terms in this chapter have very specific meanings and are defined here:

- Snap-fit capable - An organization where the snap-fit knowledge of individuals (i.e., *personal* capability) is leveraged through rational business and engineering strategies for maximum effectiveness as *organizational* capability.
- Developer - Anyone, a designer or an engineer, who is responsible for conceiving and executing product designs, including snap-fit applications.
- Manager or management - Leadership of an engineering organization, including engineering supervisors, managers, and executives.
- Organization - A product engineering entity. It may be an independent company or it may be an activity, department, or organization within a larger company. In this chapter, *organization* refers to all of them.

- Development - The *entire* product development process, from the initial concept stage through pre-production prototype and final production-ready parts.
- Design - The stage of the development process where part geometry is finalized by adding dimensions and specifications to create part drawings and/or math files.
- Problems - Problems with attachments go beyond breakage or unintended separation, which are only the most dramatic and visible problems. Other common problems include difficult assembly, squeaks and rattles, part distortion, the need for close (costly) tolerances, and a lack of robustness. All of these problems cost money and/or will result in customer dissatisfaction. Some may require costly engineering study and redesign.
- Demand - The manufacturability, assembly, performance, and reliability expectations for the application. It is a continuum, with *low* and *high* representing the extremes.
- Complexity - The geometric and manufacturing complications of the part to part interface. Like demand, complexity is a continuum, with *low* and *high* representing the extremes of the scale.
- Impact - The effect of snap-fit problems on specific (critical) application requirements as defined by management. Demand and complexity as well as other factors will influence judgments about resources and the level of attention required for a given snap-fit application. Impact is the *cumulative effect* of demand, complexity, and other factors. Depending on the number of factors involved, the matrix may have more than 2 or 3 dimensions. The word *impact* as used in this chapter does **not** refer to force related impacts on an application

15.2 Managing Expectations

These are good practices to follow for all technical specialties, not just snap-fits.

When attempting a cultural or technical change, minor or major, managing organizational and individual participants' expectations help smooth the transition and reduce uncertainty.

- In larger engineering organizations, do not expect one snap-fit technology leader or expert to do it all. Every product developer should become *capable* in snap-fit technology and some will become experts. Imagine a scenario where the technology expert suddenly disappeared. What would happen to your company's snap-fit capability?
- In any case, all fastening decisions, snap-fit and others, should be closely integrated with the product development process. It is essential that every product developer have a basic working knowledge of the subjects. At the very least, they must know enough to be aware of their *knowledge limitations* and know when to ask for help. In other words, *be mindful of what you don't know!*
- Do not expect overnight success. Study some well executed snap-fit attachments. You will begin to appreciate the complexity of the problem. Remember that snap-fit devel-

opment is iterative, especially when the attachment is complex. Very few attachments will be perfect the first time around. The goal is to minimize the design iterations needed to get to a final product design.

- Play it safe and be conservative. You are still saving money over conventional fasteners. With snap-fits, *over-design* often means simply using thicker or wider sections for more strength in some features and this is very inexpensive. Even adding locking features for more strength can be relatively inexpensive if done during the concept stage of development.
- Allow additional time for snap-fit development. Recognize that the time and effort spent in developing reliable and robust snap-fit attachments will likely exceed the time spent on a threaded fastener based attachment for the same application. This is especially true during the early learning phase. However, benefits that far exceed the initial engineering costs will be realized when that design is assembled into thousands of products without the cost of using assembly tools or loose fasteners.
- Recognize that permanent cultural change will not happen overnight. Depending on an organization's size and the type of change required (in this case, a technical change) permanent change can take years. It will require constant nurturing through management attention and support.

15.3 Harmful Beliefs

Individuals and organizations can hold these beliefs.

While not necessarily *fatal*, these commonly held beliefs can seriously interfere with developing true snap-fit expertise and they will keep an organization *of any size* from becoming the best it can be.

Expect to find these beliefs in yourself, your management team, the product design community, customers, and suppliers. They may be hidden or unspoken, but they will be there and will interfere with your organization's attempts to become snap-fit capable. Address them directly through awareness and constant reminders. Even after you think you have conquered them, be aware that they could creep back in over time.

A very easy first step in changing the culture is to print the following harmful beliefs on wall posters and put them on display.

- **The battery cover syndrome:**

  Most people are familiar with some applications of snap-fit technology thanks to their usage on simple applications like battery access covers on remote controls and on toys. This familiarity can cause two erroneous beliefs:

  Snap-fits are only appropriate for trivial applications. This is not true. Snap-fits are used in such diverse and important applications as medical devices, automobiles, consumer electronics, and structures.

  Snap-fits are easy to design. There are certainly some applications that are easier than others, but many applications are complex, requiring a thorough understanding of the systemic nature of snap-fits. Furthermore, even the so-called *trivial* applications can be difficult to design properly.

Attachment problems have been found on the simplest applications imaginable. Every snap-fit principle discussed in this book, regardless of how trivial or obvious it may seem, is based on observation of real problems in real products.

- **Snap-fits are a materials technology.**
  Because snap-fits are mostly found in products made from polymers, there has been a long-held belief that polymer experts, (including resin suppliers) should be the go-to design resource for snap-fit applications.

  This belief probably has its roots in the traditional feature level approach to snap-fits. While polymers experts and suppliers are certainly necessary for assisting with the material strength and behavior aspects of an application, they should not be expected to design the attachment interface.

  Snap-fits are a mechanical engineering technology and only the product developer has access to the skills, information, and resources needed to develop world-class snap-fit attachments. Besides, it's their job.
- **Cantilever hooks represent snap-fit technology.**
  This belief is related to the battery-cover syndrome: when one sees cantilever hook style locking features everywhere, it is easy to believe they are the only or the best snap-fit option.

  When asked to create a snap-fit attachment, many designers will default to the cantilever hook because of its familiarity, whether or not it is appropriate for the application. There are many other lock styles available as options for the snap-fit developer.

  Cantilever hooks have inherent shortcomings and, as a rule, are only appropriate for low-demand applications. The decoupling discussion in Chapter 7 explains these shortcomings and the technical advantages of other lock styles over cantilever hooks.
- **All I need to do is design the locking feature.**
  This is related to two other harmful beliefs, the *battery-cover syndrome* and *snap-fits are a materials technology.* A snap-fit attachment is an interface system and it must be developed as such. Many well-designed features fail to perform as expected because the systemic aspects of the attachment are ignored.
- **Experience in other fastening methods transfers to snap-fits.**
  A common managerial mistake is to simply tell an engineering organization to *start doing snap-fits.* The assumption may be that product developers will apply their knowledge of threaded fastening methods to snap-fits.

  However, threaded fastener knowledge does not transfer to snap-fits. One result when this scenario plays out is that snap-fit locking features (often cantilever hooks) are simply substituted at threaded fastener sites in the product design and the transition to snap-fit technology is considered done. This approach may work in some applications (creating a false sense of security) but it will also result in attachments with poor assembly behavior, improper constraint and ultimately, product problems. Sometimes the result is outright attachment failure. In virtually all cases, the design is not optimized for performance and cost.

Surprisingly, while threaded fastener knowledge does not transfer to snap-fits, *the reverse is not true.* A basic understanding of snap-fit principles, when applied to a threaded fastener attachment, (or most other mechanical attachment methods) can result in a more cost-effective, robust and assembly-friendly joint.

Snap-fit design principles can improve other fastening applications.

A strong case can be made for learning about snap-fit design principles even if actual application of the technology to your particular product is limited.

- **Every snap-fit application is a new invention.**
  For most potential snap-fit applications, there are already fundamentally good design concepts that can be found, studied, and used as a starting point.

  With snap-fits, the same fundamental rules of design are true for parts of *any size part* in any snap-fit application. These rules are the *key requirements* discussed in Chapter 2 and the minimum requirements discussed later in this chapter. Follow them. One common reaction to some of these requirements is: *Why should this small part have those features? It's not that big.* Positioning and locking one part to another using snap-fits is analogous to properly fixturing a part for a machining operation or for dimensional checking. It does not matter what size the part is, it must be constrained according to the same principles. A larger part may require *more* of some features, but the fundamental rules for their usage and arrangement still apply.

  Many design rules are also shared by families of application defined by the generic attributes *basic shapes* and *assembly motions*. Benchmark good product designs and refer to a library of good interface concepts to begin a new application.

- **I can create the attachment after I do everything else.**
  Do not wait to the end of the product development process to think about the attachment interface. The attachment concept must be developed simultaneously with the rest of the product. Attachment *details* can wait until later in the development process, but getting the basic concept right is critical to the attachment's success.

15.4 The Demand-Complexity Matrix

The Demand-Complexity Matrix, shown in a very simplified form in Fig. 15.2 can help organizations learn about snap-fit application possibilities and understand the resources that might be required to develop a snap-fit attachment.

The matrix leads to conclusions about the impact of snap-fit problems on critical application requirements and is entirely subjective. Every organization will have its own levels of comfort and judgment on this subject.

Application *demand* and interface *complexity* are primary decision factors in the development process.

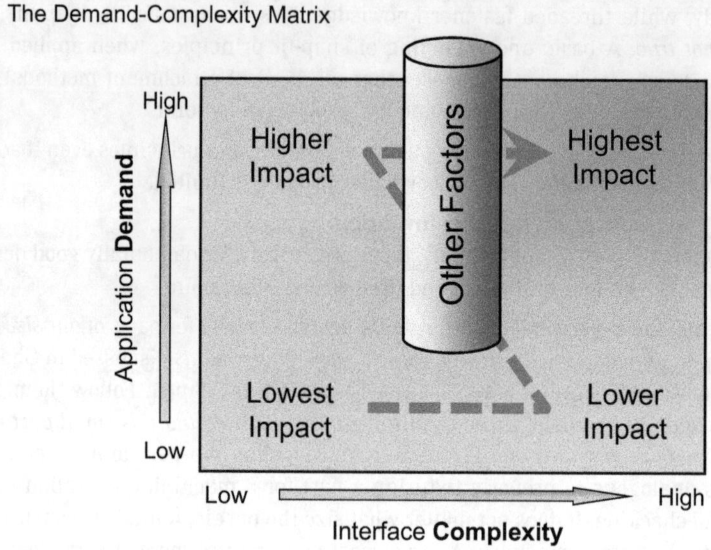

The dotted arrow is a recommended learning path for individuals or organizations beginning to use snap-fit technology.

- **Demand** – The application's quality, reliability, and durability (QRD) expectations.
- **Complexity** – The parts' geometric attributes that influence the design difficulty itself and may increase manufacturing difficulty.
- **Other Factors** – There can also be more factors than demand and complexity which can also influence impact.
- **Impact** – A judgment about the application that guides decisions about development resources, testing, and validation.

Figure 15.2 Simplified demand-complexity matrix

Impact is a qualitative way of thinking about how to approach a snap-fit application's development program. It is the expected effect of a snap-fit problem or issue on *X* where *X* represents performance requirements defined as *critical* for the application by management or government regulations. *X* might be one or more of the following:

- Product safety requirements
- Part replacement cost
- Warranty considerations
- Customer satisfaction and/or expectations
- The organization's reputation
- Ease of assembly
- Ergonomic requirements or limitations
- Serviceability

While demand and complexity are the primary factors that influence resource decisions and impact, the third dimension of the matrix, *other factors*, is also important and may include:

- Manufacturing capability - The mold-maker and part manufacturer's expertise.
- Development capability - The availability and quality of expertise.
- Uncertainty - Unavailability of information about application and material requirements.
- Application ownership - When one individual or organizational entity is responsible for both the mating and the base part (the interface), cost sharing and responsibility conflicts between the two parts are minimized. The greater the separation between ownership, the more difficult the development becomes and the possibility of mistakes increases. (An extreme case would be separate companies in geographically distant countries with a language barrier.)
- The nature of any applied force(s) and possible effects. The magnitude of the applied force(s) is included in *demand*. Other force considerations are listed here:
  - Lowest effect: Forces at the interface are compression only. Locators carry all forces.
  - Low effect: Forces at the interface are shear only and may include rotation in the interface plane. Again, locators carry all forces.
  - Higher effect: Tensile forces at the interface try to separate the parts, but locators carry those forces.
  - Highest force related effects:
    - Lock features must carry some forces.
    - Applied forces are in several directions.
- Any other factors judged to be critical.

Before proceeding, go back to Fig. 15.1 at the beginning of this chapter and consider where each of those applications might fall in the demand-complexity matrix.

Because of its size, the matrix is divided into two figures below. Figure 15.3 shows the interior area of a detailed demand-complexity matrix. It contains activities and resources that an engineering organization might consider for different degrees of interface complexity and application demand. The concepts *complexity* and *demand* are integrated into an outcome called *impact.*

This matrix will appear to be an over-complication of the subject. It is simply an attempt to model the big picture for a snap-fit technology and development process, the details of which have been largely ignored. Similar development, analysis, and resource allocation concepts and issues are already understood and embedded in the product design culture for other disciplines. These activities and their positioning in the matrix should be treated as thought-starters.

Figure numbers of applications that *could* fall into the four categories shown in Fig. 15.3 are included as examples.

Figure 15.4 suggests details of the complexity and demand axes for the matrix. Both axes are a continuum and this figure includes intermediate stages between the lowest and highest extremes of demand and complexity.

Figures 15.3 and 15.4 can be merged to construct one large illustration of the demand-complexity matrix.

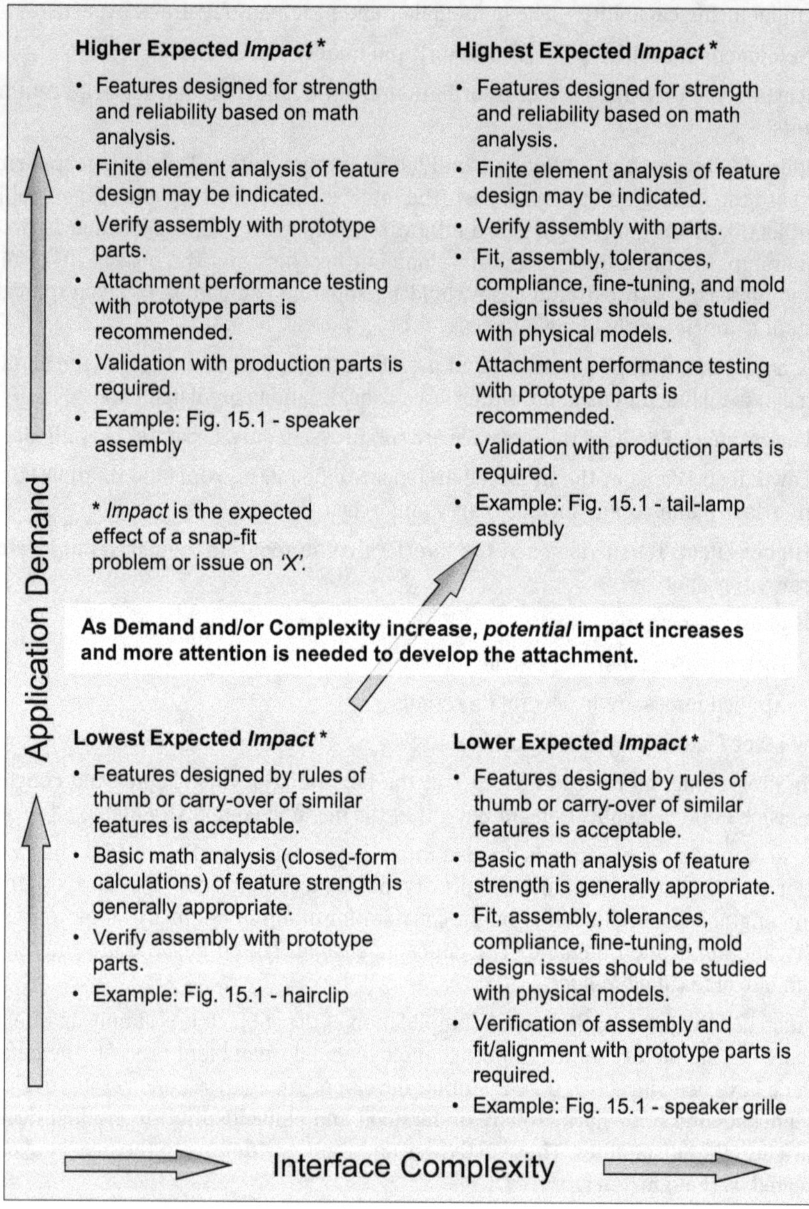

Figure 15.3 Demand, complexity, and impact influence on development activities and resources

Application Demand

Highest Demand

- Attachment must be 100% reliable.
- Feature failure or release results in safety concerns, high cost product damage or failure to meet Government Standards.
- Applied forces may be low or high.
- Tolerance and alignment requirements may be low or high.

Intermediate Demand

- Feature failure or release may cause some loss of performance (squeak and rattle), repair cost or part replacement.
- Significant forces across the attachment must be resisted by constraint features.
- Close tolerance or alignment is required.
- Customer will be operating the snap-fit.

Lowest Demand

- Feature failure or release is not a significant event.
- Separated parts can be re-attached by customer without special effort or tools and little or no degradation of attachment performance.
- Forces across the interface are low.
- Tolerance or alignment requirements are low.

Interface Complexity

Lowest Complexity

- Parts can be treated as rigid bodies for constraint purposes.
- Mating and base parts are simple basic shapes, (panel, enclosure, solid to opening or surface).
- Part geometry does not interfere with feature selection or orientation.
- A simple assembly motion is possible.
- Attachment occurs in one plane.

Intermediate Complexity

- Parts may not be rigid for constraint purposes.
- Mating and/or base parts are a more complex shape (cavity).
- Mating or base part is not a simple shape, but has strong similarities to defined basic shapes.
- Interface has multiple parallel planes.

Highest Complexity

- One or both parts are complex shapes and may not be a defined basic shape.
- Part geometry may interfere with feature selection or orientation.
- A complex or compound assembly motion may be required.
- Multiple interface planes are not parallel to each other.

Figure 15.4 Example of possible content on the demand and complexity axes

15.5 The Snap-Fit Capability Plan

The goal is world-class capability in snap-fit attachments.

One component of *organizational* capability is *individual* capability, and it is possible to have the latter without the former. The true competitive advantage lies in having both. The balance of this chapter describes a detailed plan, summarized in Fig. 15.5 that goes beyond simply training individuals about snap-fits. It should be adapted to reflect an organization's particular needs, culture, resources, and business environment. A few *must do* items are identified, but the reader is generally free to choose how to adapt the plan to their organization.

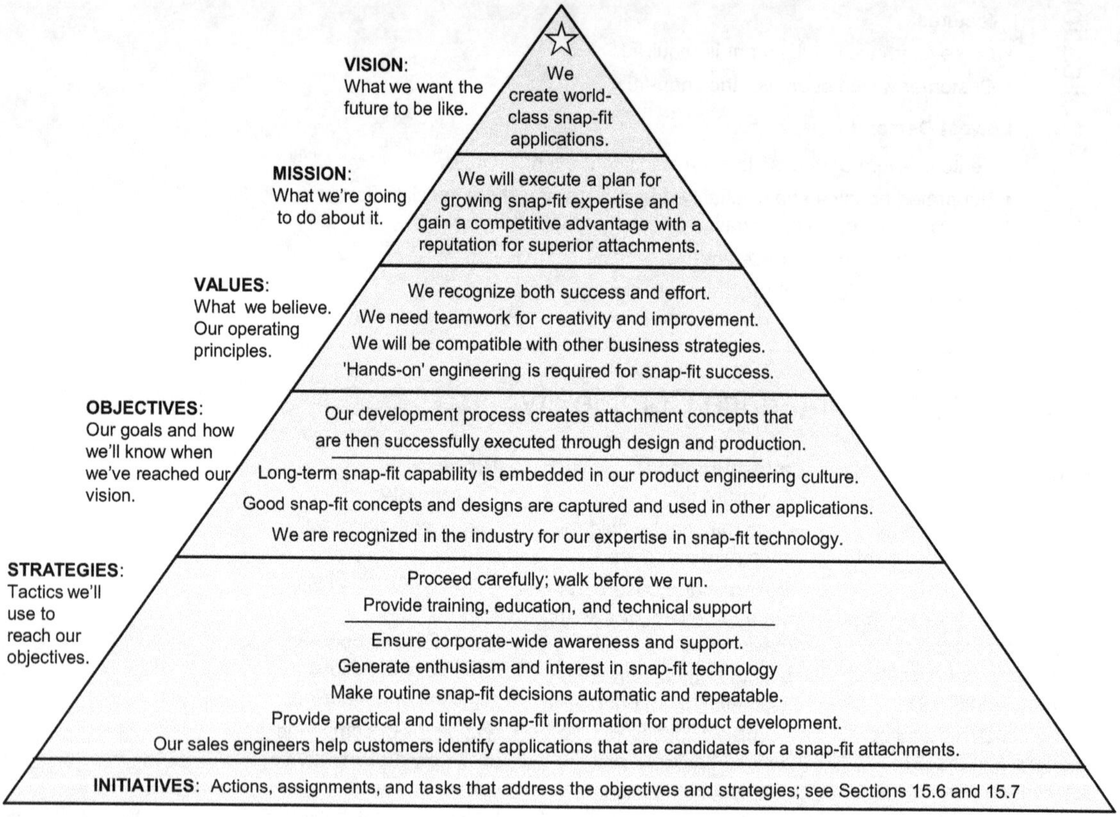

Table 15.1 shows a simplified version of this plan.

Figure 15.5 Snap-fit capability plan for an organization

15.5.1 Vision, Mission, and Values

The *vision* and *mission* statements should be adapted to reflect the organization's own needs and culture.

Some of the statements in the *values* area reflect generally recognized good personnel practices, *teamwork* and *recognition* for example. Others can be developed by the organization.

The value *hands-on engineering is essential to understanding and creativity* should be included in every organization's plan for snap-fit competence. Because of the creative and visual aspects of snap-fit attachments and the spatial-reasoning required for good concept development, it is essential that product developers have access to real parts and models.

Get your hands on real parts!

15.5.2 Objectives

Objectives are also *goals*. We are now moving from intangibles to more concrete elements of the plan. All the objectives are observable outcomes; they can be seen and measured. When we see them, we know we are doing the right things to reach our corporate vision. By measuring them, we can ensure steady progress toward that vision. All strategies must be realistic and targeted to ensure meeting these objectives.

One objective reflects personal or individual snap-fit expertise and is essential if you simply wish to ensure your developers can create reliable snap-fit applications.

- Our development process consistently creates sound attachment concepts which are then successfully executed through design and production.

Some companies may choose to address this objective only and go no farther. However, it does not resolve any *long-term* capability issues.

Three more objectives are recommended if your organization is to become *snap-fit capable*. They will move the organization's engineering culture toward a higher level of snap-fit expertise and ensure a long-term competitive advantage.

- Long-term snap-fit capability is embedded in our product engineering culture.
- Good snap-fit concepts and designs are captured and used in other applications.
- We are recognized in the industry for our expertise in snap-fit technology.

15.5.3 Strategies

Strategies are *tactics* used to reach the objectives. Strategies are where an organization can identify unique strengths or opportunities to gain an advantage over the competition. Consider using those listed here and others developed within the organization. Each strategy must be supported by specific initiatives.

Two near-term strategies will get individual product developers started on snap-fits. Both are highly recommended. As with the essential objectives described above, an organization may choose to address these strategies and forgo the larger corporate effort.

- *Proceed carefully: walk before we run*
  It is important to avoid bad experiences with any new technology so it is not rejected before it has a chance to take hold. Manage the transition to snap-fits carefully and start your designers on low risk applications. With experience, they will be comfortable taking on applications that are more difficult. A careful, managed approach will also allow other parts of the organization with a stake in snap-fits to get up to speed.
- *Provide training, education and technical resources*
  Training and education will help designers move quickly up the learning curve, avoiding many common mistakes made by beginners. Of course, training and education should be on-going and although it starts out as a near-term strategy, it should remain in place for new designers. Development of in-house advanced training specific to your products is also possible. Access to technical resources, including materials and manufacturing subject matter experts, literature, and software is also important. Refer to the appendix for more information.

Longer-term strategies build on the near-term strategies and are intended to embed a high level of snap-fit capability into the organization's culture.

- *Ensure corporate-wide awareness and support*
  Snap-fit decisions will affect other parts of the organization. Make sure all stakeholders are involved.
- *Generate enthusiasm and interest in snap-fit technology* This is a common human resources and motivation based strategy.
- *Make routine snap-fit decisions automatic and repeatable*
  Most snap-fit decisions will be of the routine variety. Prioritize and capture them first in the preferred concepts library. A logical place to start is with the basic shape combinations that appear most frequently in your products. Attachment concepts for these applications can be standardized to reduce the possibility of problems and save time and effort in future product development work.

  Once less time is spent *reinventing* these routine attachments, the less common applications can be addressed.
- *Provide practical and timely snap-fit information for product development*
  This strategy has aspects of the near-term "Provide Technical Support" strategy, but it goes far beyond passive or reactive support from other experts.
- *Sales engineers can identify applications that are candidates for snap-fit attachments*
  Helping customers reduce cost is a great way to gain business and good results will build credibility.

Once the strategies are established, initiatives to support those strategies can be identified. Each initiative must support at least one strategy and one objective. The following sections discuss the initiatives in detail.

The capability plan shown in Fig. 15.5 and described in the following sections is intentionally large and detailed in order to capture as many important topics as possible. Table 15.1 shows the most important parts of the plan in a very simplified form and is probably a more realistic starting point for most organizations. The initiatives are selected for their importance and ease of implementation.

Table 15.1 Simplified Capability Plan

Values	▪ Hands-on engineering with real parts and models is required for snap-fit success.
Objectives	▪ Long-term snap-fit capability is embedded in our product development culture. ▪ Good snap-fit concepts and designs are captured and used as a starting point for future applications.
Strategies	▪ Make routine snap-fit decisions automatic and repeatable.
	Start with the first four initiatives, shown in bold.
Initiatives	▪ **Make display posters of the harmful beliefs.** ▪ **Make display posters of the snap-fit technical domain, the ALC or your own.** ▪ **Make education, training, and technical resources available.** ▪ **Create and maintain a library of preferred snap-fit concepts.** ▪ Make snap-fit technology visible in the organization ▪ Provide parts, physical models, and other products for study and benchmarking. ▪ Include specific snap-fit requirements in your bidding and purchasing process.

15.6 Initiatives for Getting Started

Initiatives are practical *working level* activities expressed as actions, assignments, and tasks. The results or outcomes of each initiative should be observable and measurable.

A total of 15 initiatives are proposed; think of them as a wish list. All are important, but reality may dictate that some be excluded. Some are more critical to success than others and the author's recommendations will be shown in Table 15.4. A manager may choose to implement some of them as stated, ignore some, modify others, and perhaps create new ones.

The first seven initiatives focus on developing and supporting individual expertise and are also the starting point on a path to becoming a snap-fit capable organization. They are:

- *Provide education and training.*
- *Provide technical resources.*
- *Identify low-impact applications as a starting point.*

- *Use physical models.*
- *Provide benchmarking opportunities.*
- *Include snap-fit technical requirements in your bidding and purchasing processes.*
- *Identify intermediate applications.*

15.6.1 Provide Education and Training

Designers need to learn how to think about snap-fits as an interactive system. This book is a beginning. Live, in-depth instruction is also available.

A change to snap-fits requires widespread and continuous support from the entire management team. Short, awareness level presentations on the subject to the engineering management team and other organizational entities such as purchasing, sales, service, and manufacturing are an important part of this initiative.

15.6.2 Provide Technical Resources

Access to manufacturing, material and analysis information will help designers understand related technologies and communicate snap-fit design issues to subject matter experts in those areas. For feature level analysis, calculations using manual methods, closed-form software tools, and finite-element analysis will sometimes be necessary. One of the Technology Leader's duties should be to assemble these resources and make their availability known to everyone. If a snap-fit technical display center is available, reference material can be collected there. The appendix lists some reference materials that may be useful in a technical resource center.

15.6.3 Identify Low-Impact Applications as a Starting Point

There will be certain applications that more readily lend themselves than others to snap-fits. For a beginner (individual or organization), these applications are a reasonable starting point for gaining experience and confidence with minimum impact. They also represent the easiest way to begin realizing savings.

Applications for which your company has design responsibility for both sides of the interface are also much easier to develop. Some snap-fit decisions will drive cost into one or the other of the joined parts. When one organization is responsible for the mating part and another is responsible for the base part, neither organization may want to assume the cost of additional features on their side of the interface, regardless of the technical need.

Tables 3.1 through 3.4 in Chapter 3 will help you determine which applications are most appropriate for venturing into snap-fits. Pick an application with a high number of favorable responses. In Chapter 10, Tables 10.4, 10.5, and 10.12 are lists of things to remember when doing a snap-fit application. Table 7.14 is a feasibility checklist for early screening of potential applications.

Low-demand applications are also a good starting point for implementing snap-fit technology. Some applications that are often low-demand are decorative trim, bezels, access doors, close-out panels, text and instruction signs, and electronic module covers.

The term *impact* as used here has two dimensions, demand and complexity. Figure 15.2 summarizes the concept of impact as used in this discussion. The arrows show the suggested learning/experience path, starting with the lowest impact applications and as confidence and capability increase, moving toward higher impact applications.

15.6.4 Use Physical Models

Make parts and models available during product development. The spatial and creative aspects of snap-fits cannot be represented or understood on paper or on a computer screen, especially by novices. Handling and seeing parts and models in three-dimensions is essential for learning and success. Parts can come from many sources, including benchmarking samples, existing products, toys, scrapped computer printers, other electronic devices, and small household appliances.

Models that represent the application will help the designer visualize details of part geometry and behavior during assembly and removal. A model will help in understanding reactions to applied loads, constraint requirements, and lock and locator placement. They are also useful when discussing potential molding and manufacturing issues with the plastic supplier and part manufacturer. Early models can be crude handmade cardboard and Styrofoam constructs. When math information is available, models can be made using rapid-prototyping methods.

15.6.5 Provide Benchmarking Opportunities

Particularly in the more complex applications, snap-fit development has a significant component of creativity. When first learning about snap-fits, beginners can benefit from studying existing snap-fit applications. Provide snap-fit example applications and encourage study and discussion to support the initial learning phase and enable creativity. Having parts to study is, of course, closely related to the use of models as previously discussed.

Benchmark snap-fit usage in your own products as well as other products. Use attachment level understanding of the snap-fit *key requirements and elements* (see Chapter 2) to guide your analysis of what is right and what is wrong with the applications studied. Look at lots of products, including toys, small appliances, electronics, cameras, electronic devices, and automobiles. It is nice if you can benchmark products similar to yours, *but it is not critical.* Many applications can be classified according to their basic shapes. Any applications with the same basic shapes will provide benchmarking opportunities.

Benchmarking is discussed in detail in Chapter 10, Section 10.3.

15.6.6 Include Snap-Fit Technical Requirements in the Bidding and Purchasing Processes

Three snap-fit key requirements and four enhancements should appear in every snap-fit application. Table 15.2, summarizes these *minimum requirements.*

Whether your organization is a product buyer or supplier, ensure these requirements are stated explicitly and in detail in any product proposal.

Table 15.2 Minimum Requirements for all Snap-Fit Applications

Minimum snap-fit application requirements	Comments
The snap-fit interface should provide proper **constraint** between the mating parts in all degrees-of-motion (DOM).	Minimize the DOM removed by lock features. Maximize the DOM removed by locators. Lock features should only provide constraint in the separation direction.
Snap-fit interface features must be **compatible** with assembly motions and the part shapes.	Assembly (and separation) motions must not create unintended deflections or high strains on the interface features.
The lock and locator features must provide **strength** against assembly damage and failure or unintended release under applied forces.	Verify with feature level analysis or end-use testing.
Assembly guides must be provided to direct locking features to the mating features during assembly.	For ease of assembly and prevention of feature damage, the first features to make contact should be guides. Use selected locators as guides when possible.
Clearance must be designed into all constraint pairs and all potential interference corners must have relief (radii or bevels).	For ease of assembly.
All features must be manufacturing **process-friendly**.	Follow common rules of good mold design.
The attachment must provide **feedback** to the assembly operator of proper engagement.	Feedback may be tactile (preferred), audible, or visual.

Table 15.3 Additional Suggestions for Product Proposals

Snap-fit application attributes that should be addressed in product proposals when appropriate.	
All interface features must have a radius called out at all strain sites. No sharp internal corners are permitted.	Follow common injection molding guidelines for determining minimum allowable radii.
Where feasible, the tip, slide, twist, and pivot assembly motions are preferred over a push motion.	The push motion is least preferred because it maximizes degrees-of-motion that must be removed by the lock features.

Table 15.3 Additional Suggestions for Product Proposals *(Continuation)*

Snap-fit application attributes that should be addressed in product proposals when appropriate.	
Cantilever hook style locking features should be used in low-demand applications only. Consider other lock styles for applications that are moderate or high demand.	The cantilever hook style has the lowest strength capability and robustness of the available beam-based locking features.
Cantilever hook style locking features should not be used in short grip length applications.	As a general rule of thumb, the minimum grip length for a cantilever hook lock must be greater than 5x the beam thickness: 7x to 10x is preferred.
Interface feature mold tolerances should be loose or normal. Fine and close tolerances should not be necessary.	Fine and close tolerances may indicate a lack of robustness in the design. Proper lock and locator selection and constraint management will enable loose or normal tolerances.

As a manufacturer/supplier, when bidding to produce and sell a product containing snap-fits, the proposal should always include the minimum snap-fit requirements. If meeting the minimum requirements adds cost, be prepared to make the business case supporting those requirements. You will have to convince your customer that there are benefits in it for them and that awareness of the minimum requirements reflects your company's level of expertise. Ensure that you can justify, with technical reasons, why your bid may be higher than others. Understand how other enhancement features can help your product meet or exceed the customer's expectations and include them in the proposal.

Use the bidding process to demonstrate your company's expertise

If, on the other hand, you buy parts from a supplier and then assemble them, when soliciting bids for products that contain snap-fits, be certain that each proposal meets the minimum snap-fit requirements.

Your purchasing department should play a role in this initiative. Also, understand which enhancements are needed to satisfy specific requirements of the application and ensure they too are included in the bid process. If the lowest bid does not reflect the minimum snap-fit requirements, consider that you will likely be paying for them eventually in one form or another.

Protect your products from snap-fit incapable companies

You need to convince a supplier company to invest the effort and cost to produce parts with snap-fits. You may have to be willing to pay a higher piece-price to realize the significant assembly savings and you must have enough volume to recover these costs. You will also ensure that the company selling you the part understands how to design snap-fit attachments.

Enhancements were mentioned several times in the above discussion. Enhancements are an important part of the snap-fit interface and are discussed in Chapter 9. They represent the attention to detail that will make your snap-fits world-class. In Section 9.5, Tables 9.2, 9.3, and 9.4 summarize snap-fit enhancements. Those enhancements required in every snap-fit are included in Table 15.2 above.

Other information useful for writing performance requirements can be found in:

- Chapter 3, Section 3.3, Table 3.1
- Chapter 10, Section10.1, discussion of *red-flag* issues
- Chapter 10, Section10.2, Tables 10.4 and 10.5
- Chapter 10, Section 10.5, Table 10.12

15.6.7 Identify Intermediate Applications

Some applications lend themselves to a *transition technology*, where a move from a loose threaded fastener to a snap-fit can be preceded by an intermediate step using loose fasteners that provide no clamp load. This approach can sometimes offer immediate savings without the risk of a commitment to a snap-fit.

The most likely transition technology is plastic push-in fasteners. These are loose plastic fasteners installed by hand. They provide no clamp load and, once installed, behave very much like an integral or snap-fit lock feature. They can often use the same pilot and clearance holes used by threaded fasteners, making a simple substitution possible. If a hole diameter must be changed, it is often a relatively easy change.

Identifying low-demand applications that use threaded fasteners and replacing them with push-in fasteners is a relatively *low-impact* proposition. If testing indicates the new attachment method works, immediate savings are realized because threaded fasteners and power tool operations are eliminated. Once a history of successful use in the field has increased confidence, a future generation of the product may then use integral snap-fit locks features to eliminate the loose push-in fastener. Beware that *if* testing indicates the new attachment does not work, then a return to threaded fasteners is easy.

Table 15.4 Summary of Initiatives for Getting Started

Initiatives for getting started	Importance	Comments
15.6.1 Provide education and training.	Required	Technical for developers, awareness for others.
15.6.2 Provide technical resources.	Required	Improve communication and analysis capability.
15.6.3 Identify low-risk applications as a starting point.	Highly recommended	Gain experience and confidence on the 'easy' ones.
15.6.4 Use physical models.	Required	Hands-on is critical to success.
15.6.5 Provide benchmarking opportunities.	Highly recommended	Will encourage learning and creativity.
15.6.6 Include snap-fit technical requirements in your bidding and purchasing processes.	Highly recommended	Use your expertise as a selling point and protect yourself from snap-fit 'incapable' companies.
15.6.7 Identify intermediate applications.	Suggested	Potential cost-savings even if you never go all the way to snap-fits.

15.7 Initiatives for Organizational Capability

The last eight initiatives build on individual expertise to create an organization capable of sustaining and growing snap-fit expertise long after the original Champion, Technical Leader, and expert designers are gone. These initiatives will ensure that snap-fit capability becomes embedded in the organization's engineering culture.

- *Identify and empower a snap-fit Champion.*
- *Identify and empower a snap-fit Technology Leader.*
- *Make snap-fits visible in the organization.*
- *Link snap-fits to other business strategies.*
- *Create and maintain a library of preferred concepts.*
- *Have a model of the snap-fit technical domain*
- *Reward teamwork and make snap-fits interesting.*
- *Identify supportive customers and suppliers.*

Don't wait, some of these *organizational* initiatives can begin while the *getting started* initiatives are underway. They are summarized in Table 15.3.

15.7.1 Identify and Empower a Snap-Fit Champion

This individual should be an executive with the rank, credibility and personality to be a salesman and enabler for leading the transition to snap-fit capability. As with any change, there will be roadblocks and frustrations on the road to true snap-fit capability. The Champion can also provide protection for and motivation to the effort.

15.7.2 Identify and Empower a Snap-Fit Technical Leader

This individual should have the technical ability to understand snap-fit development and design as well as the desire and ability to generate enthusiasm for the subject among their peers. They will be the working-level driver for the technology and the cultural change. The Technology Leader should be prepared to work with the Champion and to use the Champion's leverage to ensure the effort stays on track. The Technology Leader will be responsible for actually executing many of the initiatives.

This is much more than a technical responsibility, it has many elements of training and *knowledge management*. Select this individual carefully; many very good technical people are not necessarily good at the relatively intangible task of managing knowledge.

15.7.3 Make Snap-Fit Technology Visible in the Organization

One of the requirements for successful change is to keep the object of the change in the minds of the participants.

The spatial and creative aspects of snap-fits make them visually interesting, particularly to technical people. Use hardware displays, posters, pictures, and illustrated summaries to generate and maintain that interest. Show unusual and interesting applications as well as successes and lessons-learned from solving problems. If you have the space, set up a dedicated snap-fit display area or center where technical information can be displayed, technical references collected and success stories told.

If you decide to create display posters, some of the things to include are listed here in a suggested order of importance:

- The harmful beliefs – To get attention.
- The attachment level construct (ALC) – To help organize understanding.
- Pictures, sketches, and parts with explanations of good/bad attributes – To make it real.
- The minimum snap-fit requirements – To get started.
- Decoupling and the limitations of cantilever hooks – To reduce the improper use of hooks.
- Other lock feature styles as alternatives to cantilever hooks – To provide alternatives to hooks.
- The concept of basic shapes – To enable more effective benchmarking.
- Using assembly motion to generate different concepts – To encourage creative thinking.
- Enhancements – Practical additions to improve snap-fits.

15.7.4 Link Snap-Fits to Other Business Strategies

Quality, cost reduction, workplace ergonomics, and design for assembly/manufacturing (DFA and DFM) are a few business strategies that can be leveraged to support snap-fit technology.

Demonstrate and spread awareness of how implementation of effective snap-fit technology and technical capability strategies can support and enable these and other business goals.

15.7.5 Create and Maintain a Library of Preferred Concepts

The author believes in the high value of a simple and practical corporate memory activity.

Create and maintain a library of standard attachments. Identify the most common basic shapes used in your products and create a set of fundamentally sound attachment concepts.

This is one of the most powerful and important initiatives for becoming a snap-fit capable organization. Create a set of fundamentally sound and proven attachment concepts for use as a starting point for all future designs. Also keep a benchmarking or reference collection of the final products that evolved from those concepts.

A preferred concept library has many advantages including:

- It is a valuable repository of corporate technical knowledge for everyone, particularly inexperienced designers.
- Developers will not waste time reinventing the same attachment concepts. They will be able to quickly identify the *routine* applications and create sound concepts. More time will be available for completing detailed product designs and creating solutions to the *nonroutine* new or unique applications.
- Problems and issues associated with new, untried concepts will be avoided.
- All design, performance, and manufacturing issues associated with each application will be understood and captured over time.
- The manufacturing issues and cost drivers of each attachment will be better understood, leading to more accurate product pricing and estimates.

Preferred concepts are arrangements of constraint features and enhancements that are desirable and generic (or common) *starting points* for interface design.

A preferred concept is *not* a detailed design. It is a fundamentally sound (technically correct and robust) arrangement of constraint features and certain enhancements. In its most basic form, the preferred concept satisfies the minimum snap-fit requirements discussed in Section 9.6.5. A preferred concept is adaptable to multiple applications having the same basic shape combination.

When a new application (product) is proposed, an appropriate preferred interface concept is identified. Then, using that concept as a starting point, the designers exercise their creativity and expertise to

- Select specific constraint feature styles.
- Design those features, doing feature level analysis if necessary and determining dimensions and tolerances.
- Add appropriate enhancements for the application.

A suggested approach to this initiative is to classify your existing snap-fit applications according to their *basic shapes*. You may find that many different applications actually fall into a limited number of shape combinations. For each of those combinations, a limited number of *best* concepts that satisfy all the rules of good snap-fits can be identified and then used on all similar applications, see Chapter 4, Section 4.2. Your own particular product line may have a different set of common basic shape combinations.

Tables 4.2 and 4.3 in Chapter 4 summarize the available basic shape combinations and show the relative frequency of these combinations in automotive applications. Of all the high-frequency combinations observed by the author, the *panel-opening* basic shape combination was the most common. This suggests high value in investigating and identifying a limited set of preferred interface concepts for panel-opening applications.

15.7.5.1 Example of a Preferred Concepts Initiative

A survey of automotive applications from many manufacturers reveals that fuse doors, speaker grilles, access panels, radiator grilles, radiators, closeouts, lamp lenses, reflectors, and a great many miscellaneous trim pieces fall into the panel-opening basic shape combination.

Studying the snap-fit attachment details of these applications shows that *each one is different*! Not just in the interface feature dimensions, but in the attachment's basic concept and/or lock feature selection. The range of differences is as great within each manufacturer's products as it is between manufacturers. For example, for 10 different speaker grilles, there are 9 different attachment methods. For fuse access door applications, of all manufacturers studied, there are 52 variations. Within one manufacturer there are 17 variations!

Technical benchmarking of these applications reveals that they also vary widely in ease of assembly, retention capability, probable durability, serviceability, perceived quality, and actual quality. A few are excellent or good, many have one or more shortcomings and some are poor designs.

The big question is *What is so unique about each of these applications that a unique attachment method is needed for each one?*

The answer is *NOTHING! Most of the applications could have been developed using the same Preferred Concept as a starting point.*

Figure 15.6 illustrates how the idea of a library of preferred concepts might be executed and used. It is explained in the following points:

- Studying a number of panel-opening applications using attachment knowledge and understanding leads to a very limited group of preferred concepts for panel-opening applications.
- These are generic concepts and do not include dimensions or details. Each concept description may include details of lessons-learned, potential pitfalls, and other issues that have occurred or may occur in similar applications.
- This group can be categorized by basic shapes, assembly motions and any other parameters needed to fully define the organization's product line, see Fig. 15.7. Note the use of complexity, demand, grip-length, and action as distinguishing subcategories for this panel-opening group.
- A product developer reviews the appropriate preferred concepts and selects one as a starting point for developing and designing a new product, but does not invent a new attachment.
- Any lessons-learned during development, manufacturing, assembly, or customer usage of this new application are fed back into the preferred concept library.

Figure 15.7 shows how one basic shape combination, a panel to an opening, might be organized in a library of preferred concepts.

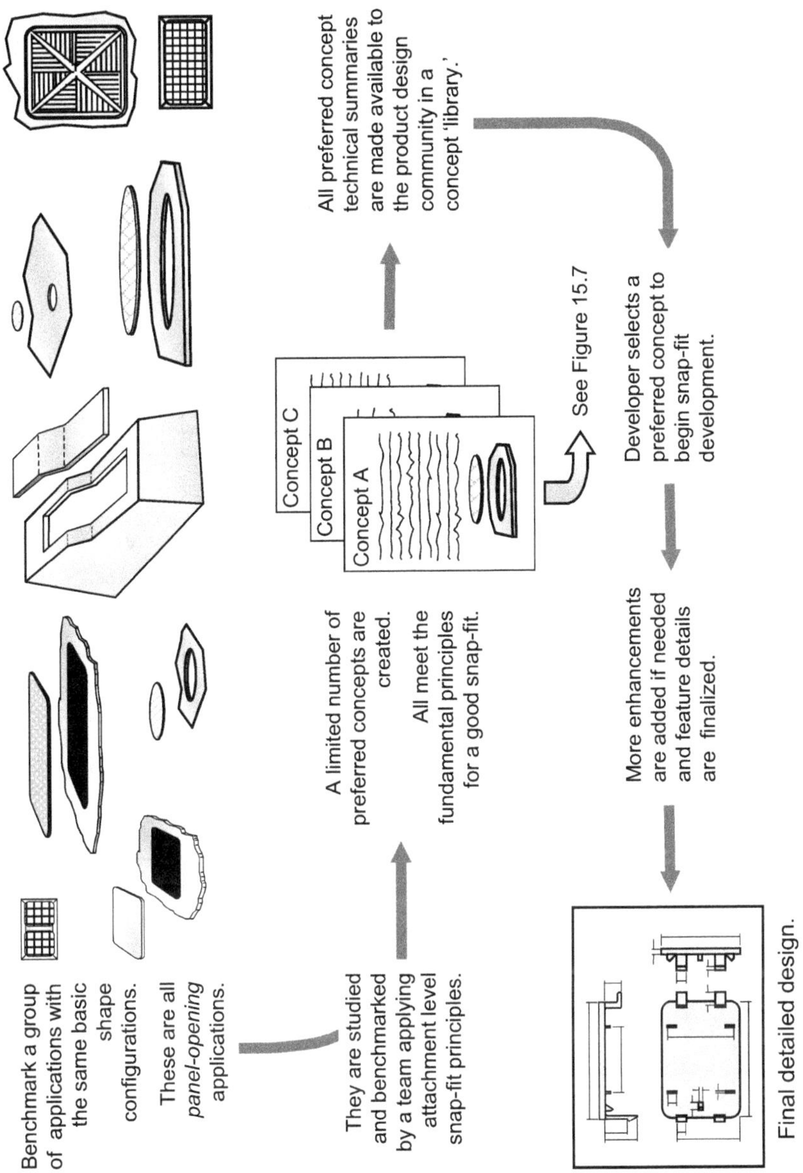

Figure 15.6 Applying the preferred concept principles

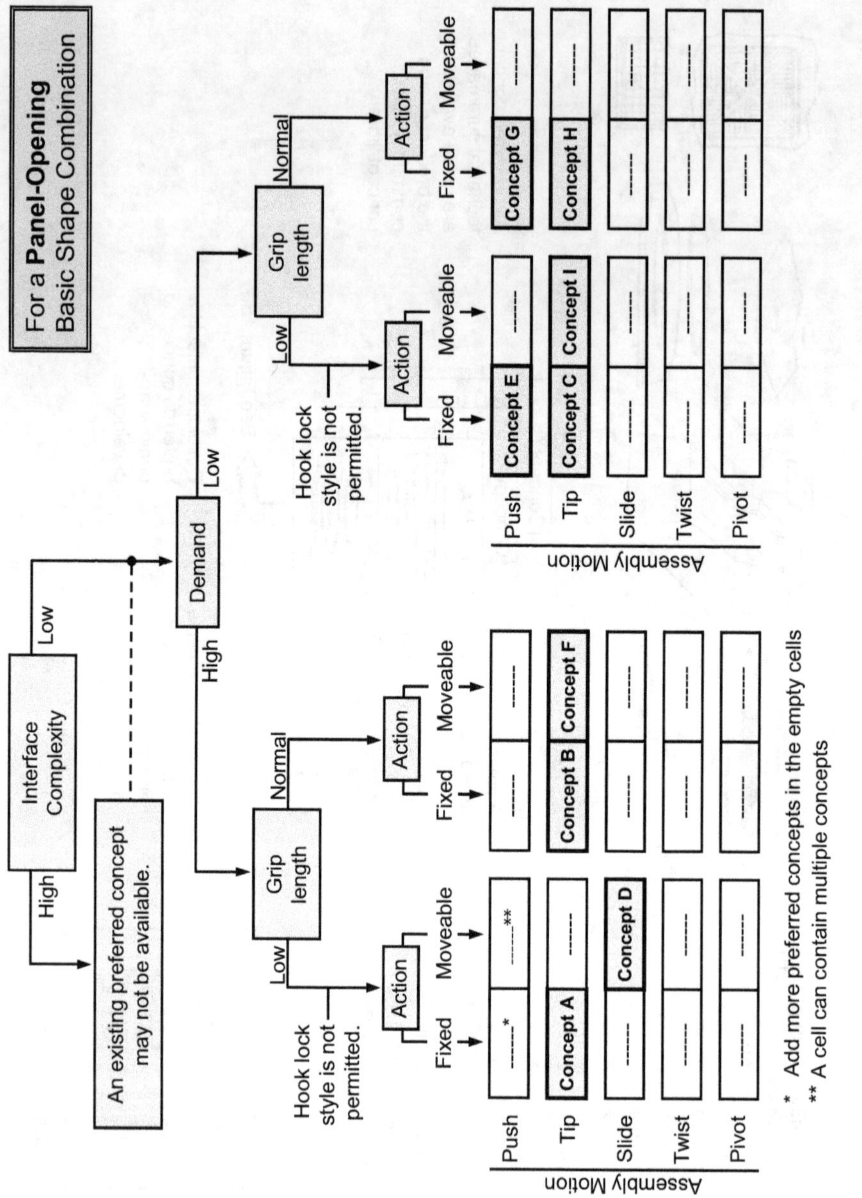

Figure 15.7 Example: The *Panel-Opening* basic shape combination organized into preferred concepts

When an application cannot be readily classified by basic shape, a preferred attachment concept that satisfies the rules of good snap-fits can still be defined and included in the library for future generations of that product.

Establishing and maintaining a technical memory takes time and effort. Technical memory will not happen or remain viable unless there is a long-term commitment to it.

Establishing the library, helping designers make contributions to it, and ensuring its use as a technical resource should be the job of the snap-fit Technology Leader. The Champion should show interest in the library and encourage, recognize, and reward contributions to the library.

Finally, make the library visible in a *preferred interfaces display* with illustrations and product examples of some common applications and the resulting product successes.

Make all of the library's contents available in an online resource if possible. Establish a process to ensure that all new knowledge gained is captured in the library.

15.7.6 Have a Model of the Snap-Fit Technical Domain

Use the Attachment Level Construct described in this book or create your own. Make the model visible and continuously refer to it until it becomes second nature. Use the model as a tool to capture and organize knowledge gathered during benchmarking. Adopting the terminology of the model and the feature and enhancement definitions is very important for creating a common language for communicating about snap-fits.

Even if you plan to create your own model, it is suggested that your organization live with the ALC model in this book for a while until sufficient expertise and understanding exist to make meaningful modifications. The model as it exists has proven to be fairly robust and adaptable to most organizational needs.

15.7.7 Reward Teamwork and Make Snap-Fits Interesting

Benchmarking and studying parts and models are activities that lend themselves particularly well to the synergy of team involvement. The learning and creativity that results will not only improve the product, but will help make snap-fits interesting. Recognize and reward clever solutions to design problems and capture them in the snap-fit library so they are available in the future.

- Have regular technical updates at departmental and staff meetings.
- Maintain a display of snap-fit successes and interesting or unique applications.
- Encourage writing and presenting technical papers and website articles.
- Have a snap-fit technical display and keep it up to date with new parts and information.
- Encourage brainstorming sessions.

15.7.8 Identify Supportive Customers and Suppliers

Close cooperation with organizations both upstream and downstream in the product development/manufacturing process will help avoid some problems as your organization develops snap-fit capability. A strategic partnership with certain customers or suppliers for developing mutual capability may also be advantageous.

When your company must coordinate development of a part with another company's part, (because they are to be attached to one another), the interface becomes of common interest. Some snap-fit decisions will drive cost into one or the other of the joined parts. One such decision is choosing which part will carry the locking features. (Locking features will tend to add cost and complexity to that part.) Having a common technical basis for making decisions can help overcome non-technical cost issues that may become areas of contention.

Table 15.5 Summary of Initiatives for Long-Term Capability

Initiatives for long-term capability	Importance	Comments
15.7.1 Identify and empower a snap-fit Champion.	Required	Required to overcome road-blocks and cause permanent change.
15.7.2 Identify and empower a snap-fit Technology Leader.	Required	Not a technical leader, a *technology* leader.
15.7.3 Make snap-fits visible in the organization.	Highly recommended	Keeps everyone thinking about snap-fits.
15.7.4 Link snap-fits to other business strategies.	Recommended	Can help 'sell' the idea in your company.
15.7.5 Create and maintain a library of preferred snap-fit concepts.	Required	Once you have a good attachment concept, why change it? Work on other things.
15.7.6 Have a model of the snap-fit technical domain.	Required	Use the one in this book to start.
15.7.7 Reward teamwork and make snap-fits interesting.	Recommended	Will help drive creative synergies and maintain interest.
15.7.8 Identify supportive customers and suppliers.	Recommended	Cooperation and understanding on both sides of the supply chain.

15.8 Summary

Individual expertise is important for developing reliable and efficient snap-fit applications. But individuals come and go and expert-level knowledge will be lost. Creating new expertise will always involve a learning-curve and mistakes will be made.

To maintain a long-term competitive advantage, an organization must sustain and grow its snap-fit expertise by embedding it into the product development culture. The organization must become *snap-fit capable*. This chapter presented a plan for doing that by integrating certain snap-fit concepts with corporate memory and organizational development principles.

Important points in Chapter 15:

- Snap-fit expertise should be managed at both the individual and the organizational levels to ensure good snap-fit designs and long-term technical excellence.
- There will be a learning curve for the new technology. Do not leap into snap-fit technology on high-complexity or high-demand attachments.
- Snap-fits are a true paradigm shift from threaded fastening methods because snap-fits do not rely on clamp load.
- Threaded fastener knowledge does not transfer to snap-fits. However, many principles of good snap-fit design do transfer to threaded fastener attachments.
- Certain harmful beliefs can interfere with developing snap-fit capability. Be proactive and repetitive in addressing them; many times, they are unspoken.
- Piece cost will be higher with snap-fits; the savings are in ease of assembly and part reduction.
- Require that minimum snap-fit requirements be addressed on product proposals and bids.
- The initiative to *create and maintain a library of preferred concepts* is so important that it bears repeating here. The author feels strongly that this is one of the most important things an organization can do to become snap-fit capable and to maintain that capability over the long term.

References

Some of the ideas discussed in this chapter first appeared in *A Management Strategy for Implementing Snap-Fit Technology*, an article by the author published in Business Briefing: Global Automotive Manufacturing and Technology (2003)

Bibliography

Argyris, Chris, *Management and Organizational Development*, McGraw-Hill, New York (1971)

Cummings, Thomas G., Worley, Christopher G., *Organization Development and Change*, 6th ed., South-Western College Publishing, Cincinnati, OH (1997)

Davenport, Thomas H., Prusak, Laurence, *Working Knowledge*, Harvard Business School Press, Boston, MA (1998)

Drucker, Peter F., *Managing in a Time of Great Change*, Truman Talley Books/Dutton, New York (1995)

Kotter, John P., *Leading Change*, Harvard Business School Press, Boston, MA (1996)

Lindsay, William M., Petrick, Joseph A., *Total Quality and Organization Development*, St. Lucie Press, Boca Raton, FL (1997)

Senge, Peter M., Kleiner, Art, Roberts, Charlotte, Ross, Richard B., Smith, Bryan J., *The Fifth Discipline Fieldbook*, Doubleday, New York (1994)

Tichy, Noel M., *Managing Strategic Change*, John Wiley and Sons, New York (1983)

Walton, Mary, *The Deming Management Method*, The Berkeley Publishing Group, New York (1986)

Zyngier, Suzanne, Burstein, Frada, McKay, Judy, *Knowledge Management Governance: A Multifaceted Approach to Organizational Decision and Innovation Support*, The IFIP TC8/WG8.3 International Conference (2004)

Appendix – Resources

Some chapter references and bibliographies cite material used when the chapter was originally written. If the original source of that material or information has changed names or URL's, the updates will appear here.

This list of plastic part design, materials and analysis resources was correct as of April 2016. Companies change names, merge, split and remerge, and URL's change. If you cannot find an item, do some searches on the old names and you are likely to find the current location or at least a path you can follow to the resource.

This list is far from complete. These are resources the author has used or is aware of. Some were used for this book and are referenced within the chapters. Resources the author has used and found particularly useful for a *nonexpert* in polymers (myself) are preceded by an asterisk. Inclusion of a resource here is not an endorsement nor is exclusion to be taken as nonendorsement.

Books listed in the chapter references, because their names do not change, are not included here.

The list is generally in alphabetical order except when related resources are grouped together.

Analysis Information and Tools

- ANSYS (nonlinear finite element analysis)
  http://www.ansys.com/
- CAE Associates (ANSYS Consultant)
  https://caeai.com/
- CAE Report (Analyzing Plastic Parts with Finite Element Analysis)
  https://caeai.com/sites/default/files/CAEA_Plastics_Demo.pdf
- BASF (snap-fit design workspace)
  http://snapfit4.cmg.net/SnapFit/workspace.jsp
- COVESTRO – FEMSnap (Finite element analysis limited to COVESTRO products: Apec, Bayblend, Makroblend, and Makrolon.)
  http://www.plastics.covestro.com/Engineering/Tools/FEMSnap-tool/FemSnap-Tool
- FEMSnap MANUAL
  https://techcenter.lanxess.com/scp/emea/en/docguard/FEMSnap_-_manual.pdf?docId=7959239
- EFUNDA (plastic part design and tapered beams)
  http://www.efunda.com/DesignStandards/design_home.cfm
- ENGINEERS EDGE (beam bending equations and calculator)
  http://www.engineersedge.com/beam_bending/tapered-snap-fit-beam.htm

- MOLDFLOW-AUTODESK (injection molding simulation software)
  http://www.autodesk.com/products/moldflow/overview#
- PROTO LABS (injection molding prototypes)
  https://www.protolabs.com/injection-molding
- TRIBOLOGY ABC (beam and torsional member calculators)
  http://www.tribology-abc.com/calculators/t14_9.htm
- XCP (useful plastic design tools)
  http://xcp.x-castro.com/useful-design-tools/

Design Guidelines

- BASF SNAP-FIT DESIGN MANUAL
  http://web.mit.edu/2.75/resources/random/Snap-Fit%20Design%20Manual.pdf
- * DESIGNING WITH PLASTIC – THE FUNDAMENTALS
  http://www.polymerhouse.com/datasheets/Designing%20with%20Plastics%20-%20Ticona.pdf
- ENGINEERS EDGE (resources for plastic part design)
  http://www.engineersedge.com/plastic_design_menu.shtml
- * SNAP-FIT JOINTS FOR PLASTICS – A DESIGN GUIDE
  http://fab.cba.mit.edu/classes/S62.12/people/vernelle.noel/Plastic_Snap_fit_design.pdf
  or
  http://www.x-castro.com/docs/Plastic_Snap_fit_design.pdf
- PROTO LABS (design resources)
  https://www.protolabs.com/resources
- * STANDARDS AND PRACTICES OF PLASTIC MOLDERS – Guidelines for Molders and Their Customers (SPI)
  https://www.pct.edu/academics/library/secure/pdf/Standards_and_Practices_of_Plastic_Molding.pdf

Consulting and Training

- ETS PLASTICS CONSULTING
  http://www.ets-corp.com/
- FASTENINGSMART, INC.
  http://www.fasteningsmart.net/
- INJECTION MOLDING HELP
  http://www.moldinghelp.com/dreamteam.html
- GLENN BEALL PLASTICS, LTD., Libertyville IL, 847-549-9970
  URL not available
- PENN STATE PLASTICS TRAINING ACADEMY
  http://psbehrend.psu.edu/community-and-workforce-programs/business-industry/plastics

- POLYMERS CENTER OF EXCELLENCE
  http://polymers-center.org/
- UNIVERSITY OF WISCONSIN - MILWAUKEE
  http://uwm.edu/sce/program_area/engineering/

General Design Information

- CPC PLASTICS (technical services)
  http://www.theplasticsconsultant.com/index.html
- PARALLEL DESIGN (technical services)
  http://www.paralleldesign.com/
- PROSPECTOR (plastics industry search engine)
  http://www2.ulprospector.com/press/060316.asp

Loose Metal Fasteners

- ITW CIP
  http://www.itwcip.com/
- SALES SITES
  http://www.alibaba.com/showroom/metal-clips-fasteners.html
- SHAKEPROOF GROUP
  http://www.shakeproof.com/
- STANLEY ENGINEERED FASTENING - (Metal and Plastic)
  http://www.stanleyengineeredfastening.com/

Loose Plastic Fasteners

- ITW AUTOMOTIVE GROUP
  http://www.itwautomotive.com/
- ITW DELTAR
  http://www.itwef.com/
- ITW FASTEX
  http://www.itw-fastex.com/index.html
- SALES SITES
  http://www.alibaba.com/cache/Push-in-fastener-plastic-injection-parts-_410174796.html
- STANLEY ENGINEERED FASTENING (Metal and Plastic)
  http://www.stanleyengineeredfastening.com/
- TRW-EFC
  http://www.trw-efc.com/en/products/fastening_systems
- WCL
  https://www.wclco.com/plastic-components/panel-fasteners/

Special Screws for Plastics

- REMINC
  http://www.taptite.net/taptite/plastite.asp
- SHAKEPROOF GROUP
  http://www.shakeproof.com/products/threaded/bosscrew.html
  http://www.shakeproof.com/products/threaded/hi-lo.html
- STANLEY ENGINEERED FASTENING
  http://www.stanleyengineeredfastening.com/sites/www.emhartamericas.com/files/downloads/Threaded_Fasteners_for_Plastics_EN.pdf

Materials Databases

- CAMPUS
  http://www.campusplastics.com/
- * MATWEB (Information and Database)
  http://www.matls.com/index.asp?ckck=1
- * PROSPECTOR (Free)
  https://www.ulprospector.com/en/na
- PROSPECTOR (Subscription)
  http://www2.ulprospector.com/prospector/
- SOCIETY OF PLASTICS ENGINEERS (SPE)
  http://www.4spe.org/contentfullscreen.aspx?ItemNumber=22772&navItemNumber=23281

Organizations

- SOCIETY OF PLASTICS ENGINEERS (SPE)
  http://www.4spe.org/
- SOCIETY OF THE PLASTICS INDUSTRY, INC. (SPI)
  http://www.plasticsindustry.org/

Supplier Sites

- BASF
  http://www2.basf.us/businesses/plasticportal/pp_home_en.html
- CELANESE
  http://www.celanese.com/engineered-materials/engineered-materials.aspx
- COVESTRO
  http://www.covestro.com/
- DUPONT
  http://www.dupont.com/
- SABIC (Formerly GE Plastics.)
  https://www.sabic-ip.com/gep/en/Home/Home/home.html

About the Author

Paul Bonenberger has experience in final assembly, product test and development, engineering standards, technical training development, and instruction. As an engineer with the General Motors North America Engineering Center, he was involved with mechanical attachments for 30 years, and is recognized as a threaded fastener and snap-fit expert. He has an Industrial Engineering Degree from General Motors Institute, a Master of Engineering Management Degree from the University of Detroit and a Master of Training and Development Degree from Oakland University. In response to a lack of documented snap-fit attachment design knowledge, he created the *Attachment Level Construct©* described in this book. He teaches technical classes independently and through various professional organizations, and can be contacted at paulrb@fasteningsmart.net or paulrb@ameritech.net.

Index

D

E

F

G

H

I

J

K

L

M

N

O

P

Q

R

S

www.ingramcontent.com/pod-product-compliance
Lightning Source LLC
LaVergne TN
LVHW081930200926
848484LV00009B/2399